AF361859

# Nano-Engineering Solutions for Dental Implant Applications

# Nano-Engineering Solutions for Dental Implant Applications

Editor

**Karan Gulati**

MDPI • Basel • Beijing • Wuhan • Barcelona • Belgrade • Manchester • Tokyo • Cluj • Tianjin

*Editor*
Karan Gulati
The University of Queensland
Australia

*Editorial Office*
MDPI
St. Alban-Anlage 66
4052 Basel, Switzerland

This is a reprint of articles from the Special Issue published online in the open access journal *Nanomaterials* (ISSN 2079-4991) (available at: https://www.mdpi.com/journal/nanomaterials/special_issues/Dental_Implant_nano).

For citation purposes, cite each article independently as indicated on the article page online and as indicated below:

LastName, A.A.; LastName, B.B.; LastName, C.C. Article Title. *Journal Name* **Year**, *Volume Number*, Page Range.

**ISBN 978-3-0365-3144-1 (Hbk)**
**ISBN 978-3-0365-3145-8 (PDF)**

Cover image courtesy of Dr Karan Gulati

# Contents

# About the Editor

**Karan Gulati** (Dr) is a Research Group Leader and the Deputy Director of Research at the School of Dentistry, The University of Queensland.   Dr Gulati is a pioneer in electrochemically nano-engineered implants with over 11 years of extensive research experience using nano-engineering towards various bioactive and therapeutic applications.  He completed his PhD from the University of Adelaide in 2015 and was awarded the Dean's Commendation for Doctoral Thesis Excellence. His career has been supported by prestigious fellowships from NHMRC (National Health and Medical Research Council, Australia), JSPS (Japan Society for the Promotion of Science, Japan), Erasmus+ (Germany) and the University of Queensland.

*nanomaterials*

*Editorial*

# Nano-Engineering Solutions for Dental Implant Applications

**Karan Gulati**

School of Dentistry, The University of Queensland, Herston, QLD 4006, Australia; k.gulati@uq.edu.au

**Citation:** Gulati, K. Nano-Engineering Solutions for Dental Implant Applications. *Nanomaterials* **2022**, *12*, 272. https://doi.org/10.3390/nano12020272

Received: 29 December 2021
Accepted: 13 January 2022
Published: 15 January 2022

**Publisher's Note:** MDPI stays neutral with regard to jurisdictional claims in published maps and institutional affiliations.

This Special Issue of *Nanomaterials* explores the recent advances and trends with respect to nano-engineered strategies towards dental implant applications. A dental implant microenvironment is complex, and an implantation surgery results in a local trauma [1]. Further, exacerbated by the ongoing patient conditions (age, osteoporosis, diabetes or smoking), long-term dental implant success may be compromised due to inappropriate integration (both soft-tissue and osseo-integration), inflammation and bacterial infection [2,3]. As a result, surface modification of dental implants to fabricate desirable topographical and chemical features towards enhancing osseo- and soft-tissue integration (STI), has been well documented [4]. Various physical, chemical and biological modifications have been investigated across the macro-, micro- and nano-scales to find the most optimum dental implant surface features [5].

The goal of this Special Issue is to shine light on the recent nano-engineering advances that revolutionize the dental implant technology, with a focus on the next generation of implants capable of providing maximum local therapy to drastically reduce implant failures. This Special Issue will inform the readers of the latest nano-engineering developments in the domain of dental implants, aiming to bridge the gap between research and clinical translation, from lab to clinics. This Special Issue contains a blend of eight original research, communication-style research and review papers from leading scientists across the world with expertise in nano-engineered dental implant technology.

Titanium (Ti) is the most popular choice for the fabrication of dental implants and hence several articles were focussed on surface modification of Ti-based dental implants to augment their bioactivity or therapeutic potential, as reviewed by Zhang et al. [6]. The review summarizes key progress, challenges and research gaps relating to nano-engineered dental implants, spanning across the use of nano-engineered Ti and therapeutic nanoparticle (NP) modification of Ti dental implants. Similarly, the importance of nanoscale surface modification with respect to achieving desirable microbial decontamination and antibacterial efficacy is reviewed by Hosseinpour et al. [7]. While metallic and non-metallic NPs have shown great promise in both bioactivity and antimicrobial functions, natural micro-/nanoparticles such as extracellular vesicles (EVs, membrane bound lipid particles secreted by all cell types) possess considerable therapeutic potential. Hua et al. reviewed the current status of periodontal and dental pulp cell derived small EVs towards anti-inflammatory, osteo/odontogenic, angiogenic and immunomodulatory functions, suitable as effective therapeutic molecules for alleviating dental implant challenges [8]. Next, Alali et al. investigated the soft-tissue integration and antibacterial performance of Lithium (Li)-doped Ti implants [9]. Briefly, chemically modified Ti doped with Li presented an extracellular matrix (ECM) mimicking nanowire network that enhanced collagen-I and fibronectin gene expression (of cultured human gingival fibroblasts) and reduced bacterial metabolic activity (of *Staphylococcus aureus*), confirming the suitability for dental implant applications.

Electrochemical anodization of Ti-based dental implants has been utilized to fabricate controlled titania ($TiO_2$)-based nanotopographies including nanotubes or nanopores to augment cellular functions towards soft- and osseo-integration and enable loading and release of potent therapeutics (antibiotics or proteins) [10,11]. Briefly, anodization involves immersion of metal electrode/implant (anode) and a counter metal electrode (cathode) in an appropriate electrolyte containing water and fluoride ions and supply of optimized voltage and current, which facilitates self-ordering of various metal-oxide nanostructures on the implant (anode) surface [12]. It is known that nanoscale implant surface can influence blood coagulation that can modulate cellular functions and early osseointegration. Further, long non-coding (Lnc) RNAs regulate various processes within the skeletal system, however, the interdependence between LncRNAs (derives from clot cells) and osseointegration remains unexplored. Bai et al. bridged this research gap and investigated the correlation between LncRNAs and $TiO_2$ nanotube (TNT) modified Ti implants towards osseointegration [13]. Briefly, the sequence analysis (detailed Gene Ontology and Kyoto Encyclopedia of Genes and Genomes pathway investigation) of LncRNAs (expressed within the clot formed) on TNTs of various diameters (15, 60 and 120 nm) indicated that implant nanotopography can influence the clot-derived LncRNAs expression profile, which dictates the *de novo* bone formation.

Besides Ti, Zirconium (Zr) or Zirconia ($ZrO_2$) is emerging as a popular dental implant material choice attributed to its reduced affinity to bacterial plaque, appropriate mechanical properties, white colour and non-magnetic nature [14]. In a pioneering study, Chopra et al. reported nano-engineering of curved and micro-rough Zr surfaces via electrochemical anodization to fabricate various nanotopographies [15]. Briefly, by optimizing anodization conditions, dental implant/abutment relevant surfaces were modified with $ZrO_2$ nanotubes, nanocrystals or nanopores, bringing anodization of dental implants closer to clinical translation.

Peri-implantitis is characterized by peri-implant mucosa inflammation and progressive destruction of the supporting bone attributed to biofilm formation [1,2]. Due to the high prevalence of peri-implantitis, various debridement techniques including mechanical treatment, chemical disinfection, antibiotic treatment, lasers and their combinations have been explored. Among these, the use of various lasers like erbium-doped: yttrium, aluminum and garnet (Er:YAG); and erbium, chromium-doped: yttrium, scandium, gallium and garnet (Er, Cr:YSGG) lasers have been proposed for implant debridement. Advancing this domain, Secgin-Atar et al. investigated the use of erbium lasers (Er:YAG and Er, Cr:YSGG) and mechanical methods (curette, ultrasonic device) on implant debridement (of implants lost to peri-implantitis) to obtain implant characteristics similar to virgin implants [16]. In total, 28 failed implants (4 failed implants in each group: titanium curette; ultrasonic scaler; Er:YAG very short pulse; Er:YAG short-pulse; Er:YAG long-pulse; Er, Cr:YSGG1; Er, Cr:YSGG2) were debrided for 120s and compared with two virgin implants (as controls) using SEM, EDX and profilometry characterizations. The results indicated that ultrasonic and Er:YAG long pulse groups were most effective debridement techniques.

Next, Casarrubios et al. studied the influence of Ipriflavone (IP) incorporated $SiO_2$–CaO mesoporous bioactive glasse based hollow nanospheres (nanoMBGs) as an alternative to bioactive glasses for treating periodontal defects [17]. The authors reported that nanoMBG–IPs entered pre-osteoblasts and enabled their differentiation into mature osteoblast phenotype and enhanced the alkaline phosphatase activity, demonstrating the osteogenic potential of the nanoMBGs, which can be used towards periodontal augmentation.

In summary, this Special Issue in *Nanomaterials* entitled "Nano-Engineering Solutions for Dental Implant Applications" compiles a series of cutting-edge research and extensive review articles demonstrating the potential of advance nano-engineering towards fabrication of the next-generation of bioactive and therapeutic dental implants that overcome challenges associated with conventional implants, while maintaining clinical translatability. The Special Issue also informs the readers of the current challenges and future directions in this domain. The Editor would like to thank all contributing authors for the success of the Special Issue. This Special Issue would not have been of such quality without the constructive criticism of the Reviewers.

**Funding:** Karan Gulati is supported by the National Health and Medical Research Council (NHMRC) Early Career Fellowship (APP1140699).

**Acknowledgments:** The Editor would like to thank all contributing authors for the success of the Special Issue. This Special Issue would not have been of such quality without the constructive criticism of the Reviewers.

**Conflicts of Interest:** The author declare no conflict of interest.

# References


1. Guo, T.; Gulati, K.; Arora, H.; Han, P.; Fournier, B.; Ivanovski, S. Race to invade: Understanding soft tissue integration at the transmucosal region of titanium dental implants. *Dent. Mater.* **2021**, *37*, 816–831. [CrossRef] [PubMed]
2. Guo, T.; Gulati, K.; Arora, H.; Han, P.; Fournier, B.; Ivanovski, S. Orchestrating soft tissue integration at the transmucosal region of titanium implants. *Acta Biomater.* **2021**, *124*, 33–49. [CrossRef] [PubMed]
3. Chopra, D.; Gulati, K.; Ivanovski, S. Understanding and optimizing the antibacterial functions of anodized nano-engineered titanium implants. *Acta Biomater.* **2021**, *127*, 80–101. [CrossRef] [PubMed]
4. Gulati, K.; Moon, H.-J.; Kumar, P.T.S.; Han, P.; Ivanovski, S. Anodized anisotropic titanium surfaces for enhanced guidance of gingival fibroblasts. *Mater. Sci. Eng. C* **2020**, *112*, 110860. [CrossRef] [PubMed]
5. Gulati, K.; Kogawa, M.; Maher, S.; Atkins, G.; Findlay, D.; Losic, D. Titania Nanotubes for Local Drug Delivery from Implant Surfaces. In *Electrochemically Engineered Nanoporous Materials: Methods, Properties and Applications*; Losic, D., Santos, A., Eds.; Springer International Publishing: Cham, Switzerland, 2015; pp. 307–355.
6. Zhang, Y.; Gulati, K.; Li, Z.; Di, P.; Liu, Y. Dental Implant Nano-Engineering: Advances, Limitations and Future Directions. *Nanomaterials* **2021**, *11*, 2489. [CrossRef] [PubMed]
7. Hosseinpour, S.; Nanda, A.; Walsh, L.J.; Xu, C. Microbial Decontamination and Antibacterial Activity of Nanostructured Titanium Dental Implants: A Narrative Review. *Nanomaterials* **2021**, *11*, 2336. [CrossRef] [PubMed]
8. Hua, S.; Bartold, P.M.; Gulati, K.; Moran, C.S.; Ivanovski, S.; Han, P. Periodontal and Dental Pulp Cell-Derived Small Extracellular Vesicles: A Review of the Current Status. *Nanomaterials* **2021**, *11*, 1858. [CrossRef] [PubMed]
9. Alali, A.Q.; Abdal-hay, A.; Gulati, K.; Ivanovski, S.; Fournier, B.P.J.; Lee, R.S.B. Influence of Bioinspired Lithium-Doped Titanium Implants on Gingival Fibroblast Bioactivity and Biofilm Adhesion. *Nanomaterials* **2021**, *11*, 2799. [CrossRef] [PubMed]
10. Jayasree, A.; Ivanovski, S.; Gulati, K. ON or OFF: Triggered therapies from anodized nano-engineered titanium implants. *J. Control. Release* **2021**, *333*, 521–535. [CrossRef] [PubMed]
11. Rahman, S.; Gulati, K.; Kogawa, M.; Atkins, G.J.; Pivonka, P.; Findlay, D.M.; Losic, D. Drug diffusion, integration, and stability of nanoengineered drug-releasing implants in bone ex-vivo. *J. Biomed. Mater. Res. Part A* **2016**, *104A*, 714–725. [CrossRef] [PubMed]
12. Gulati, K.; Li, T.; Ivanovski, S. Consume or Conserve: Microroughness of Titanium Implants toward Fabrication of Dual Micro–Nanotopography. *ACS Biomater. Sci. Eng.* **2018**, *4*, 3125–3131. [CrossRef] [PubMed]
13. Bai, L.; Chen, P.; Tang, B.; Hang, R.; Xiao, Y. Correlation between LncRNA Profiles in the Blood Clot Formed on Nano-Scaled Implant Surfaces and Osseointegration. *Nanomaterials* **2021**, *11*, 674. [CrossRef] [PubMed]
14. Chopra, D.; Jayasree, A.; Guo, T.; Gulati, K.; Ivanovski, S. Advancing dental implants: Bioactive and therapeutic modifications of zirconia. *Bioact. Mater.* **2021**. [CrossRef]
15. Chopra, D.; Gulati, K.; Ivanovski, S. Towards Clinical Translation: Optimized Fabrication of Controlled Nanostructures on Implant-Relevant Curved Zirconium Surfaces. *Nanomaterials* **2021**, *11*, 868. [CrossRef] [PubMed]
16. Secgin-Atar, A.; Aykol-Sahin, G.; Kocak-Oztug, N.A.; Yalcin, F.; Gokbuget, A.; Baser, U. Evaluation of Surface Change and Roughness in Implants Lost Due to Peri-Implantitis Using Erbium Laser and Various Methods: An In Vitro Study. *Nanomaterials* **2021**, *11*, 2602. [CrossRef] [PubMed]
17. Casarrubios, L.; Gómez-Cerezo, N.; Feito, M.J.; Vallet-Regí, M.; Arcos, D.; Portolés, M.T. Ipriflavone-Loaded Mesoporous Nanospheres with Potential Applications for Periodontal Treatment. *Nanomaterials* **2020**, *10*, 2573. [CrossRef] [PubMed]

 nanomaterials

**MDPI**

*Review*

# Dental Implant Nano-Engineering: Advances, Limitations and Future Directions

**Yifan Zhang** [1,†], **Karan Gulati** [2,†], **Ze Li** [3], **Ping Di** [2,*] **and Yan Liu** [4,*]

1    Department of Oral Implantology, Peking University School and Hospital of Stomatology & National Clinical Research Center for Oral Diseases & National Engineering Laboratory for Digital and Material Technology of Stomatology & Beijing Key Laboratory of Digital Stomatology, Beijing 100081, China; zyfbjmu2012@163.com

2    School of Dentistry, The University of Queensland, Herston, QLD 4006, Australia; k.gulati@uq.edu.au

3    School of Stomatology, Chongqing Medical University, Chongqing 400016, China; lizhecqmu@126.com

4    Laboratory of Biomimetic Nanomaterials, Department of Orthodontics, Peking University School and Hospital of Stomatology & National Clinical Research Center for Oral Diseases & National Engineering Laboratory for Digital and Material Technology of Stomatology & Beijing Key Laboratory of Digital Stomatology, Beijing 100081, China

*    Correspondence: diping2008@163.com (P.D.); orthoyan@bjmu.edu.cn (Y.L.)

†    Shared first authorship: Yifan Zhang and Karan Gulati.

**Citation:** Zhang, Y.; Gulati, K.; Li, Z.; Di, P.; Liu, Y. Dental Implant Nano-Engineering: Advances, Limitations and Future Directions. *Nanomaterials* **2021**, *11*, 2489. https://doi.org/10.3390/nano11102489

Academic Editors: May Lei Mei and Elena Ivanova

Received: 25 June 2021
Accepted: 18 September 2021
Published: 24 September 2021

**Publisher's Note:** MDPI stays neutral with regard to jurisdictional claims in published maps and institutional affiliations.



**Abstract:** Titanium (Ti) and its alloys offer favorable biocompatibility, mechanical properties and corrosion resistance, which makes them an ideal material choice for dental implants. However, the long-term success of Ti-based dental implants may be challenged due to implant-related infections and inadequate osseointegration. With the development of nanotechnology, nanoscale modifications and the application of nanomaterials have become key areas of focus for research on dental implants. Surface modifications and the use of various coatings, as well as the development of the controlled release of antibiotics or proteins, have improved the osseointegration and soft-tissue integration of dental implants, as well as their antibacterial and immunomodulatory functions. This review introduces recent nano-engineering technologies and materials used in topographical modifications and surface coatings of Ti-based dental implants. These advances are discussed and detailed, including an evaluation of the evidence of their biocompatibility, toxicity, antimicrobial activities and in-vivo performances. The comparison between these attempts at nano-engineering reveals that there are still research gaps that must be addressed towards their clinical translation. For instance, customized three-dimensional printing technology and stimuli-responsive, multi-functional and time-programmable implant surfaces holds great promise to advance this field. Furthermore, long-term in vivo studies under physiological conditions are required to ensure the clinical application of nanomaterial-modified dental implants.

**Keywords:** dental implants; osseointegration; $TiO_2$ nanotubes; surface modification; nanoparticles; antibacterial

## 1. Introduction

### 1.1. Dental Implants: History, Survival Rates and Related Complications

In the 1960s, the first preclinical and clinical studies revealed that implants made of commercially pure titanium (Ti) could achieve anchorage in bone, which shifted the paradigm in implant dentistry [1]. Direct bone-to-implant contact, known as osseointegration, formed the foundation of oral implantology [2]. In the next two decades, other materials and different shapes of implants were clinically tested, such as ceramic implants made of aluminum oxide [3], non-threaded implants with a Ti plasma-sprayed surface [4], and Ti-aluminum-vanadium implants [5]. By the end of the 1980s, commercially pure Ti became the preferred material choice of implants [6]. In the 1990s, research findings reported that significantly stronger bone response and higher bone-to-implant contact were achieved in moderately rough or microrough implant surfaces [7]. Next, sandblasted and acid-etched surfaces, as well as microporous surfaces produced by anodic oxidation, were marketed [8,9]. In the past 10 years, zirconium dioxide implants showed comparable preclinical and clinical outcomes as those of moderately rough Ti implants [10]. Currently, microrough implant surfaces are the *'gold standard'* in implant dentistry.

Dental implant treatment is highly predictable, with a survival rate of around 95% according to 10-year clinical observations [11–13]. Despite the favorable clinical results, there are still implant-related mechanical, biological and functional complications [14,15]. One major complication is peri-implantitis, which can cause bone loss around the implant, eventually leading to implant failure. According to several reviews, more than 20% of patients and 10% of implants will be affected by peri-implantitis 5–10 years after implantation [16,17].

*1.2. Current Ti Surfaces and Their Physicochemical Modifications*

Various implant characteristics influence the osseointegration of dental implants, such as implant geometry (parallel-walled, root-form, conical), thread design and implant-abutment connection (trichannel, external hexagon, internal conical hexagon), as well as implant surfaces [18,19]. Among them, the surface characteristics of dental implants are important determinants of short-term and long-term clinical performance [20–22]. Various attempts have been made to optimize implants' bioactivity by increasing their surface roughness and performing physicochemical modifications, which have reduced the incidence of implant failures and peri-implantitis.

In the 1980s, the majority of marketed implants featured turned or machined surfaces, with an estimated average roughness (Ra) of 0.5 μm to 0.8 μm. Later, a much rougher surface, Ti plasma sprayed surface (TPS), as well as surfaces coated with hydroxyapatite (HAp) and calcium phosphate (CaP), emerged, with an Ra value of >2 μm [14]. However, these TPS implants coated with HAp soon disappeared from the market, owing to the delamination of the HAp-coating, which can cause severe marginal bone resorption and even implant failure. Next, moderately rough surfaces manufactured by blasting, etching, and oxidation techniques were introduced to the market during the 1990s and early 2000s. One of the most successful surfaces in current clinical implant dentistry is the sandblasted, large-grit, acid-etched (or SLA) surface. Smooth titanium implant surfaces are formed into a primary mechanical cavity of about 200 μm using sand blasting technology, and subsequently cleaned by acid etching to form a secondary cavity of 20μm, resulting in a multi-level rough implant surface, which is conducive to bone bonding. It is worth noting that the 10-year survival rate of SLA Ti implants was reported to be 95–97% [11,23,24]. As a mainstream dental implant surface treatment technology, SLA surfaces have been frequently applied in clinical practice.

Another comparable surface is produced by using the anodic oxidation technique, which uses a Ti implant as an anode to form a thickened and roughened $TiO_2$ layer upon electrochemical treatment. This surface is characterized as isotropic, with an Ra value between 1 μm and 1.5 μm [25]. A recently published meta-analysis comparing the 10-year clinical outcomes of different dental implant surfaces (machined, blasted, acid-etched, sandblasted and acid-etched, anodized, Ti plasma-sprayed, sintered porous and micro-textured) demonstrated that the anodized implants had the lowest failure rate (1.3%, 0.2–2.4%) and minor peri-implantitis rate (1–2%) [14]. According to this article, in addition to a moderate microroughness that increases surface area and oxide thickness, anodized implants also provide additional adhesion points for proteins and cells, which contributes to the augmentation of osseointegration [26].

It is well established that osteogenic cells prefer and respond to microrough Ti surfaces, as compared to the machined surfaces [7,27]. However, additional investigations are needed to find the most optimized implant surface topography (SLA or anodized) that enhances bioactivity and osteogenesis [28]. Currently, both SLA and anodized implants present a suitable topography for clinical use. However, SLA surfaces remain the preferred choice in clinical dentistry, with many manufacturers opting for SLA over anodized implants.

*1.3. Nano-Scale Modifications and Coatings of Ti Implant Surfaces*

While micro-roughness is regarded as the 'gold standard' towards establishment of appropriate implant-bone bonding, nano-engineering is emerging as a new platform for further enhancement of the dental implant bioactivity. It has been established by several studies, in both in vitro and in vivo settings, that the nano-scale surface modification of Ti implants offers enhanced bioactivity, outperforming the clinical micro-roughness [29]. To fabricate nano-engineered Ti implant surfaces, various strategies have been employed, including plasma treatment, micro-machining, polishing/grinding, particle blasting, chemical etching and electrochemical anodization [30]. The following summarizes the various techniques utilized in the fabrication nano-engineered dental implants.

### 1.3.1. Mechanical Modification

While techniques including grinding, machining, blasting and polishing have been used in the production of rough/smooth surfaces, attrition can be used to produce nano-scale layers in order to improve mechanical characteristics, such as hardness and wettability. Machining, polishing and grid-blasting involve the shaping or removal of material surfaces and have been extensively utilized in the fabrication of controlled micro-scale surface topographies on dental implants. Further, machining can result in the deformations of crystalline grains, which increases the surface hardness. The polishing of implant surfaces has also been utilized to obtain smoother finishes. The blasting of abrasive particles against the implant surface can enhance surface reactivity. It is noteworthy that micro-scale surface texturing may be inadequate for the early establishment and subsequent maintenance of osseointegration, especially in compromised conditions. Attrition can enable nanoscale surfaces on implants, which can improve tensile properties, surface hardness and hydrophilicity.

### 1.3.2. Chemical Modification

Changing surface chemistry enables the alteration of topography, as well as the incorporation of chemical moieties that can augment bioactivity and corrosion resistance and offer surface decontamination. A simple acid or alkaline immersion can impart unique surface chemistries/topographies, which have shown promising outcomes. A few tens of nanometers or few micrometers of surface oxide have enhanced the osteogenic potential of implants. Similarly, sol-gel and chemical vapour deposition (CVD) have also been utilized to promote the bioactivity of conventional dental implants.

### 1.3.3. Physical Modification

Processes such as thermal treatment, physical vapour deposition (PVD), ion implantation and plasma treatments are included in the physical modification of implants. The involvement of either thermal, kinetic or electrical energy drives the deposition of specific molecules or ions on the implants' surface. For instance, thermal or plasma spraying has been used in the coating of hydroxyapatite, calcium silicate, alumina, zirconia and titania on Ti implants, to augment their wear and corrosion resistance and bioactivity. Further, PVD and sputtering also enable favorable biocompatibility and wear/corrosion resistance. Besides, glow discharge plasma can be used not only for surface oxidation, but also for its sterilization.

### 1.3.4. Electrochemical Modification

Anodic oxidation enables the growth of 10 nm to 40 μm of $TiO_2$ oxide layer and can also allow the adsorption and incorporation of ions from the electrolyte. Through anodic oxidation, controlled topographies can be fabricated on implants, which also offers corrosion resistance and augmented bioactivity. Alternatively, in electrochemical anodization

(EA), fluoride and water in electrolytes drive the self-ordering of controlled metal oxide nanostructures when the implant (anode) and counter electrode (cathode) are immersed, and appropriate current/voltage is supplied [31].

### 1.3.5. Biomolecule Modification

The coating of bioactive molecules, such as collagen or peptides, has been performed on dental implants to enhance bone-implant contact and peri-implant bone formation [32,33]. Additionally, inherently bioactive and antibacterial polymers such as chitosan have also been used in order to modify implant surfaces [34]. Bioactive modifications can induce specific cell and tissue responses, as well as biomimetic precipitation of CaP, via their immersion in simulated body fluid.

## 2. Nanoscale Dental Implant Modifications

### 2.1. Titania Nanotubes

### 2.1.1. Fabrication Optimization

Titania ($TiO_2$) nanotubes (TNTs) can be fabricated on Ti or its alloys via electrochemical anodization (EA) [35]. Briefly, EA involves the immersion of a Ti implant as an anode and a bare Ti/Pt electrode (cathode) inside an electrolyte (containing fluoride and water), with the supply of adequate current/voltage [31]. Under controlled and optimized conditions and the attainment of an equilibrium (characterized by metal oxide formation and dissolution), the self-ordering of $TiO_2$ nanotubes (like test-tubes, open at the top and closed at the bottom) or nanopores (nanotubes fused together, with no distance between them) on the entire surface of the implant occurs [36]. It is noteworthy that EA represents a cost-effective and scalable Ti implant surface modification strategy. Recent attempts to optimize EA to enable clinical translation include fabrication of controlled nanostructures on clinical dental implants [36], superior mechanical stability (nanopores > nanotubes) [37], and fabrication of dual micro-nano structures [38] by preserving the underlying 'gold standard' micro-roughness of dental implants [39]. It is worth noting that EA is a versatile technique that can be used to nano-engineer controlled topographies on various biomedical implants, spanning various metals and alloys, including Ti [40], Ti alloys [41], Zr [42] and Al [43]. A schematic representation of TNTs and their various characteristics and research challenges is shown in Figure 1.

### 2.1.2. Osseointegration

Attributed to improved bioactivity and the ability to load and release proteins/growth factors, TNTs are a promising surface modification strategy for orchestrating osteogenesis, as established by various in vivo investigations [29,45]. The incorporation of fluoride ions into TNTs during anodization and the mechanical stimulation of osteoblasts also contribute towards the enhancement of osseointegration [46]. Further, to ensure the successful establishment and maintenance of osseointegration, TNTs on Ti implants have loaded with various orthobiologics, including bone morphogenetic protein-2 (BMP-2) [47], platelet-derived growth factor-BB [48], alendronate [49], ibandronate [50], N-acetyl cysteine (NAC) [51], and parathyroid hormone (PTH) [52]. Lee et al. loaded TNT-modified dental mini-screws with N-acetyl cysteine [NAC, a reactive oxygen species (ROS) scavenger with anti-inflammatory and osteogenic properties], implanted them in rat mandibles in vivo and, at 4 weeks, observed significantly enhanced osteointegration at the NAC–TNT sites [51]. In another study, machined dental implant screws were modified with HF etching and EA to fabricate dual micro- and nanotubular structures, which, upon implantation in ovariectomized sheep in vivo for 12 weeks, showed significantly increased pull-out force and bone-implant contact [53]. Further, various nanoparticles, ions or coatings of Sr [54], Ta [55], La [56], and Zn [57] onto/inside TNTs have also shown upregulated osteogenic outcomes.

It is worth noting that various ions or NPs have exhibited favorable osseointegration through their use in in vitro and in vivo investigations; however, these may illicit immunotoxic reactions in a dose-dependent manner and remain the subject of active research. Further, with respect to bone-forming proteins, future investigations into the estimation of the local need for bioactive agents and the evaluation of their release inside the bone micro-environment are needed.

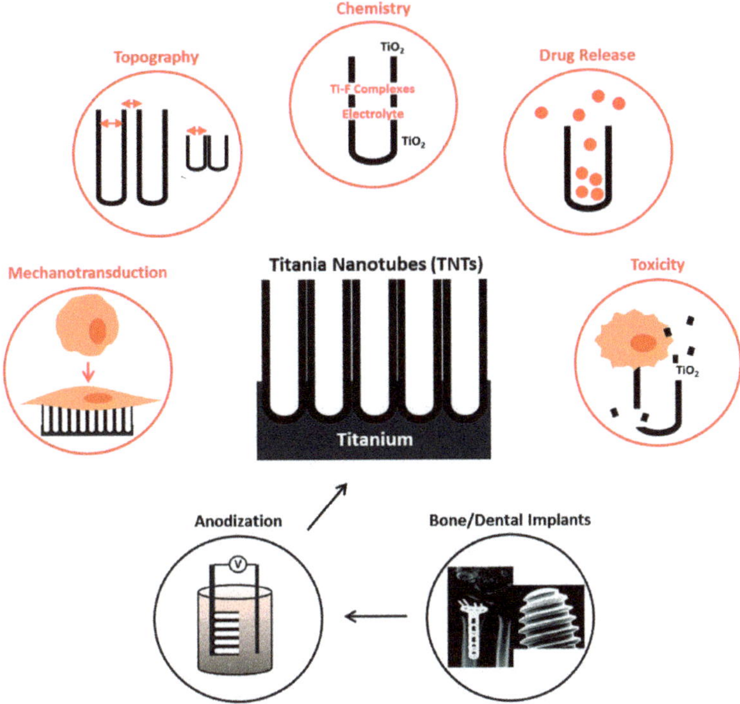


**Figure 1.** Electrochemically anodized dental implants with titania nanotubes (TNTs) for the purpose of enhanced bioactivity and local therapy. Adapted with permission from [44].

### 2.1.3. Soft-Tissue Integration (STI)

Studies relating to the use of TNTs for enhancing STI for dental implants are very limited, as reviewed elsewhere [58]. Recently, Gulati et al. reported the enhanced proliferation and adhesion of human gingival fibroblasts (HGFs) on dual-micro-nano anisotropic $TiO_2$ nanopores [38]. Further, beginning at 1 day of culture, the HGFs started to align parallel to the nanopores; and the gene expression analysis (type I collagen, type III collagen and integrin β1) indicated a wound-healing profile that promoted substrate–cell and cell–cell interactions [59]. Further, anodization combined with heat treatment has also been used to upregulate fibroblast activity. Briefly, the proliferation and adhesion of gingival epithelial cells were enhanced on heat-treated anodized Ti surfaces, which was attributed to hydrothermal treatment precipitation of hydroxyapatite crystals [60]. Alternatively, hydrothermally treated TNTs have been reported to upregulate the integrin α5 and β4 expressions of gingival epithelial cells [61], the adhesion of murine fibroblast-like NIH/3T3 cells and the expression of adhesion kinase [62], as compared to unmodified TNTs.

The biofunctionalization of TNTs has also been explored in order to enhance the functions of fibroblasts and epithelial cells towards augmenting STI. For instance, Xu et al. reported that the inhibition of human gingival epithelial cells on TNTs was reversed when the electrochemical deposition of CaP was performed on TNTs, which was attributed to the local elution of Ca and P ions [63]. Next, Liu et al. investigated the influence of bovine serum albumin (BSA) loading inside TNTs on HGF functions [64]. Unmodified TNTs promoted early HGF adhesion and COL-1 secretion; however, BSA-TNTs enhanced early HGF adhesion, while suppressing late proliferation and COL-1 secretion. It is interesting that contradictory behaviors among bioactive coatings on TNTs have been reported and further in-depth investigation into the influence of these modifications on the STI performance is needed. Furthermore, the local elution of fibroblast growth factor-2 (FGF-2, immobilized on Ag nanoparticles) from TNTs effectively enhanced the proliferation, adhesion and extracellular matrix formation in the cultured HGFs [65]. Augmented proliferation, adhesion, and expression of VEGF and LAMA1 genes in vitro was observed, which were pronounced after the loading of 500 ng/mL of FGF-2.

### 2.1.4. Antibacterial Functions

The local release of therapeutics from TNTs has been widely explored towards optimizing the loading and local elution of potent antibacterial agents [30]. It is noteworthy that within minutes of implantation, saliva proteins adhere to the dental implant, forming a pellicle, and early colonizers such as *Streptococci* adhere to these pellicles within 48h [66]. This can be followed by secondary colonizers, including *Fusobacterium nucleatum*, *Aggregatibacter actinomycetemcomitans* and *Porphyromonas gingivalis* [67]. These bacteria can further lead to peri-implantitis [68]. Once a biofilm is established, the routine administration of antibiotics is insufficient and, hence, local therapy using dental implants has been proposed. Further, TNTs can enhance bacterial adhesion due to their nano-scale roughness, increased number of dead bacteria and amorphous nature. Hence, the synergistic antibacterial functions of TNT-modified dental implants are needed to prevent bacterial colonization and implant failure. Further, the size and crystal structure of TNTs influences bacterial adhesion properties. Ercan et al. investigated the influence of the size and the heat treatment of TNTs on their antibacterial effect and reported that heat-treated and 80 nm diameter TNTs exhibit strong antibacterial effects [69]. Similarly, when comparing 15, 50 and 100 nm diameter TNTs, the lowest number of adherent bacteria were reported on the smallest-diameter TNTs [70]. Further, annealed TNTs show the best bactericidal response, as reported by Mazare et al. [71] and Podporska-Carroll et al. [72].

Various commonly prescribed antibiotics including Gentamicin [73], Vancomycin [74], Minocycline, Amoxicillin, Cephalothin [75], Cefuroxime [76] and Cecropin B [77] have been incorporated inside TNT-modified Ti implants to enable local antibacterial functions. Further, to target methicillin-resistant *Staphylococcus aureus* (MRSA), antimicrobial peptides (AMPs) such as HHC-36 have been loaded inside TNTs to achieve a bactericidal effect of almost 99.9% against MRSA [78]. Biopolymer coatings have also been applied to antibiotic-loaded TNTs to: (a) control drug release, (b) promote bioactivity, and (c) harness the inherent antibacterial property of biopolymers in order to provide long-term antibacterial functions. As a result, bare/drug-loaded TNTs have been modified with chitosan [79], polydopamine [80], silk fibroin [81] and PLGA (poly(lactic-co-glycolic acid)), which exhibited synergistic bioactivity and antibacterial enhancements. In addition, various antibacterial ions and nanoparticles (NPs), such as Ag [82], Au [83], Cu [84,85], B, P, Ca [86], Ga [87], Mg [88], ZnO [89], etc., have also been immobilized on or incorporated inside TNTs, with or without the use of hydroxyapatite or biopolymers, using techniques such as micro-arc oxidation, chemical reduction, photo-irradiation, spin-coating, and sputtering.

Multiple synergistic therapies, including osseointegration, immunomodulation, soft-tissue integration and antibacterial functions can also be enabled using nano-engineered Ti with TNTs. For instance, TNTs modified by Ag via plasma immersion ion implantation (PIII) showed excellent antibacterial effects against *P. gingivalis* and *A. actinomycetemcomitans*, while enhancing the bioactivity of epithelial cells and fibroblasts in vitro and reducing inflammatory responses in vivo [90]. Similarly, hydrothermally doped Mg-TNTs exhibited up-regulated osteoprogenitor cell adhesion and proliferation (without cytotoxicity) and suppressed osteoclastogenesis, while showing long-lasting antimicrobial effects against methicillin-susceptible *S. aureus* (MSSA), methicillin-resistant *S. aureus* (MRSA) and *E. coli* [88].

### 2.1.5. Immuno-Modulation

The modulation of the host immuno-inflammatory response is crucial to the timely establishment of osseointegration. Hence, attempts have been made to obtain immunomodulatory functions from modified TNTs [44]. These include the influence of physical/chemical characteristics and the local elution of anti-inflammatory drugs from TNTs. The influence of Ti nanotopography on immune cells, including macrophages, monocytes and neutrophils, has supported the attenuation of inflammation [91,92]. Clearly, the presence of nano-scale cues controls macrophage adhesion and inflammatory cytokine production. Similarly, in vitro cultures of such cells on TNTs have also established the influence TNTs nanotopography on immuno-inflammatory responses [38].

Smith et al. reported reduced functions (viability, adhesion, proliferation and spreading) of immune cells on TNTs, as compared with bare Ti [93]. Alternatively, other studies have shown enhanced nitric oxide and the absence of foreign-body giant cells on TNTs [93,94]. With respect to the nanotube diameters, inconsistent results have been obtained, with some studies indicating 60–70 nm diameters as the most immuno-compatible [93,95]. Further, Ma et al. compared the functions of monocytes/macrophages on nanotubes and polished Ti, and reported post-attachment stretching inhibition (repulsed adhesion), enhanced M2 phenotype (wound healing) and suppressed M1 phenotype (pro-inflammatory) polarization for TNTs anodized at 5V [96]. Furthermore, to understand the mechanism behind selective immunomodulation due to TNTs, Neacsu et al. reported that this effect is attributed to the suppression of the phosphorylation of MAPK (mitogen-activated protein kinase) signaling molecules (p38, ERK1/2, and JNK) on TNTs [97]. More recently, using 50 and 70 nm diameter anodized anisotropic $TiO_2$ nanopores, we showed that macrophage proliferation was significantly reduced on the 70 nm nanopores [38]. Further, the spread of macrophage on nanopores indicated an oval morphology, which was suggestive of an inactivated state.

The local elution of potent drugs, such as non-steroidal anti-inflammatory drugs (NSAIDs), bypasses the limitations associated with systemic administration (delayed bone healing and toxicity). These drugs have been loaded inside TNTs for the purpose of local release. Briefly, Ibuprofen [98], Indomethacin [99], Dexamethasone [100], Aspirin [101], Sodium naproxen [102], Quercetin [103], Enrofloxacin [104], Propolis [105] and immunomodulatory cytokines [106] have been successfully loaded and locally eluted from TNTs in vitro. Further, to achieve substantial loading and the delayed/controlled release of anti-inflammatory drugs, approaches including biopolymer coating on drug-loaded TNTs [107,108], polymeric micelle encapsulation of drugs prior to loading [109], the periodic tailoring of TNTs [110], the chemical intercalation of drugs inside TNTs [111] and trigger-based release [112,113] have been reported for TNT-based Ti implants. Additionally, metal ions and nanoparticles (NPs), including Au [83], Ag [114] and Zn [115] have also been incorporated on/inside TNTs to impart synergistic immunomodulatory functions with antibacterial or osteogenic activity. More recently, super-hydrophilic TNTs were fabricated via anodization and hydrogenation, and significantly reduced macrophage proliferation; upregulated M2 and downregulated M1 surface markers were exhibited on the modified TNTs, translating into effective immunomodulation and wound healing functionality [116]. It is also noteworthy that various in vivo tests of TNT modifications intended for use in various therapies, including antibacterial [30], osteogenic [29] or anti-cancer [117] applications, have established the immuno-compatibility of TNTs.

### 2.2. Nanoparticles

NPs can enable multiple therapies at the surface of dental implants, including antibiofouling, osseo- and soft-tissue integration and immunomodulation [118,119]. While NPs have been utilized towards controlled therapies for periodontal, orthodontic, endodontic and restorative treatments, this section will primarily focus on the uses of NP-modified Ti dental implants in implant-based local therapy [120]. As reported in the previous section, NP-doped TNTs have also been widely explored in the context of the controlled release of NPs, which aims to strike a balance between therapy and toxicity.

### 2.2.1. Silver

Ag NPs are one of the most widely used dentistry restoration and dental implant doping choices due to their outstanding antimicrobial properties [121]. Ag adheres to the bacterial cell wall and the cytoplasmic membrane electrostatically, which causes structural disruption [120]. This results in extensive damage to bacterial DNA, proteins and lipids, resulting in the inhibition of bacterial growth/viability and effective bactericidal action. Besides, Ag NPs can also stimulate osteogenesis and soft-tissue integration, making them an ideal choice for dental implant surface modification [122]. For instance, dental abutments modified with Ag NP suspension prevented *C. albicans* contamination, in comparison with the controls of unmodified abutments [123]. Further, citrate-capped Ag NPs offered bactericidal effects against *S. aureus* and *P. aeruginosa* [124]. Ti implants deposited with Ag NPs using anodic spark deposition have also been co-doped with Si, Ca, P and Na ions, to offer synergistic antibacterial (*S. epidermidis*, *S. mutans* and *E. coli*) and osteogenic (human osteoblast-like cells, SAOS-2) functions [125]. Similarly, to confirm that the used dosage of Ag NPs is safe, a culture of HGFs on Ag NPs/Ti was performed in vitro and the results confirmed no adverse effects [126]. Further, Ag NPs have also been immobilized on Ti implants pre-modified with hydroxyapatite [127], hydrogen titanate [128], chitosan/hyaluronic acid multilayer [129], nanoporous silica coatings [130], Pt and Au [131], and sandblasting and acid-etching [132] in order to achieve superior antibacterial and bioactivity effects. However, while Ag NPs offer effective antimicrobial action, they may cause cytotoxicity via the release of free Ag+ ions, ROS production, transport across blood-brain-barrier, and inflammation [120]. In a manner that is also applicable to other NPs discussed below, the toxicity of NPs depends on their chemical composition, surface charge, size and shape [133]

### 2.2.2. Zinc

Like Ag NPs, Zn/ZnO NPs are not only antimicrobial but also osteogenic, hence their use in the modification of dental implants [118]. Zn is an essential element in all biological tissues and offers antibacterial effects against a wide range of microbes; however, its aggregation can cause cytotoxicity in mammalian cells [134]. To demonstrate its effectiveness against oral biofilms, Kulshrestha et al. reported that graphene/zinc oxide nanocomposite showed a significant reduction in biofilm formation [135]. Further, Hu et al. incorporated Zn into $TiO_2$ coatings on Ti implants through plasma electrolytic oxidation and observed superior bactericidal and bone-forming effects [136]. In 2017, Li et al. synthesized N-halamine labeled Silica/ZnO hybrid nanoparticles to functionalize Ti implants to enable antibacterial functions [137]. The hybrid NP-modified Ti exhibited excellent antibacterial activity against *P. aeruginosa*, *E. coli* and *S. aureus*, without any cytotoxicity against MC3T3-E1 preosteoblast in vitro. Recently, selective laser-melted porous Ti was biofunctionalized using Ag and Zn NPs via plasma electrolytic oxidation and tested against methicillin-resistant *Staphylococcus aureus* (MRSA) [138]. The results confirmed that 75% Ag and 25% Zn fully eradicated both adherent and planktonic bacteria in vitro and ex vivo. Further, Zn-modified Ti (0% Ag) enhanced the metabolic activity of preosteoblasts, indicating its suitability for dual osteogenic and antibacterial implant modification. Further, it is worth noting that ZnO NPs may cause cell apoptosis or necrosis and DNA damage [139]

### 2.2.3. Copper

CuO NPs offer advantages over Ag NPs, including cost-effectiveness, chemical stability and ease of combining with polymers, which makes them an attractive choice for biomaterial applications [140]. Further, Cu NPs have antibacterial, osteogenic and angiogenic properties [141], and have been applied towards the enhancement of both the bioactivity and the antimicrobial properties of Ti dental implants [142]. More recently, van Hengel et al. incorporated varying amounts of Ag and Cu NPs into $TiO_2$ coating on additively manufactured Ti–6Al–4V porous implants via plasma electrolytic oxidation [143]. Further, 75% Ag and 25% Cu caused the eradication of all bacteria in a murine femora model ex vivo, while only Cu NP-modified implants (0% Ag) augmented the metabolic activity of pre-osteoblastic MC3T3-E1 cells in vitro. Alternatively, Ti-6Al-7Nb alloy dental implants were coated with Cu NPs and cultured with *P. gingivalis* in vitro, and the findings suggested that Cu NPs can aid in local infection control around implants [144]. In 2020, Xia et al. reported the use of plasma immersion ion implantation and deposition (PIIID) technology to modify Ti implants with C/Cu NPs co-implantation [145]. The modified implants displayed superior mechanical and corrosion resistance properties and enhanced the antibacterial performance of Ti implants (against *S. aureus* and *E. coli*) without causing cytotoxicity (to mouse osteoblast cells) in vitro. In a more dental implant setting, Cu-deposited (micro-/nanoparticles) commercially pure (cp) grade 4 Ti discs (via spark-assisted anodization) were shown to exhibit dose-dependent antibacterial effects against peri-implantitis-associated strain *P. gingivalis* [146]. Similarly, micro-arc oxidation Cu NP-doped $TiO_2$ coatings showed excellent antibacterial activity, while augmenting the proliferation and adhesion of osteoblast and endothelial cells in vitro [85]. The interaction of Cu NPs with microbes and the bioactivity and toxicity evaluations of Cu NPs can be found elsewhere [147].

### 2.2.4. Zirconia

Zirconium (Zr) and zirconia ($ZrO_2$) are rising as dental implant material choices due to their biocompatibility, corrosion resistance and superior mechanical properties [42]. It is established that $Zr^{4+}$ ions can interact with negatively charged bacterial membranes and cause cell damage and death [148]. Furthermore, Zr-based implants have been electrochemically anodized in order to fabricate controlled $ZrO_2$ nanostructures, including nanotubes and nanopores, which can augment implant bioactivity due to their nanoscale roughness [42,149,150]. For instance, anodized Zr cylinders were placed in rat femur osteotomy models in vivo, and accelerated bone formation was obtained, in comparison with the controls of unmodified Zr [151]. Further, Indira et al. reported the dip coating of Zr ions into anodized TNTs to form $ZrTiO_4$ over the nanotubes, which exhibited enhanced bioactivity (HAp formation in Hank's solution in vitro) and corrosion resistance [152]. Similarly, the application of a Zr film on a TiNi alloy via plasma immersion ion implantation and deposition (PIIID) augmented its corrosion resistance [153]. Nanotube formation has also been extended to TiZr alloys. For instance, Grigorescu et al. used two-step EA to fabricate nanotubes of varied diameters and observed an increase in hydrophilicity with reduction in diameter [154]. Further, the smallest nanotube diameters exhibited the highest antibacterial effects against *E. coli*. While $ZrO_2$/Zr is extensively used as a dental implant material, the leaching of Zr NPs may initiate cytotoxicity. For instance, the application of both Zr and $TiO_2$ NPs in a dose-dependent fashion could lead to osteoblast morphology changes and apoptosis, affecting both osteoblast differentiation and osteogenesis at high dosages [155].

### 2.2.5. Silica

Si/$SiO_2$ NPs have been utilized in biomedical applications, including biosensing and drug delivery [156]. In dentistry, Si NPs have been used as dental filler, for tooth polishing and in hypersensitivity treatments [157]. Varied concentrations of $SiO_2$ NPs within HAp fabricated on Ti hydrothermally were analyzed for bioactivity and cytotoxicity [158]. The results confirmed homogenous distribution of $SiO_2$ NPs on hexagonal HAp crystals and favourable biocompatibility with human osteoblast-like cells in vitro. Furthermore, in order to achieve superior bioactivity, a protein-based Si NP coating (via the genetic fusion of recombinant MAP with the R5 peptide derived from a marine diatom *Cylindrotheca fusiformis*) was performed on Ti implants to explore their osteogenic potential [159]. Briefly, the assembly of Si NPs augmented the in vitro osteogenic cellular behaviors of preosteoblasts and bone tissue formation in vivo (calvarial defect model). Further, Si NP coatings were performed on Ti-based implants to enable the local elution of potent therapeutics in order to achieve antibacterial [160] and osseointegration [161] functions. For example, the enhancement of osteogenesis via immunomodulation by Si NP-doped chitosan-modified TNTs has been reported [162]. Furthermore, 100 nm mesoporous Si NPs were loaded with dexamethasone (an anti-inflammatory drug) in order to achieve its local elution, which demonstrated favourable macrophage cytocompatibility. Additionally, local release of dexamethasone modulated M2 macrophage polarization which supported osteogenesis. Immuno-toxicity evaluations of Si NPs have been reviewed elsewhere [163].

*2.3. Hydroxyapatite*

HAp is biocompatible, non-toxic and non-immunogenic, and has been widely used as a coating material in the modification of dental implants [164]. In the late 1980s, plasma spray coating of HAp became obsolete due to the delamination of the HAp-coat, which could cause severe marginal bone resorption and incompatibility with antibiotic incorporation [165,166]. Later, some alternative coating techniques, such as electrochemical deposition [167], electrophoretic deposition and electrospray deposition [168] made it possible to combine HAp coatings with antibiotics to achieve both enhanced bioactivity and antibacterial effects. Moreover, due to their special crystalline structure and positive-charged surface, the ability of substituted HAp to immobilize proteins and growth factors through noncovalent interactions has offered new possibilities for the preparation of hybrid coatings that accelerate bone healing [169].

Geuli et al. reported the use of drug-loaded HAp nanoparticles on Ti implants through single-step electrophoretic deposition. The release profiles of the gentamicin sulfate (Gs)-HAp and ciprofloxacin (Cip)-HAp coatings demonstrated a prolonged release of up to 10 and 25 days, respectively. In vitro antibacterial tests of the Cip and Gs-HAp coatings showed the efficient inhibition of *P. aeruginosa* [170]. Liu et al. applied a nano-silver-loaded HAp (Ag-HAp) nanocomposite coating to a Ti6Al4V surface by laser melting. They found that the coating containing 2% Ag showed excellent biocompatibility and antibacterial ability, which was conducive to the deposition of apatite on the implant's surface [171]. Further, Zhao et al. compared the application of magnesium (Mg)-substituted and pure HAp coatings in the osseointegration of dental implants in vitro and in vivo [172]. They observed increased cell proliferation, higher alkaline phosphatase activity and enhanced osteocalcin production in the Mg-HAp group in vitro. In vivo testing using a rabbit femur model revealed a slightly higher bone implant contact for the Mg-Hap-coated implants at 2 weeks post-implantation, whereas no significant differences were seen after 4 and 8 weeks. Recently, Vu et al. coated Ti and Ti6Al4V implants with a ternary dopant coating, which used commercial HA powder doping with 0.25 wt% ZnO to induce osteogenesis, 0.5 wt% $SiO_2$ to induce angiogenesis, and 2 wt% $Ag_2O$ to control infection [173]. The Zn/Si/Ag-HAp coatings resulted in better antibacterial properties in vitro against *E. coli* and *S. aureus*. Meanwhile, the Zn/Si/Ag-HAp implants with higher shear modulus augmented bone mineralization and total bone formation compared to pure HAp implants in rats by week 5, while no evidence of angiogenesis or antibacterial properties, as demonstrated in vivo (Figure 2). Other ionic substitutes, such as Si, $F^-$, $Sr^{2+}$ have also been utilized in combination with HAp to accelerate bone healing and improve bioactivity [164].

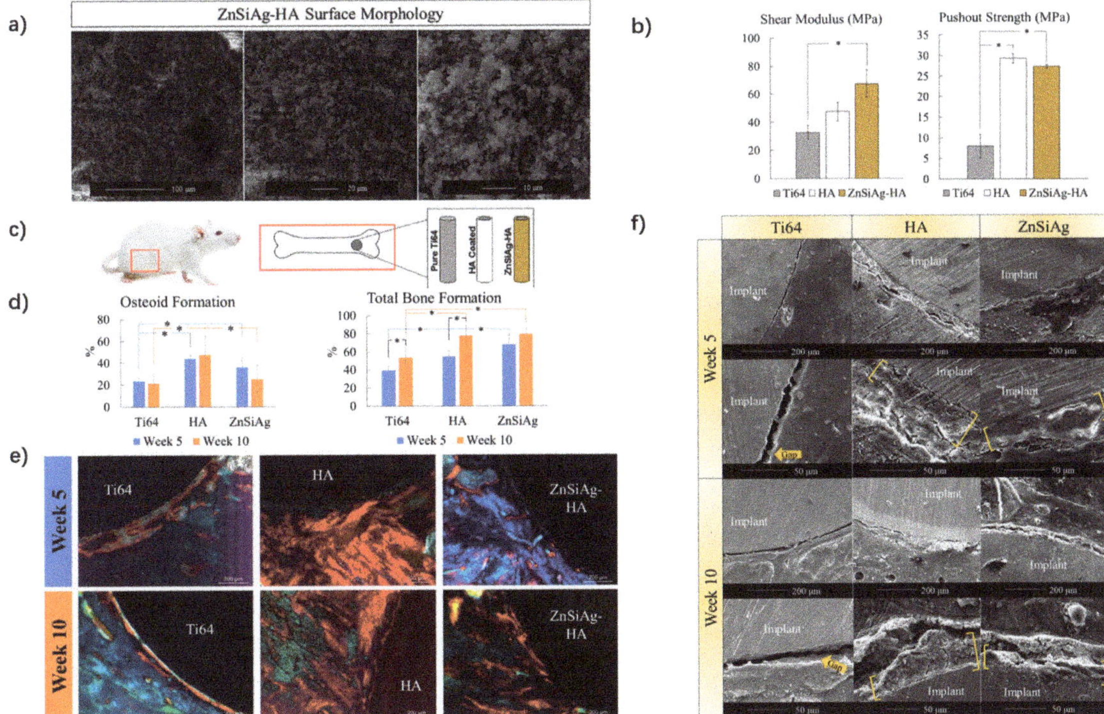

**Figure 2.** Mechanical and biological properties of Zn/Si/Ag-HAp coating implants. (**a**) Scanning electron microscopy (SEM) showing roughness and porosity of the implant surface. (**b**) Shear modulus and pushout test of implants from harvested femurs. (* $p < 0.05$). (**c**) The in-vivo bilateral model rat's distal femur. (**d**) Total osteoid formation and bone formation in % around implant at week 5 and week 10. (* $p < 0.05$). (**e**) Modified Masson-Goldner trichrome staining 5 weeks and 10 weeks after implantation. (**f**) SEM images of implant interface for all compositions at week 5 and week 10. Adapted with permission from Vu, et al. Coatings are outlined with yellow brackets [173].

### 2.4. Biopolymers

Polymeric layers are a promising strategy for the enhancement of bioactivity and controlled release of potent drugs. The use of biopolymers, such as chitosan, cellulose and silk fibroin-based nanomaterials provide the synthetic implant surface coatings with superior bioactivity and antibacterial functions. A combination of implant surface treatment with polymer-incorporated antibiotics, drugs or biomolecular delivery systems has shown promising results when compared to polymers and drugs alone. Here, we discuss and detail the application of the two most commonly utilized polymers in dental implants.

Chitosan is an inherently antibacterial and non-toxic polysaccharide that is widely applied in wound healing, tissue engineering and drug delivery [174]. Moreover, nanofibrous chitosan provides a more favorable microenvironment for cellular activity than bulk chitosan, which can be attributed to the way its unique morphological characteristics mimic extracellular matrices [175,176]. Benefiting from its positive surface charge, chitosan is also antibacterial and ruptures negatively charged bacterial cells [177]. Many studies have been conducted on the use of chitosan for the fabrication of antibacterial medical implants [178–180]. Furthermore, when chitosan is incorporated in the form of nanoparticles on the implant surface, it shows a high loading rate and the capacity for sustained drug release. Chitosan [181], chitosan/gelatin [182], chitosan/alginate [107] and chitosan/graphene oxide [183] have also been utilized in the coating of implants.

Song et al. used chitosan to wrap Semaphorin 3A (Sema 3A), a proven osteoprotection molecule, and to immobilize oxidized Ti surface. A burst release of Sema 3A was maintained for more than 2 weeks [184] (Figure 3a–e). Further, Mattioli-Belmonte et al. prepared a ciprofloxacin-loaded chitosan nanoparticle-based coating on Ti substrates for the in situ release of the antibiotic for post-operative infections. According to the in vitro results, this coating inhibited the growth of *Staphylococci aureus* and did not impair the viability, adhesion or expression of MG63 osteoblast-like cells [185]. Ma et al. applied chitosan-gelatin (CS/G) coatings to a Ti surface and evaluated its biological *performance* in vitro and in vivo [186]. The CS/G coatings supported MC3T3-E1 cell attachment, migration and proliferation. In addition, Micro-CT and histomorphometrical analysis revealed new bone formation around CS/G implants at 8 and 12 weeks, while the majority of the coatings were degraded at 12 weeks (Figure 3f,g).

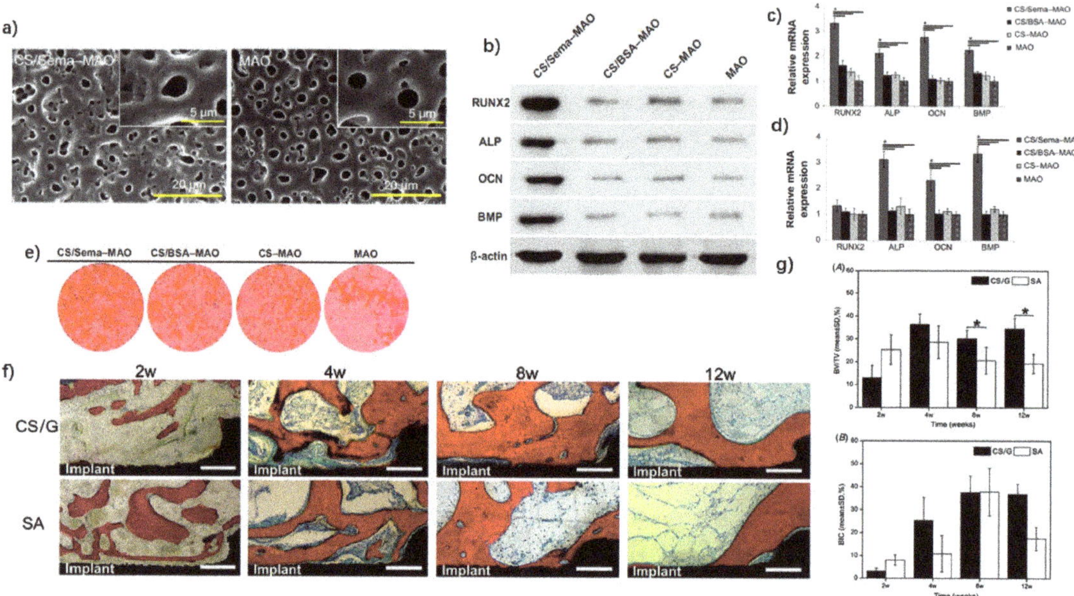

**Figure 3.** Applications of nano-chitosan in implant surface coatings. (**a**) SEM images showing surface morphology of chitosan–semaphorin 3A-microarc oxidation (CS/Sema-MAO) and MAO. (**b–e**) Osteogenic-related gene and protein expression of MG63 cells cultured on CS/Sema-MAO surface and control surface at day 3 (**b,c**), day 7 (**d**) and day 21. * $p < 0.05$ vs CS/Sema–MAO, CS–MAO, and MAO. (**e**) Alizarin red study. (**f**) Basic fuchsin and methylene blue staining showing histological appearance around chitosan–gelatin (CS/G) coatings and sandblasted/acid-etched (SA) implants at weeks 2, 4, 8, and 12. (**g**) Histomorphometrical variables evaluating the gap level of CS/G and SA implants within the region of interest: (A) the percentage of bone volume fraction (BV/TV) around the implant within 300 μm; and (B) the ratio of BIC. * $p < 0.05$. Adapted with permission from [184,186].

A recent review focused on the toxicity/safety concerns in zebrafish models, and described the toxicity of different chitosan nanocomposites [187]. According to Hu et al., 200 nm chitosan nanoparticles (Ch NPs) were able to cause 100% mortality to the embryos and severe teratogenic deformities at 40 mg/L, compared to the 340 nm particles [188]. By contrast, both Wang et al. [189] using 200 mg/L Ch NPs, and Abou-Saleh et al. [190], using 100–150 nm Ch NPs, failed to induce significant mortality or teratogenic phenotypes, even at 200 mg/L. The contradictory results suggest that more cytotoxicity and toxicity investigations of Ch NPs are required to advance this field.

### 2.5. Carbon Composites

Graphene, obtained through the physicochemical exfoliation of graphite, provides several advantages, such as its low cost and safe preparation. There are several derivative forms of graphene, such as graphene oxide (GO), which is highly oxidative, and reduced GO (rGO), which is prepared via the chemical or thermal reduction of GO. It has been reported that pure graphene shows a certain degree of cytotoxicity [191]. However, whether GO causes cytotoxicity remains controversial, as some studies have shown that GO does not initiate cytotoxicity [192–194]. However, others have revealed that micro-sized GO (and not nano-sized GO) can induce high levels of cytotoxicity [195]. It is worth noting that the main difference among all the carbon-based materials is the hybridization type of their carbon atoms [196,197]. In a study by Wang et al., the hybridization type of carbon atoms (sp2 or sp3) was the critical point in determining their biological properties. The larger amount and smaller size of dispersing sp2 domains regulated the behavior of cells by affecting the amount and properties of the adsorbed proteins [198].

Recently, Gu et al., attempted to improve the adhesion strength of graphene on the surface of Ti substrate through a thermal treatment and observed enhanced antibacterial effects (*E. coli* and *S. aureus*), cell adhesion, proliferation and osteogenesis in vitro (human adipose-derived stem cells and human bone marrow mesenchymal stem cells) and in vivo (dorsal subcutaneous area of eight-week-old male BALB/c nude mice) on graphene-coated Ti implant surfaces after dry heating treatment [199]. More recently, Wei et al. synthesized a new Ti biomaterial containing graphene (Ti-0.125G) by using the spark plasma sintering technique. Bioactivity (human gingival fibroblasts) and antimicrobial (*Streptococci mutans*, *Fusobacterium nucleatum*, and *Porphyromonas gingivalis*) findings revealed that graphene modification upregulated both functions [200].

Unlike hydrophobic graphene, GO is a hydrophilic derivative form with the addition of bounded oxygen atoms. Due to the large number of carboxyl and hydroxyl groups containing active functional groups on its surface, it is easy to perform biomaterial functionalization using GO [201,202]. Wang et al. assembled GO coatings on a laser microgroove Ti alloy [203]. The in vitro bioactivity results showed superior adhesion, proliferation, differentiation and osteogenic capability compared with bare Ti implant, due to the wettability and apatite formation induced by the GO coating. Based on the findings of Li et al., the FAK/P38 signaling pathways were proven to be involved in the enhanced osteogenic differentiation of bone marrow mesenchymal stem cells, accompanied by the upregulated expression of focal adhesion (vinculin) on the GO-coated surface [204]. It is noteworthy that GO can cause direct damage to bacterial cell membranes through its sharp structure and its destructive extraction of lipid molecules, together with ROS reactions [205]. Besides, the extensive two-dimensional (2D) honeycomb structure can be loaded with biomolecules or drugs in order to enable local therapy [195].

Compared to GO, rGO possesses structural defects that enhance molecular interactions. For instance, Kang et al. fabricated rGO-coated Ti substrates through meniscus-dragging deposition and investigated their biological behaviors [206]. They cultured human mesenchymal stem cells on the Rgo-Ti substrates and found superior bioactivity and osteogenic potential via a cell counting kit-8 assay, an alkaline phosphatase activity assay and alizarin red S staining, suggesting that these graphene derivatives had potent applications in dental implants. Further, Rahnamaee et al. assembled both chitosan nanofibers (CH) and reduced graphene oxide (rGO) onto TNTs [183]. This multifunctional coating offered the synergistic effects of CH and rGO against both long-term and short-term antibacterial activity, promoted osteoblast cell viability, prolonged antibiotic release profile and inhibited bacterial biofilm formation.

Another 2D carbon-based nanomaterial is graphdiyne (GDY), which has been predicted to become the most stable carbon derivative form [207]. Compared with graphene, the particular sp and sp2 hybridized carbon atoms of GDY offer superior electrical conductivity and enhanced catalytic effects, and exhibited enhanced biocompatibility and stability in some in vivo studies [208,209]. Further, Wang et al. successfully assembled

GDY onto TiO$_2$ to synthesize a TiO$_2$/GDY composite by using electrostatic force [210]. Its antibacterial effect against *Methicillin-resistant Staphylococcus aureus* (MRSA) was prolonged with sustained ROS release, which prevented the formation of biofilm. A mouse implant infection model further demonstrated excellent sterilization and bone regeneration effects in vivo (Figure 4). The reviewed research works on various modifications for dental implants with their main advantages and drawbacks are summarized in Table 1.

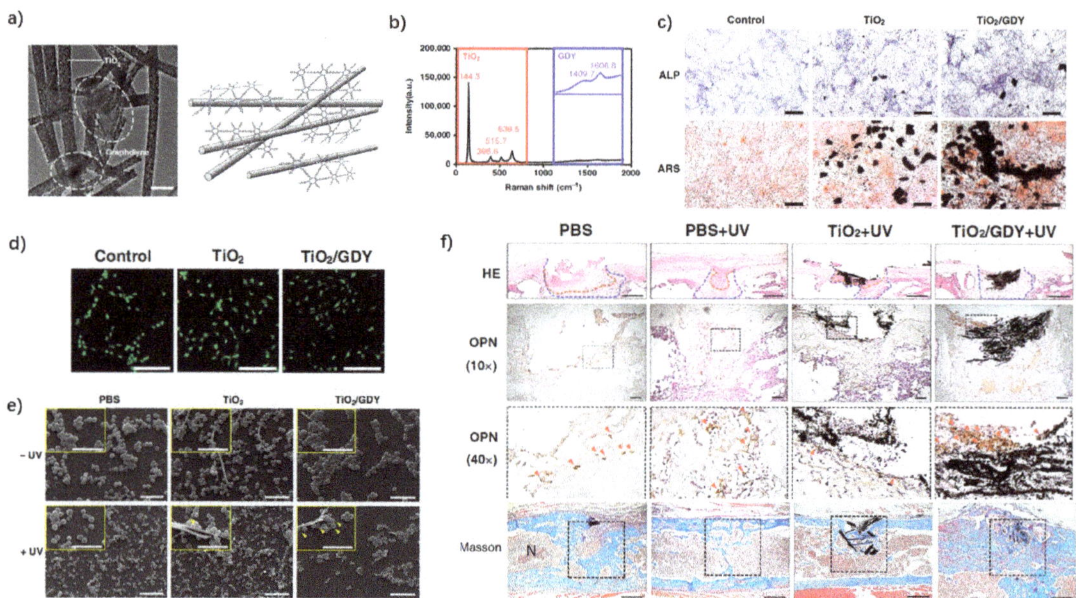

**Figure 4.** Graphdiyne (GDY)-modified TiO$_2$/Ti implants. (**a**) Transmission electron microscopy image of TiO$_2$/GDY. (**b**) Raman spectra of TiO$_2$/GDY. (**c**) Osteogenic effects of TiO$_2$/GDY and TiO$_2$ in vitro: alkaline phosphatase activity (upper labeled ALP) and alizarin red S staining (lower-labeled ARS) on day 14. (**d**) Live/dead staining for MC3T3-E1 cells cultured with nanofibers (scale bar = 50 μm). (**e**) SEM images of MRSA biofilms after exposure to different conditions. yellow arrows in the magnified inset images show holes on the bacterial surface; scale bar = 5 μm (upper), 10 μm (below). (**f**) Hematoxylin and Eosin (HE) staining and immunohistochemical staining of infected tissues after 5 days; Masson staining for bone formation after 4 weeks. In vivo implant infection model: femur bone defect with MRSA infection in 8-week-old mouse. Adapted with permission from [210].

**Table 1.** Summary of nanoscale dental implant modifications and their key features.

| Implant Modification | Fabrication | Advantages | Drawbacks | Main Reference |
|---|---|---|---|---|
| $TiO_2$ nanotubes | • Electrochemical anodization | • Enhanced osseointegration | | [29,45] |
| | | • Soft-tissue integration: enhanced proliferation and adhesion of human gingival fibroblasts | | [38] |
| | | • Local release of therapeutics | | [30] |
| | | • Immunomodulatory functions | | [94,95] |
| Ag NPs | • Anodic spark deposition | • Outstanding antimicrobial properties<br>• Stimulation of osteogenesis and soft-tissue integration | • Toxicity: via release of free Ag+ ions | [120–122] |
| Zn/ZnO NPs | • Plasma electrolytic oxidation | • Antibacterial properties<br>• Osteogenic effects | • Cytotoxicity: ZnO NPs may cause cell apoptosis or necrosis and DNA damage | [118,134,136] |
| CuO NPs | • Plasma electrolytic oxidation<br>• Plasma immersion ion implantation and deposition (PIIID)<br>• Micro-arc oxidation | • Cost-effectiveness<br>• Chemical stability<br>• Ease of mixing with polymers<br>• Antibacterial effects<br>• Osteogenic properties<br>• Angiogenic properties | • Toxicity | [140,145,147] |
| $ZrO_2$ nanostructures | • Electrochemical anodization<br>• Plasma immersion ion implantation and deposition (PIIID) | • Enhanced bioactivity<br>• Corrosion resistance<br>• Antibacterial effects | • Cytotoxicity: dose-dependent, affecting both osteoblast differentiation and osteogenesis at high dosages | [152,153,155] |
| Si/SiO$_2$ NPs | • Hydrothermal method | • Biocompatibility with human osteoblast-like cells in vitro<br>• Antibacterial properties<br>• Immunomodulation | | [158,160,162] |
| Hydroxyapatite | • Electrochemical deposition<br>• Electrophoretic deposition<br>• Electrospray deposition | • Biocompatibility<br>• Non-toxicity<br>• Non-immunogenicity<br>• Prolonged drug release | | [170,173] |
| Chitosan | • Microarc oxidized and silane glutaraldehyde coupling | • Antibacterial properties<br>• High loading rate and sustained drug release ability | | [185,188] |
| Carbon composites | • Dry heating treatment<br>• Meniscus-dragging deposition | • Low cost<br>• Safer preparation<br>• Enhanced antibacterial effects<br>• Bioactivity in vitro and in vivo<br>• Highly efficient drug loading and therapy | • Cytotoxicity: remains controversial | [184,196,200,205,211] |

## 3. Research Challenges

The use of various nano-engineering strategies to enhance the bioactivity and therapeutic performance of dental implants shows great promise; however, many research gaps remain unaddressed with respect to the clinical application of nano-engineered dental implants. Next, we take a close look at the key challenges that must be investigated in order to bridge the gap between nano-engineering dental implant research and its clinical translation.

- The key physical, chemical and mechanical characteristics of the implant and its surface modification are crucial towards the understanding and prediction of cell response and therapeutic efficacy [211]. These also include appropriate corrosion resistance and electrochemical stability. Hence, testing under masticatory loading conditions, under varied pH and physiological conditions (matching healthy and compromised conditions, such as infection and inflammation) for extended durations are essential for nano-engineered coatings of implants. Any delamination or release of nanoparticles from implant modifications can initiate a cytotoxic response, and only a few attempts have been made to ensure the successful fabrication of robust nano-engineered coatings on commercial implants with appropriate mechanical stability [36,37].

- Nano-engineered implants can enable the local elution of potent drugs, proteins or therapeutic nanoparticles/ions. While the concept of local drug release has gained attention, its investigation has largely remained restricted to proof-of-concept in vitro studies or short-term in vivo investigations without mechanical loading. Further, to enable the deep loading of drugs and a controlled initial burst release, drugs have been encapsulated in micelles prior to loading [109], or loaded in TNTs covered with biopolymers [52,108]; however, the release only lasts for a few weeks or 1–2 months. It is noteworthy that therapeutic action may be needed for prolonged periods (several months to years) in order to achieve long-term implant success, specially in compromised conditions.

- When a drug-releasing implant is placed, several cells *'race to invade'* the site [66], and often the nanotopography is immediately covered with proteins and cells, which may block the open pores [117,212]. This can impact drug release, given that the latter is dependent on a diffusion gradient that is impeded by poor perfusion inside the bone micro-environment. These conditions, especially considering that surgical placement causes trauma, even in healthy patients, may be difficult to approximate in vitro and *in silico* [213]. Hence, the performance of drug-releasing implants must be tested in real traumatized tissue in vivo, based on therapeutic needs identified ex vivo [214].

- Ideally, the implant surface modification should cater to the *three Is, integration* (both osseo- and soft-tissue integration), *inflammation* and *infection,* in order to enable early acceptance and long-term survival. While multi-therapeutic nano-engineered implants have been applied, either by combining various drugs or through the inclusion of biopolymers or metal ions/nanoparticles, their effectiveness in compromised patients conditions including advanced age, diabetes or osteoporosis, has not been investigated. It is worth noting that the success of dental implants is further challenged in these patient conditions. Further, nano-engineering attempts to augment soft-tissue integration in order to form a barrier to the ingress of oral pathogens is not explored adequately.

- To ensure clinical translation, avoiding the *'valley of death'*, nano-engineered implants must survive packaging, handling, implantation and operation inside the dental micro-environment. This also includes optimizations at all stages of product development, from the fabrication of controlled and reproducible nanostructures to bioactivity and local therapy. Further, bioactivity and cytotoxicity evaluations specifically considering initial burst release, the early consumption of drugs and dead bacteria/cells blocking the open pores of TNTs are vital. Additionally, with the use of metals ions and nanoparticles to augment the therapeutic effects of implants, it is important to determine and control their release profile to reduce cytotoxicity.

## 4. Future Perspectives

The next generation of dental implants will employ optimized nanotopography to simultaneously augment antibacterial and osseointegration functions. The following details the future directions in the domain of nano-engineered dental implants:

- The integration of new materials and technologies is the key factor in the development of new hybrid dental implants. However, it remains difficult to fabricate uniform nanostructures rapidly and on a large scale. Additive manufacturing or three-dimensional (3D) printing technology may provide customized implants to match patient needs [215]. In 2014, Dong et al. successfully fabricated a novel 3D porous scaffold by mixing anti-tuberculosis bacterium drugs, Poly-DL-lactide and nano-hydroxyapatite via additive manufacturing technology [216]. In the field of orthopedic surgery, the use of 3D printing is increasing and patient-specific implants have been produced to meet the surgical requirements [217,218]. By controlling the shape and porosity using the rapid prototyping method, 3D-printed implants enable rapid bone in-growth and reduce implant stiffness. However, the use of 3D-printed implants is limited due to high costs and time demands. Although it is still in development, 3D printing technology is the most important direction for fabricating future dental implants.
- Another direction for future dental implants is triggered drug release, whereby the therapeutic payloads are released via an internal or external stimulus, which significantly reduces the initial burst release, ensuring release *'on-demand'* [113]. The triggering mechanisms can be temperature, pH, electric or magnetic fields, or radio or ultrasonic frequencies. Further, future 'smart' dental implants could detect/sense the type of cellular attachment or tissue formation around the implant, and switch the release of a drug on or off.

Additive manufacturing, as well as biosensing and triggered drug release techniques are the future of multi-functional and customizable dental implants [219].

## 5. Conclusions

The nano-engineering of dental implants has been performed in order to augment the antibacterial and bioactivity performances of conventional implants, improving long-term treatment outcomes. This article reviewed the nano-engineering of Ti-based dental implants and evaluated modifications with titania nanotubes, nanoparticles, biopolymers and carbon-based coatings in terms of biocompatibility, antimicrobial activity, toxicity and in vivo evidence. Various nanoscale dental implant modifications and their key features have been summarized in Table 1. While in vitro and short-term in vivo studies have shown favorable outcomes, long-term in vivo investigations in compromised models (including inflammation and infection), under masticatory loading, are needed to ensure the clinical translation of nano-engineered dental implants. Clearly, the future of dental implants will include customized, patient-specific, nano-engineered implants that enable long-term therapeutic action, while augmenting implant-tissue integration, without initiating any cytotoxicity.

**Author Contributions:** P.D. and Y.L. designed, prepared and revised the manuscript; Y.Z. and K.G. wrote the manuscript and prepared the figures; Z.L. revised the manuscript. All authors have read and agreed to the published version of the manuscript.

**Funding:** The authors are grateful for the financial support from the National Natural Science Foundation of China No. 81871492 (Y.L.) and ITI Research Grant No. 1544_2020 (Y.L.). P.D. is supported by National Natural Science Foundation of China No. 81771106 (P.D.). K.G. is supported by the National Health and Medical Research Council (NHMRC) Early Career Fellowship (APP1140699).

**Institutional Review Board Statement:** Not applicable.

**Informed Consent Statement:** Not applicable.

**Data Availability Statement:** Not applicable.

**Conflicts of Interest:** The authors declare no conflict of interest.


## References

1. Branemark, P.I.; Adell, R.; Breine, U.; Hansson, B.O.; Lindstrom, J.; Ohlsson, A. Intra-osseous anchorage of dental prostheses: I. Experimental studies. *Scand. J. Plast. Reconstr. Surg.* **1969**, *3*, 81–100. [CrossRef] [PubMed]
2. Branemark, P.I.; Hansson, B.O.; Adell, R.; Breine, U.; Lindstrom, J.; Hallen, O.; Ohman, A. Osseointegrated implants in the treatment of the edentulous jaw. Experience from a 10-year period. *Scand. J. Plast. Reconstr. Surg. Suppl.* **1977**, *16*, 1–132. [PubMed]
3. Schulte, W.; Kleineikenscheidt, H.; Lindner, K.; Schareyka, R. The Tubingen immediate implant in clinical studies. *Dtsch. Zahnarztl. Z.* **1978**, *33*, 348–359. [PubMed]
4. Kirsch, A.; Ackermann, K.L. The IMZ osteointegrated implant system. *Dent. Clin. N. Am.* **1989**, *33*, 733–791.
5. Niznick, G.A. The Core-Vent implant system. *Oral Health* **1983**, *73*, 13–17.
6. Buser, D.; Sennerby, L.; De Bruyn, H. Modern implant dentistry based on osseointegration: 50 years of progress, current trends and open questions. *Periodontology 2000* **2017**, *73*, 7–21. [CrossRef]
7. Buser, D.; Schenk, R.K.; Steinemann, S.; Fiorellini, J.P.; Fox, C.H.; Stich, H. Influence of surface characteristics on bone integration of titanium implants. A histomorphometric study in miniature pigs. *J. Biomed. Mater. Res.* **1991**, *25*, 889–902. [CrossRef]
8. Buser, D.; Nydegger, T.; Hirt, H.P.; Cochran, D.L.; Nolte, L.P. Removal torque values of titanium implants in the maxilla of miniature pigs. *Int. J. Oral Maxillofac. Implant.* **1998**, *13*, 611–619.
9. Buser, D.; Nydegger, T.; Oxland, T.; Cochran, D.L.; Schenk, R.K.; Hirt, H.P.; Snetivy, D.; Nolte, L.P. Interface shear strength of titanium implants with a sandblasted and acid-etched surface: A biomechanical study in the maxilla of miniature pigs. *J. Biomed. Mater. Res.* **1999**, *45*, 75–83. [CrossRef]
10. Cionca, N.; Hashim, D.; Mombelli, A. Zirconia dental implants: Where are we now, and where are we heading? *Periodontology 2000* **2017**, *73*, 241–258. [CrossRef]
11. Buser, D.; Janner, S.F.; Wittneben, J.G.; Bragger, U.; Ramseier, C.A.; Salvi, G.E. 10-year survival and success rates of 511 titanium implants with a sandblasted and acid-etched surface: A retrospective study in 303 partially edentulous patients. *Clin. Implant. Dent. Relat. Res.* **2012**, *14*, 839–851. [CrossRef]
12. Degidi, M.; Nardi, D.; Piattelli, A. 10-year follow-up of immediately loaded implants with TiUnite porous anodized surface. *Clin. Implant. Dent. Relat. Res.* **2012**, *14*, 828–838. [CrossRef]
13. Fischer, K.; Stenberg, T. Prospective 10-year cohort study based on a randomized controlled trial (RCT) on implant-supported full-arch maxillary prostheses. Part 1: Sandblasted and acid-etched implants and mucosal tissue. *Clin. Implant. Dent. Relat. Res.* **2012**, *14*, 808–815. [CrossRef]
14. Wennerberg, A.; Albrektsson, T.; Chrcanovic, B. Long-term clinical outcome of implants with different surface modifications. *Eur. J. Oral Implantol.* **2018**, *11* (Suppl. S1), S123–S136.
15. Albrektsson, T.; Canullo, L.; Cochran, D.; De Bruyn, H. "Peri-Implantitis": A Complication of a Foreign Body or a Man-Made "Disease". Facts and Fiction. *Clin. Implant. Dent. Relat. Res.* **2016**, *18*, 840–849. [CrossRef]
16. Fu, J.H.; Wang, H.L. Breaking the wave of peri-implantitis. *Periodontology 2000* **2020**, *84*, 145–160. [CrossRef]
17. Berglundh, T.; Armitage, G.; Araujo, M.G.; Avila-Ortiz, G.; Blanco, J.; Camargo, P.M.; Chen, S.; Cochran, D.; Derks, J.; Figuero, E.; et al. Peri-implant diseases and conditions: Consensus report of workgroup 4 of the 2017 World Workshop on the Classification of Periodontal and Peri-Implant Diseases and Conditions. *J. Periodontol.* **2018**, *89* (Suppl. S1), S313–S318. [CrossRef]
18. Rompen, E.; Domken, O.; Degidi, M.; Pontes, A.E.; Piattelli, A. The effect of material characteristics, of surface topography and of implant components and connections on soft tissue integration: A literature review. *Clin. Oral Implant. Res.* **2006**, *17* (Suppl. S2), 55–67. [CrossRef] [PubMed]
19. Albrektsson, T.; Wennerberg, A. On osseointegration in relation to implant surfaces. *Clin. Implant. Dent. Relat. Res.* **2019**, *21* (Suppl. S1), 4–7. [CrossRef]
20. Chrcanovic, B.R.; Albrektsson, T.; Wennerberg, A. Bone Quality and Quantity and Dental Implant Failure: A Systematic Review and Meta-analysis. *Int. J. Prosthodont.* **2017**, *30*, 219–237. [CrossRef]
21. Milleret, V.; Lienemann, P.S.; Gasser, A.; Bauer, S.; Ehrbar, M.; Wennerberg, A. Rational design and in vitro characterization of novel dental implant and abutment surfaces for balancing clinical and biological needs. *Clin. Implant. Dent. Relat. Res.* **2019**, *21* (Suppl. S1), 15–24. [CrossRef] [PubMed]
22. Wennerberg, A.; Albrektsson, T. Effects of titanium surface topography on bone integration: A systematic review. *Clin. Oral Implants Res.* **2009**, *20* (Suppl. S4), 172–184. [CrossRef] [PubMed]
23. Roccuzzo, M.; Bonino, L.; Dalmasso, P.; Aglietta, M. Long-term results of a three arms prospective cohort study on implants in periodontally compromised patients: 10-year data around sandblasted and acid-etched (SLA) surface. *Clin. Oral Implant. Res.* **2014**, *25*, 1105–1112. [CrossRef] [PubMed]
24. Rossi, F.; Lang, N.P.; Ricci, E.; Ferraioli, L.; Baldi, N.; Botticelli, D. Long-term follow-up of single crowns supported by short, moderately rough implants-A prospective 10-year cohort study. *Clin. Oral Implant. Res.* **2018**, *29*, 1212–1219. [CrossRef]
25. Wennerberg, A.; Albrektsson, T. On implant surfaces: A review of current knowledge and opinions. *Int. J. Oral Maxillofac. Implant.* **2010**, *25*, 63–74.
26. Karl, M.; Albrektsson, T. Clinical Performance of Dental Implants with a Moderately Rough (TiUnite) Surface: A Meta-Analysis of Prospective Clinical Studies. *Int. J. Oral Maxillofac. Implant.* **2017**, *32*, 717–734. [CrossRef]

27. Klokkevold, P.R.; Johnson, P.; Dadgostari, S.; Caputo, A.; Davies, J.E.; Nishimura, R.D. Early endosseous integration enhanced by dual acid etching of titanium: A torque removal study in the rabbit. *Clin. Oral Implant. Res.* **2001**, *12*, 350–357. [CrossRef]

28. Yeo, I.L. Modifications of Dental Implant Surfaces at the Micro- and Nano-Level for Enhanced Osseointegration. *Materials* **2019**, *13*, 89. [CrossRef]

29. Gulati, K.; Maher, S.; Findlay, D.M.; Losic, D. Titania nanotubes for orchestrating osteogenesis at the bone–implant interface. *Nanomedicine* **2016**, *11*, 1847–1864. [CrossRef]

30. Chopra, D.; Gulati, K.; Ivanovski, S. Understanding and optimizing the antibacterial functions of anodized nano-engineered titanium implants. *Acta Biomater.* **2021**, *127*, 80–101. [CrossRef]

31. Gulati, K.; Santos, A.; Findlay, D.; Losic, D. Optimizing Anodization Conditions for the Growth of Titania Nanotubes on Curved Surfaces. *J. Phys. Chem. C* **2015**, *119*, 16033–16045. [CrossRef]

32. Cerqueira, A.; Romero-Gavilán, F.; Araújo-Gomes, N.; García-Arnáez, I.; Martinez-Ramos, C.; Ozturan, S.; Azkargorta, M.; Elortza, F.; Gurruchaga, M.; Suay, J.; et al. A possible use of melatonin in the dental field: Protein adsorption and in vitro cell response on coated titanium. *Mater. Sci. Eng. C* **2020**, *116*, 111262. [CrossRef]

33. Schliephake, H.; Scharnweber, D.; Dard, M.; Sewing, A.; Aref, A.; Roessler, S. Functionalization of dental implant surfaces using adhesion molecules. *J. Biomed. Mater. Res. Part B Appl. Biomater.* **2005**, *73*, 88–96. [CrossRef]

34. Ballarre, J.; Aydemir, T.; Liverani, L.; Roether, J.A.; Goldmann, W.H.; Boccaccini, A.R. Versatile bioactive and antibacterial coating system based on silica, gentamicin, and chitosan: Improving early stage performance of titanium implants. *Surf. Coat. Technol.* **2020**, *381*, 125138. [CrossRef]

35. Gulati, K.; Kogawa, M.; Maher, S.; Atkins, G.; Findlay, D.; Losic, D. Titania nanotubes for local drug delivery from implant surfaces. In *Electrochemically Engineered Nanoporous Materials*; Springer International Publishing AG: Berlin, Germany, 2015; pp. 307–355.

36. Li, T.; Gulati, K.; Wang, N.; Zhang, Z.; Ivanovski, S. Bridging the gap: Optimized fabrication of robust titania nanostructures on complex implant geometries towards clinical translation. *J. Colloid Interface Sci.* **2018**, *529*, 452–463. [CrossRef]

37. Li, T.; Gulati, K.; Wang, N.; Zhang, Z.; Ivanovski, S. Understanding and augmenting the stability of therapeutic nanotubes on anodized titanium implants. *Mater. Sci. Eng. C* **2018**, *88*, 182–195. [CrossRef]

38. Gulati, K.; Moon, H.-J.; Li, T.; Sudheesh Kumar, P.T.; Ivanovski, S. Titania nanopores with dual micro-/nano-topography for selective cellular bioactivity. *Mater. Sci. Eng. C* **2018**, *91*, 624–630. [CrossRef]

39. Gulati, K.; Li, T.; Ivanovski, S. Consume or Conserve: Microroughness of Titanium Implants toward Fabrication of Dual Micro–Nanotopography. *ACS Biomater. Sci. Eng.* **2018**, *4*, 3125–3131. [CrossRef]

40. Guo, T.; Oztug, N.A.K.; Han, P.; Ivanovski, S.; Gulati, K. Old is Gold: Electrolyte Aging Influences the Topography, Chemistry, and Bioactivity of Anodized TiO2 Nanopores. *Acs Appl. Mater. Inter.* **2021**, *13*, 7897–7912. [CrossRef]

41. Gulati, K.; Prideaux, M.; Kogawa, M.; Lima-Marques, L.; Atkins, G.J.; Findlay, D.M.; Losic, D. Anodized 3D–printed titanium implants with dual micro- and nano-scale topography promote interaction with human osteoblasts and osteocyte-like cells. *J. Tissue Eng. Regen. Med.* **2017**, *11*, 3313–3325. [CrossRef]

42. Chopra, D.; Gulati, K.; Ivanovski, S. Towards Clinical Translation: Optimized Fabrication of Controlled Nanostructures on Implant-Relevant Curved Zirconium Surfaces. *Nanomaterials* **2021**, *11*, 868. [CrossRef] [PubMed]

43. Saji, V.S.; Kumeria, T.; Gulati, K.; Prideaux, M.; Rahman, S.; Alsawat, M.; Santos, A.; Atkins, G.J.; Losic, D. Localized drug delivery of selenium (Se) using nanoporous anodic aluminium oxide for bone implants. *J. Mater. Chem. B* **2015**, *3*, 7090–7098. [CrossRef] [PubMed]

44. Gulati, K.; Hamlet, S.M.; Ivanovski, S. Tailoring the immuno-responsiveness of anodized nano-engineered titanium implants. *J. Mater. Chem. B* **2018**, *6*, 2677–2689. [CrossRef] [PubMed]

45. Gulati, K.; Ivanovski, S. Dental implants modified with drug releasing titania nanotubes: Therapeutic potential and developmental challenges. *Expert Opin. Drug Deliv.* **2017**, *14*, 1009–1024. [CrossRef] [PubMed]

46. Zhang, H.; Yang, S.; Masako, N.; Lee, D.J.; Cooper, L.F.; Ko, C.-C. Proliferation of preosteoblasts on TiO 2 nanotubes is FAK/RhoA related. *RSC Adv.* **2015**, *5*, 38117–38124. [CrossRef]

47. Balasundaram, G.; Yao, C.; Webster, T.J. TiO2 nanotubes functionalized with regions of bone morphogenetic protein-2 increases osteoblast adhesion. *J. Biomed. Mater. Res. Part A* **2008**, *84*, 447–453. [CrossRef]

48. Zhang, W.; Jin, Y.; Qian, S.; Li, J.; Chang, Q.; Ye, D.; Pan, H.; Zhang, M.; Cao, H.; Liu, X. Vacuum extraction enhances rhPDGF-BB immobilization on nanotubes to improve implant osseointegration in ovariectomized rats. *Nanomed. Nanotechnol. Biol. Med.* **2014**, *10*, 1809–1818. [CrossRef]

49. Shen, X.; Ma, P.; Hu, Y.; Xu, G.; Xu, K.; Chen, W.; Ran, Q.; Dai, L.; Yu, Y.; Mu, C. Alendronate-loaded hydroxyapatite-TiO$_2$ nanotubes for improved bone formation in osteoporotic rabbits. *J. Mater. Chem. B* **2016**, *4*, 1423–1436. [CrossRef]

50. Lee, S.J.; Oh, T.J.; Bae, T.S.; Lee, M.H.; Soh, Y.; Kim, B.I.; Kim, H.S. Effect of bisphosphonates on anodized and heat-treated titanium surfaces: An animal experimental study. *J. Periodontol.* **2011**, *82*, 1035–1042. [CrossRef]

51. Lee, Y.-H.; Bhattarai, G.; Park, I.-S.; Kim, G.-R.; Kim, G.-E.; Lee, M.-H.; Yi, H.-K. Bone regeneration around N-acetyl cysteine-loaded nanotube titanium dental implant in rat mandible. *Biomaterials* **2013**, *34*, 10199–10208. [CrossRef]

52. Gulati, K.; Kogawa, M.; Prideaux, M.; Findlay, D.M.; Atkins, G.J.; Losic, D. Drug-releasing nano-engineered titanium implants: Therapeutic efficacy in 3D cell culture model, controlled release and stability. *Mater. Sci. Eng. C* **2016**, *69*, 831–840. [CrossRef]

53. Xiao, J.; Zhou, H.; Zhao, L.; Sun, Y.; Guan, S.; Liu, B.; Kong, L. The effect of hierarchical micro/nanosurface titanium implant on osseointegration in ovariectomized sheep. *Osteoporos. Int.* **2011**, *22*, 1907–1913. [CrossRef]
54. Zhao, L.; Wang, H.; Huo, K.; Zhang, X.; Wang, W.; Zhang, Y.; Wu, Z.; Chu, P.K. The osteogenic activity of strontium loaded titania nanotube arrays on titanium substrates. *Biomaterials* **2013**, *34*, 19–29. [CrossRef]
55. Frandsen, C.J.; Brammer, K.S.; Noh, K.; Johnston, G.; Jin, S. Tantalum coating on TiO2 nanotubes induces superior rate of matrix mineralization and osteofunctionality in human osteoblasts. *Mater. Sci. Eng. C* **2014**, *37*, 332–341. [CrossRef]
56. Zhang, X.; Zhang, X.; Wang, B.; Lan, J.; Yang, H.; Wang, Z.; Chang, X.; Wang, S.; Ma, X.; Qiao, H.; et al. Synergistic effects of lanthanum and strontium to enhance the osteogenic activity of TiO2 nanotube biological interface. *Ceram. Int.* **2020**, *46*, 13969–13979. [CrossRef]
57. Huo, K.; Zhang, X.; Wang, H.; Zhao, L.; Liu, X.; Chu, P.K. Osteogenic activity and antibacterial effects on titanium surfaces modified with Zn-incorporated nanotube arrays. *Biomaterials* **2013**, *34*, 3467–3478. [CrossRef]
58. Guo, T.Q.; Gulati, K.; Arora, H.; Han, P.P.; Fournier, B.; Ivanovski, S. Orchestrating soft tissue integration at the transmucosal region of titanium implants. *Acta Biomater.* **2021**, *124*, 33–49. [CrossRef]
59. Gulati, K.; Moon, H.J.; Kumar, P.T.S.; Han, P.P.; Ivanovski, S. Anodized anisotropic titanium surfaces for enhanced guidance of gingival fibroblasts. *Mat. Sci. Eng. C-Mater.* **2020**, *112*, 110860. [CrossRef]
60. Takebe, J.; Miyata, K.; Miura, S.; Ito, S. Effects of the nanotopographic surface structure of commercially pure titanium following anodization–hydrothermal treatment on gene expression and adhesion in gingival epithelial cells. *Mater. Sci. Eng. C* **2014**, *42*, 273–279. [CrossRef]
61. Miyata, K.; Takebe, J. Anodized-hydrothermally treated titanium with a nanotopographic surface structure regulates integrin-$\alpha6\beta4$ and laminin-5 gene expression in adherent murine gingival epithelial cells. *J. Prosthodont. Res.* **2013**, *57*, 99–108. [CrossRef]
62. Miura, S.; Takebe, J. Biological behavior of fibroblast-like cells cultured on anodized-hydrothermally treated titanium with a nanotopographic surface structure. *J. Prosthodont. Res.* **2012**, *56*, 178–186. [CrossRef]
63. Xu, R.; Hu, X.; Yu, X.; Wan, S.; Wu, F.; Ouyang, J.; Deng, F. Micro-/nano-topography of selective laser melting titanium enhances adhesion and proliferation and regulates adhesion-related gene expressions of human gingival fibroblasts and human gingival epithelial cells. *Int. J. Nanomed.* **2018**, *13*, 5045–5057. [CrossRef]
64. Liu, X.; Zhou, X.; Li, S.; Lai, R.; Zhou, Z.; Zhang, Y.; Zhou, L. Effects of titania nanotubes with or without bovine serum albumin loaded on human gingival fibroblasts. *Int. J. Nanomed.* **2014**, *9*, 1185–1198. [CrossRef]
65. Ma, Q.; Mei, S.; Ji, K.; Zhang, Y.; Chu, P.K. Immobilization of Ag nanoparticles/FGF-2 on a modified titanium implant surface and improved human gingival fibroblasts behavior. *J. Biomed. Mater. Res. Part. A* **2011**, *98A*, 274–286. [CrossRef]
66. Guo, T.; Gulati, K.; Arora, H.; Han, P.; Fournier, B.; Ivanovski, S. Race to invade: Understanding soft tissue integration at the transmucosal region of titanium dental implants. *Dent. Mater.* **2021**, *37*, 816–831. [CrossRef]
67. Hao, Y.; Huang, X.; Zhou, X.; Li, M.; Ren, B.; Peng, X.; Cheng, L. Influence of Dental Prosthesis and Restorative Materials Interface on Oral Biofilms. *Int. J. Mol. Sci.* **2018**, *19*, 3157. [CrossRef]
68. Shibli, J.A.; Melo, L.; Ferrari, D.S.; Figueiredo, L.C.; Faveri, M.; Feres, M. Composition of supra- and subgingival biofilm of subjects with healthy and diseased implants. *Clin. Oral Implant. Res.* **2008**, *19*, 975–982. [CrossRef] [PubMed]
69. Ercan, B.; Taylor, E.; Alpaslan, E.; Webster, T.J. Diameter of titanium nanotubes influences anti-bacterial efficacy. *Nanotechnology* **2011**, *22*, 295102. [CrossRef]
70. Narendrakumar, K.; Kulkarni, M.; Addison, O.; Mazare, A.; Junkar, I.; Schmuki, P.; Sammons, R.; Iglič, A. Adherence of oral streptococci to nanostructured titanium surfaces. *Dent. Mater.* **2015**, *31*, 1460–1468. [CrossRef]
71. Mazare, A.; Totea, G.; Burnei, C.; Schmuki, P.; Demetrescu, I.; Ionita, D. Corrosion, antibacterial activity and haemocompatibility of TiO2 nanotubes as a function of their annealing temperature. *Corros. Sci.* **2016**, *103*, 215–222. [CrossRef]
72. Podporska-Carroll, J.; Panaitescu, E.; Quilty, B.; Wang, L.; Menon, L.; Pillai, S.C. Antimicrobial properties of highly efficient photocatalytic TiO2 nanotubes. *Appl. Catal. B Environ.* **2015**, *176–177*, 70–75. [CrossRef]
73. Pawlik, A.; Jarosz, M.; Syrek, K.; Sulka, G.D. Co-delivery of ibuprofen and gentamicin from nanoporous anodic titanium dioxide layers. *Colloids Surf. B Biointerfaces* **2017**, *152*, 95–102. [CrossRef] [PubMed]
74. Ionita, D.; Bajenaru-Georgescu, D.; Totea, G.; Mazare, A.; Schmuki, P.; Demetrescu, I. Activity of vancomycin release from bioinspired coatings of hydroxyapatite or TiO2 nanotubes. *Int. J. Pharm.* **2017**, *517*, 296–302. [CrossRef] [PubMed]
75. Park, S.W.; Lee, D.; Choi, Y.S.; Jeon, H.B.; Lee, C.-H.; Moon, J.-H.; Kwon, I.K. Mesoporous TiO2 implants for loading high dosage of antibacterial agent. *Appl. Surf. Sci.* **2014**, *303*, 140–146. [CrossRef]
76. Chennell, P.; Feschet-Chassot, E.; Devers, T.; Awitor, K.; Descamps, S.; Sautou, V. In vitro evaluation of TiO2 nanotubes as cefuroxime carriers on orthopaedic implants for the prevention of periprosthetic joint infections. *Int. J. Pharm.* **2013**, *455*, 298–305. [CrossRef]
77. Shen, X.; Zhang, F.; Li, K.; Qin, C.; Ma, P.; Dai, L.; Cai, K. Cecropin B loaded TiO2 nanotubes coated with hyaluronidase sensitive multilayers for reducing bacterial adhesion. *Mater. Des.* **2016**, *92*, 1007–1017. [CrossRef]
78. Ma, M.; Kazemzadeh-Narbat, M.; Hui, Y.; Lu, S.; Ding, C.; Chen, D.D.; Hancock, R.E.; Wang, R. Local delivery of antimicrobial peptides using self-organized TiO2 nanotube arrays for peri-implant infections. *J. Biomed. Mater. Res. Part A* **2012**, *100*, 278–285. [CrossRef]

79. Kumeria, T.; Mon, H.; Aw, M.S.; Gulati, K.; Santos, A.; Griesser, H.J.; Losic, D. Advanced biopolymer-coated drug-releasing titania nanotubes (TNTs) implants with simultaneously enhanced osteoblast adhesion and antibacterial properties. *Colloids Surf. B Biointerfaces* **2015**, *130*, 255–263. [CrossRef]

80. Ding, X.; Zhang, Y.; Ling, J.; Lin, C. Rapid mussel-inspired synthesis of PDA-Zn-Ag nanofilms on TiO2 nanotubes for optimizing the antibacterial activity and biocompatibility by doping polydopamine with zinc at a higher temperature. *Colloids Surf. B Biointerfaces* **2018**, *171*, 101–109. [CrossRef]

81. Fathi, M.; Akbari, B.; Taheriazam, A. Antibiotics drug release controlling and osteoblast adhesion from titania nanotubes arrays using silk fibroin coating. *Mater. Sci. Eng. C* **2019**, *103*, 109743. [CrossRef]

82. Gao, A.; Hang, R.; Huang, X.; Zhao, L.; Zhang, X.; Wang, L.; Tang, B.; Ma, S.; Chu, P.K. The effects of titania nanotubes with embedded silver oxide nanoparticles on bacteria and osteoblasts. *Biomaterials* **2014**, *35*, 4223–4235. [CrossRef]

83. Xu, W.; Qi, M.; Li, X.; Liu, X.; Wang, L.; Yu, W.; Liu, M.; Lan, A.; Zhou, Y.; Song, Y. TiO2 nanotubes modified with Au nanoparticles for visible-light enhanced antibacterial and anti-inflammatory capabilities. *J. Electroanal. Chem.* **2019**, *842*, 66–73. [CrossRef]

84. Zong, M.; Bai, L.; Liu, Y.; Wang, X.; Zhang, X.; Huang, X.; Hang, R.; Tang, B. Antibacterial ability and angiogenic activity of Cu-Ti-O nanotube arrays. *Mater. Sci. Eng. C* **2017**, *71*, 93–99. [CrossRef]

85. Zhang, X.; Li, J.; Wang, X.; Wang, Y.; Hang, R.; Huang, X.; Tang, B.; Chu, P.K. Effects of copper nanoparticles in porous TiO2 coatings on bacterial resistance and cytocompatibility of osteoblasts and endothelial cells. *Mater. Sci. Eng. C* **2018**, *82*, 110–120. [CrossRef]

86. Sopchenski, L.; Cogo, S.; Dias-Ntipanyj, M.; Elifio-Espósito, S.; Popat, K.; Soares, P. Bioactive and antibacterial boron doped TiO2 coating obtained by PEO. *Appl. Surf. Sci.* **2018**, *458*, 49–58. [CrossRef]

87. Dong, J.; Fang, D.; Zhang, L.; Shan, Q.; Huang, Y. Gallium-doped titania nanotubes elicit anti-bacterial efficacy in vivo against Escherichia coli and Staphylococcus aureus biofilm. *Materialia* **2019**, *5*, 100209. [CrossRef]

88. Yang, Y.; Liu, L.; Luo, H.; Zhang, D.; Lei, S.; Zhou, K. Dual-purpose magnesium-incorporated titanium nanotubes for combating bacterial infection and ameliorating osteolysis to realize better osseointegration. *ACS Biomater. Sci. Eng.* **2019**, *5*, 5368–5383. [CrossRef]

89. Xiang, Y.; Liu, X.; Mao, C.; Liu, X.; Cui, Z.; Yang, X.; Yeung, K.W.; Zheng, Y.; Wu, S. Infection-prevention on Ti implants by controlled drug release from folic acid/ZnO quantum dots sealed titania nanotubes. *Mater. Sci. Eng. C* **2018**, *85*, 214–224. [CrossRef]

90. Mei, S.; Wang, H.; Wang, W.; Tong, L.; Pan, H.; Ruan, C.; Ma, Q.; Liu, M.; Yang, H.; Zhang, L. Antibacterial effects and biocompatibility of titanium surfaces with graded silver incorporation in titania nanotubes. *Biomaterials* **2014**, *35*, 4255–4265. [CrossRef]

91. Alfarsi, M.A.; Hamlet, S.M.; Ivanovski, S. Titanium surface hydrophilicity modulates the human macrophage inflammatory cytokine response. *J. Biomed. Mater. Res. Part A* **2014**, *102*, 60–67. [CrossRef]

92. Hamlet, S.; Ivanovski, S. Inflammatory cytokine response to titanium chemical composition and nanoscale calcium phosphate surface modification. *Acta Biomater.* **2011**, *7*, 2345–2353. [CrossRef] [PubMed]

93. Neacsu, P.; Mazare, A.; Cimpean, A.; Park, J.; Costache, M.; Schmuki, P.; Demetrescu, I. Reduced inflammatory activity of RAW 264.7 macrophages on titania nanotube modified Ti surface. *Int. J. Biochem. Cell Biol.* **2014**, *55*, 187–195. [CrossRef] [PubMed]

94. Smith, B.S.; Capellato, P.; Kelley, S.; Gonzalez-Juarrero, M.; Popat, K.C. Reduced in vitro immune response on titania nanotube arrays compared to titanium surface. *Biomater. Sci.* **2013**, *1*, 322–332. [CrossRef] [PubMed]

95. Rajyalakshmi, A.; Ercan, B.; Balasubramanian, K.; Webster, T.J. Reduced adhesion of macrophages on anodized titanium with select nanotube surface features. *Int. J. Nanomed.* **2011**, *6*, 1765–1771. [CrossRef]

96. Ma, Q.-L.; Zhao, L.-Z.; Liu, R.-R.; Jin, B.-Q.; Song, W.; Wang, Y.; Zhang, Y.-S.; Chen, L.-H.; Zhang, Y.-M. Improved implant osseointegration of a nanostructured titanium surface via mediation of macrophage polarization. *Biomaterials* **2014**, *35*, 9853–9867. [CrossRef] [PubMed]

97. Neacsu, P.; Mazare, A.; Schmuki, P.; Cimpean, A. Attenuation of the macrophage inflammatory activity by TiO$_2$ nanotubes via inhibition of MAPK and NF-κB pathways. *Int. J. Nanomed.* **2015**, *10*, 6455–6467. [CrossRef]

98. Doadrio, A.L.; Conde, A.; Arenas, M.A.; Hernández-López, J.M.; de Damborenea, J.J.; Pérez-Jorge, C.; Esteban, J.; Vallet-Regí, M. Use of anodized titanium alloy as drug carrier: Ibuprofen as model of drug releasing. *Int. J. Pharm.* **2015**, *492*, 207–212. [CrossRef]

99. Karan, G.; Gerald, J.A.; David, M.F.; Dusan, L. Nano-engineered titanium for enhanced bone therapy. In Proceedings of the SPIE, San Diego, CA, USA, 11 September 2013.

100. Shen, K.; Tang, Q.; Fang, X.; Zhang, C.; Zhu, Z.; Hou, Y.; Lai, M. The sustained release of dexamethasone from TiO2 nanotubes reinforced by chitosan to enhance osteoblast function and anti-inflammation activity. *Mater. Sci. Eng. C* **2020**, *116*, 111241. [CrossRef]

101. Ma, A.; You, Y.; Chen, B.; Wang, W.; Liu, J.; Qi, H.; Liang, Y.; Li, Y.; Li, C. Icariin/Aspirin Composite Coating on TiO2 Nanotubes Surface Induce Immunomodulatory Effect of Macrophage and Improve Osteoblast Activity. *Coatings* **2020**, *10*, 427. [CrossRef]

102. Shokuhfar, T.; Sinha-Ray, S.; Sukotjo, C.; Yarin, A.L. Intercalation of anti-inflammatory drug molecules within TiO2 nanotubes. *RSC Adv.* **2013**, *3*, 17380–17386. [CrossRef]

103. Mohan, L.; Anandan, C.; Rajendran, N. Drug release characteristics of quercetin-loaded TiO2 nanotubes coated with chitosan. *Int. J. Biol. Macromol.* **2016**, *93*, 1633–1638. [CrossRef]

104. Lai, S.; Zhang, W.; Liu, F.; Wu, C.; Zeng, D.; Sun, Y.; Xu, Y.; Fang, Y.; Zhou, W. TiO2 Nanotubes as Animal Drug Delivery System and In vitro Controlled Release. *J. Nanosci. Nanotechnol.* **2013**, *13*, 91–97. [CrossRef]
105. Somsanith, N.; Kim, Y.-K.; Jang, Y.-S.; Lee, Y.-H.; Yi, H.-K.; Jang, J.-H.; Kim, K.-A.; Bae, T.-S.; Lee, M.-H. Enhancing of Osseointegration with Propolis-Loaded TiO2 Nanotubes in Rat Mandible for Dental Implants. *Materials* **2018**, *11*, 61. [CrossRef]
106. Gao, L.; Li, M.; Yin, L.; Zhao, C.; Chen, J.; Zhou, J.; Duan, K.; Feng, B. Dual-inflammatory cytokines on TiO2 nanotube-coated surfaces used for regulating macrophage polarization in bone implants. *J. Biomed. Mater. Res. Part A* **2018**, *106*, 1878–1886. [CrossRef]
107. Yin, X.; Li, Y.; Yang, C.; Weng, J.; Wang, J.; Zhou, J.; Feng, B. Alginate/chitosan multilayer films coated on IL-4-loaded TiO2 nanotubes for modulation of macrophage phenotype. *Int. J. Biol. Macromol.* **2019**, *133*, 503–513. [CrossRef]
108. Gulati, K.; Ramakrishnan, S.; Aw, M.S.; Atkins, G.J.; Findlay, D.M.; Losic, D. Biocompatible polymer coating of titania nanotube arrays for improved drug elution and osteoblast adhesion. *Acta Biomater.* **2012**, *8*, 449–456. [CrossRef]
109. Aw, M.S.; Gulati, K.; Losic, D. Controlling Drug Release from Titania Nanotube Arrays Using Polymer Nanocarriers and Biopolymer Coating. *J. Biomater. Nanobiotechnology* **2011**, *2*, 8. [CrossRef]
110. Gulati, K.; Kant, K.; Findlay, D.; Losic, D. Periodically tailored titania nanotubes for enhanced drug loading and releasing performances. *J. Mater. Chem. B* **2015**, *3*, 2553–2559. [CrossRef]
111. Mandal, S.S.; Jose, D.; Bhattacharyya, A.J. Role of surface chemistry in modulating drug release kinetics in titania nanotubes. *Mater. Chem. Phys.* **2014**, *147*, 247–253. [CrossRef]
112. Gulati, K.; Maher, S.; Chandrasekaran, S.; Findlay, D.M.; Losic, D. Conversion of titania (TiO2) into conductive titanium (Ti) nanotube arrays for combined drug-delivery and electrical stimulation therapy. *J. Mater. Chem. B* **2016**, *4*, 371–375. [CrossRef]
113. Jayasree, A.; Ivanovski, S.; Gulati, K. ON or OFF: Triggered therapies from anodized nano-engineered titanium implants. *J. Control. Release* **2021**, *333*, 521–535. [CrossRef] [PubMed]
114. Chen, J.; Dai, S.; Liu, L.; Maitz, M.F.; Liao, Y.; Cui, J.; Zhao, A.; Yang, P.; Huang, N.; Wang, Y. Photo-functionalized TiO2 nanotubes decorated with multifunctional Ag nanoparticles for enhanced vascular biocompatibility. *Bioact. Mater.* **2021**, *6*, 45–54. [CrossRef] [PubMed]
115. Yao, S.; Feng, X.; Lu, J.; Zheng, Y.; Wang, X.; Volinsky, A.A.; Wang, L.-N. Antibacterial activity and inflammation inhibition of ZnO nanoparticles embedded TiO2 nanotubes. *Nanotechnology* **2018**, *29*, 244003. [CrossRef] [PubMed]
116. Gao, S.; Lu, R.; Wang, X.; Chou, J.; Wang, N.; Huai, X.; Wang, C.; Zhao, Y.; Chen, S. Immune response of macrophages on super-hydrophilic TiO2 nanotube arrays. *J. Biomater. Appl.* **2020**, *34*, 1239–1253. [CrossRef] [PubMed]
117. Kaur, G.; Willsmore, T.; Gulati, K.; Zinonos, I.; Wang, Y.; Kurian, M.; Hay, S.; Losic, D.; Evdokiou, A. Titanium wire implants with nanotube arrays: A study model for localized cancer treatment. *Biomaterials* **2016**, *101*, 176–188. [CrossRef] [PubMed]
118. Priyadarsini, S.; Mukherjee, S.; Mishra, M. Nanoparticles used in dentistry: A review. *J. Oral Biol. Craniofacial Res.* **2018**, *8*, 58–67. [CrossRef] [PubMed]
119. Jandt, K.D.; Watts, D.C. Nanotechnology in dentistry: Present and future perspectives on dental nanomaterials. *Dent. Mater.* **2020**, *36*, 1365–1378. [CrossRef]
120. Noronha, V.T.; Paula, A.J.; Durán, G.; Galembeck, A.; Cogo-Müller, K.; Franz-Montan, M.; Durán, N. Silver nanoparticles in dentistry. *Dent. Mater.* **2017**, *33*, 1110–1126. [CrossRef]
121. Bapat, R.A.; Chaubal, T.V.; Joshi, C.P.; Bapat, P.R.; Choudhury, H.; Pandey, M.; Gorain, B.; Kesharwani, P. An overview of application of silver nanoparticles for biomaterials in dentistry. *Mater. Sci. Eng. C* **2018**, *91*, 881–898. [CrossRef]
122. Cao, H.; Liu, X.; Meng, F.; Chu, P.K. Biological actions of silver nanoparticles embedded in titanium controlled by micro-galvanic effects. *Biomaterials* **2011**, *32*, 693–705. [CrossRef]
123. Matsubara, V.H.; Igai, F.; Tamaki, R.; Tortamano Neto, P.; Nakamae, A.E.M.; Mori, M. Use of silver nanoparticles reduces internal contamination of external hexagon implants by Candida albicans. *Braz. Dent. J.* **2015**, *26*, 458–462. [CrossRef] [PubMed]
124. Flores, C.Y.; Miñán, A.G.; Grillo, C.A.; Salvarezza, R.C.; Vericat, C.; Schilardi, P.L. Citrate-capped silver nanoparticles showing good bactericidal effect against both planktonic and sessile bacteria and a low cytotoxicity to osteoblastic cells. *ACS Appl. Mater. Interfaces* **2013**, *5*, 3149–3159. [CrossRef] [PubMed]
125. Della Valle, C.; Visai, L.; Santin, M.; Cigada, A.; Candiani, G.; Pezzoli, D.; Arciola, C.R.; Imbriani, M.; Chiesa, R. A novel antibacterial modification treatment of titanium capable to improve osseointegration. *Int. J. Artif. Organs* **2012**, *35*, 864–875. [CrossRef] [PubMed]
126. Cochis, A.; Azzimonti, B.; Della Valle, C.; Chiesa, R.; Arciola, C.R.; Rimondini, L. Biofilm formation on titanium implants counteracted by grafting gallium and silver ions. *J. Biomed. Mater. Res. Part A* **2015**, *103*, 1176–1187. [CrossRef]
127. Fu, C.; Zhang, X.; Savino, K.; Gabrys, P.; Gao, Y.; Chaimayo, W.; Miller, B.L.; Yates, M.Z. Antimicrobial silver-hydroxyapatite composite coatings through two-stage electrochemical synthesis. *Surf. Coat. Technol.* **2016**, *301*, 13–19. [CrossRef]
128. Wang, Z.; Sun, Y.; Wang, D.; Liu, H.; Boughton, R.I. In situ fabrication of silver nanoparticle-filled hydrogen titanate nanotube layer on metallic titanium surface for bacteriostatic and biocompatible implantation. *Int. J. Nanomed.* **2013**, *8*, 2903.
129. Zhong, X.; Song, Y.; Yang, P.; Wang, Y.; Jiang, S.; Zhang, X.; Li, C. Titanium surface priming with phase-transited lysozyme to establish a silver nanoparticle-loaded chitosan/hyaluronic acid antibacterial multilayer via layer-by-layer self-assembly. *PLoS ONE* **2016**, *11*, e0146957. [CrossRef]

130. Massa, M.A.; Covarrubias, C.; Bittner, M.; Fuentevilla, I.A.; Capetillo, P.; Von Marttens, A.; Carvajal, J.C. Synthesis of new antibacterial composite coating for titanium based on highly ordered nanoporous silica and silver nanoparticles. *Mater. Sci. Eng. C* **2014**, *45*, 146–153. [CrossRef]
131. Svensson, S.; Suska, F.; Emanuelsson, L.; Palmquist, A.; Norlindh, B.; Trobos, M.; Bäckros, H.; Persson, L.; Rydja, G.; Ohrlander, M. Osseointegration of titanium with an antimicrobial nanostructured noble metal coating. *Nanomed. Nanotechnol. Biol. Med.* **2013**, *9*, 1048–1056. [CrossRef]
132. Qiao, S.; Cao, H.; Zhao, X.; Lo, H.; Zhuang, L.; Gu, Y.; Shi, J.; Liu, X.; Lai, H. Ag-plasma modification enhances bone apposition around titanium dental implants: An animal study in Labrador dogs. *Int. J. Nanomed.* **2015**, *10*, 653.
133. Sukhanova, A.; Bozrova, S.; Sokolov, P.; Berestovoy, M.; Karaulov, A.; Nabiev, I. Dependence of nanoparticle toxicity on their physical and chemical properties. *Nanoscale Res. Lett.* **2018**, *13*, 1–21. [CrossRef]
134. Yuan, J.-H.; Chen, Y.; Zha, H.-X.; Song, L.-J.; Li, C.-Y.; Li, J.-Q.; Xia, X.-H. Determination, characterization and cytotoxicity on HELF cells of ZnO nanoparticles. *Colloids Surf. B Biointerfaces* **2010**, *76*, 145–150. [CrossRef]
135. Kulshrestha, S.; Khan, S.; Meena, R.; Singh, B.R.; Khan, A.U. A graphene/zinc oxide nanocomposite film protects dental implant surfaces against cariogenic Streptococcus mutans. *Biofouling* **2014**, *30*, 1281–1294. [CrossRef]
136. Hu, H.; Zhang, W.; Qiao, Y.; Jiang, X.; Liu, X.; Ding, C. Antibacterial activity and increased bone marrow stem cell functions of Zn-incorporated TiO2 coatings on titanium. *Acta Biomater.* **2012**, *8*, 904–915. [CrossRef]
137. Li, Y.; Liu, X.; Tan, L.; Cui, Z.; Yang, X.; Yeung, K.W.K.; Pan, H.; Wu, S. Construction of N-halamine labeled silica/zinc oxide hybrid nanoparticles for enhancing antibacterial ability of Ti implants. *Mater. Sci. Eng. C* **2017**, *76*, 50–58. [CrossRef]
138. van Hengel, I.A.J.; Putra, N.E.; Tierolf, M.W.A.M.; Minneboo, M.; Fluit, A.C.; Fratila-Apachitei, L.E.; Apachitei, I.; Zadpoor, A.A. Biofunctionalization of selective laser melted porous titanium using silver and zinc nanoparticles to prevent infections by antibiotic-resistant bacteria. *Acta Biomater.* **2020**, *107*, 325–337. [CrossRef]
139. Kononenko, V.; Repar, N.; Marušič, N.; Drašler, B.; Romih, T.; Hočevar, S.; Drobne, D. Comparative in vitro genotoxicity study of ZnO nanoparticles, ZnO macroparticles and ZnCl2 to MDCK kidney cells: Size matters. *Toxicol. Vitr.* **2017**, *40*, 256–263. [CrossRef]
140. Ren, G.; Hu, D.; Cheng, E.W.; Vargas-Reus, M.A.; Reip, P.; Allaker, R.P. Characterisation of copper oxide nanoparticles for antimicrobial applications. *Int. J. Antimicrob. Agents* **2009**, *33*, 587–590. [CrossRef]
141. Burghardt, I.; Lüthen, F.; Prinz, C.; Kreikemeyer, B.; Zietz, C.; Neumann, H.-G.; Rychly, J. A dual function of copper in designing regenerative implants. *Biomaterials* **2015**, *44*, 36–44. [CrossRef]
142. Thukkaram, M.; Vaidulych, M.; Kylián, O.; Rigole, P.; Aliakbarshirazi, S.; Asadian, M.; Nikiforov, A.; Biederman, H.; Coenye, T.; Du Laing, G.; et al. Biological activity and antimicrobial property of Cu/a-C:H nanocomposites and nanolayered coatings on titanium substrates. *Mater. Sci. Eng. C* **2021**, *119*, 111513. [CrossRef]
143. van Hengel, I.; Tierolf, M.; Valerio, V.; Minneboo, M.; Fluit, A.; Fratila-Apachitei, L.; Apachitei, I.; Zadpoor, A. Self-defending additively manufactured bone implants bearing silver and copper nanoparticles. *J. Mater. Chem. B* **2020**, *8*, 1589–1602. [CrossRef] [PubMed]
144. Hameed, H.A.; Ariffin, A.; Luddin, N.; Husein, A. Evaluation of antibacterial properties of copper nanoparticles surface coating on titanium dental implant. *J. Pharm. Sci. Res.* **2018**, *10*, 1157–1160.
145. Xia, C.; Ma, X.; Zhang, X.; Li, K.; Tan, J.; Qiao, Y.; Liu, X. Enhanced physicochemical and biological properties of C/Cu dual ions implanted medical titanium. *Bioact. Mater.* **2020**, *5*, 377–386. [CrossRef] [PubMed]
146. Astasov-Frauenhoffer, M.; Koegel, S.; Waltimo, T.; Zimmermann, A.; Walker, C.; Hauser-Gerspach, I.; Jung, C. Antimicrobial efficacy of copper-doped titanium surfaces for dental implants. *J. Mater. Sci. Mater. Med.* **2019**, *30*, 1–9. [CrossRef]
147. Ingle, A.P.; Duran, N.; Rai, M. Bioactivity, mechanism of action, and cytotoxicity of copper-based nanoparticles: A review. *Appl. Microbiol. Biotechnol.* **2014**, *98*, 1001–1009. [CrossRef]
148. Zhao, Y.; Jamesh, M.I.; Li, W.K.; Wu, G.; Wang, C.; Zheng, Y.; Yeung, K.W.; Chu, P.K. Enhanced antimicrobial properties, cytocompatibility, and corrosion resistance of plasma-modified biodegradable magnesium alloys. *Acta Biomater.* **2014**, *10*, 544–556. [CrossRef]
149. Chopra, D.; Gulati, K.; Ivanovski, S. Micro + Nano: Conserving the Gold Standard Microroughness to Nanoengineer Zirconium Dental Implants. *ACS Biomater. Sci. Eng.* **2021**, *7*, 3069–3074. [CrossRef]
150. Guo, L.; Zhao, J.; Wang, X.; Xu, R.; Lu, Z.; Li, Y. Bioactivity of zirconia nanotube arrays fabricated by electrochemical anodization. *Mater. Sci. Eng. C* **2009**, *29*, 1174–1177. [CrossRef]
151. Katunar, M.R.; Sanchez, A.G.; Coquillat, A.S.; Civantos, A.; Campos, E.M.; Ballarre, J.; Vico, T.; Baca, M.; Ramos, V.; Cere, S. In vitro and in vivo characterization of anodised zirconium as a potential material for biomedical applications. *Mater. Sci. Eng. C* **2017**, *75*, 957–968. [CrossRef]
152. Indira, K.; KamachiMudali, U.; Rajendran, N. In vitro bioactivity and corrosion resistance of Zr incorporated TiO2 nanotube arrays for orthopaedic applications. *Appl. Surf. Sci.* **2014**, *316*, 264–275. [CrossRef]
153. Zheng, Y.F.; Liu, D.; Liu, X.L.; Li, L. Enhanced corrosion resistance of Zr coating on biomedical TiNi alloy prepared by plasma immersion ion implantation and deposition. *Appl. Surf. Sci.* **2008**, *255*, 512–514. [CrossRef]
154. Grigorescu, S.; Ungureanu, C.; Kirchgeorg, R.; Schmuki, P.; Demetrescu, I. Various sized nanotubes on TiZr for antibacterial surfaces. *Appl. Surf. Sci.* **2013**, *270*, 190–196. [CrossRef]
155. Ye, M.; Shi, B. Zirconia nanoparticles-induced toxic effects in osteoblast-like 3T3-E1 cells. *Nanoscale Res. Lett.* **2018**, *13*, 1–12. [CrossRef]

156. Bitar, A.; Ahmad, N.M.; Fessi, H.; Elaissari, A. Silica-based nanoparticles for biomedical applications. *Drug Discov. Today* **2012**, *17*, 1147–1154. [CrossRef]
157. Gaikwad, R.; Sokolov, I. Silica nanoparticles to polish tooth surfaces for caries prevention. *J. Dent. Res.* **2008**, *87*, 980–983. [CrossRef]
158. Bartkowiak, A.; Suchanek, K.; Menaszek, E.; Szaraniec, B.; Lekki, J.; Perzanowski, M.; Marszałek, M. Biological effect of hydrothermally synthesized silica nanoparticles within crystalline hydroxyapatite coatings for titanium implants. *Mater. Sci. Eng. C* **2018**, *92*, 88–95. [CrossRef]
159. Jo, Y.K.; Choi, B.-H.; Kim, C.S.; Cha, H.J. Diatom-Inspired Silica Nanostructure Coatings with Controllable Microroughness Using an Engineered Mussel Protein Glue to Accelerate Bone Growth on Titanium-Based Implants. *Adv. Mater.* **2017**, *29*, 1704906. [CrossRef]
160. Xu, G.; Shen, X.; Dai, L.; Ran, Q.; Ma, P.; Cai, K. Reduced bacteria adhesion on octenidine loaded mesoporous silica nanoparticles coating on titanium substrates. *Mater. Sci. Eng. C* **2017**, *70*, 386–395. [CrossRef]
161. Janßen, H.C.; Angrisani, N.; Kalies, S.; Hansmann, F.; Kietzmann, M.; Warwas, D.P.; Behrens, P.; Reifenrath, J. Biodistribution, biocompatibility and targeted accumulation of magnetic nanoporous silica nanoparticles as drug carrier in orthopedics. *J. Nanobiotechnology* **2020**, *18*, 1–18. [CrossRef]
162. Luo, J.; Ding, X.; Song, W.; Bai, J.-Y.; Liu, J.; Li, Z.; Meng, F.-H.; Chen, F.-H.; Zhang, Y.-M. Inducing macrophages M2 polarization by dexamethasone laden mesoporous silica nanoparticles from titanium implant surface for enhanced osteogenesis. *Acta Metall. Sin. (Engl. Lett.)* **2019**, *32*, 1253–1260. [CrossRef]
163. Chen, L.; Liu, J.; Zhang, Y.; Zhang, G.; Kang, Y.; Chen, A.; Feng, X.; Shao, L. The toxicity of silica nanoparticles to the immune system. *Nanomedicine* **2018**, *13*, 1939–1962. [CrossRef] [PubMed]
164. Arcos, D.; Vallet-Regi, M. Substituted hydroxyapatite coatings of bone implants. *J. Mater. Chem. B* **2020**, *8*, 1781–1800. [CrossRef] [PubMed]
165. Mittal, M.; Nath, S.K.; Prakash, S. Improvement in mechanical properties of plasma sprayed hydroxyapatite coatings by Al2O3 reinforcement. *Mater. Sci. Eng. C Mater. Biol. Appl.* **2013**, *33*, 2838–2845. [CrossRef] [PubMed]
166. Hayashi, K.; Mashima, T.; Uenoyama, K. The effect of hydroxyapatite coating on bony ingrowth into grooved titanium implants. *Biomaterials* **1999**, *20*, 111–119. [CrossRef]
167. Liu, F.; Wang, X.; Chen, T.; Zhang, N.; Wei, Q.; Tian, J.; Wang, Y.; Ma, C.; Lu, Y. Hydroxyapatite/silver electrospun fibers for anti-infection and osteoinduction. *J. Adv. Res.* **2020**, *21*, 91–102. [CrossRef] [PubMed]
168. Bosco, R.; Iafisco, M.; Tampieri, A.; Jansen, J.A.; Beucken, J.J.J.P.v.d. Hydroxyapatite nanocrystals functionalized with alendronate as bioactive components for bone implant coatings to decrease osteoclastic activity. *Appl. Surf. Sci.* **2015**, *328*, 516–524. [CrossRef]
169. Bertoni, E.; Bigi, A.; Cojazzi, G.; Gandolfi, M.; Panzavolta, S.; Roveri, N. Nanocrystals of magnesium and fluoride substituted hydroxyapatite. *J. Inorg. Biochem.* **1998**, *72*, 29–35. [CrossRef]
170. Geuli, O.; Metoki, N.; Zada, T.; Reches, M.; Eliaz, N.; Mandler, D. Synthesis, coating, and drug-release of hydroxyapatite nanoparticles loaded with antibiotics. *J. Mater. Chem. B* **2017**, *5*, 7819–7830. [CrossRef]
171. Liu, X.; Man, H.C. Laser fabrication of Ag-HA nanocomposites on Ti6Al4V implant for enhancing bioactivity and antibacterial capability. *Mater. Sci. Eng. C Mater. Biol. Appl.* **2017**, *70*, 1–8. [CrossRef]
172. Zhao, S.F.; Jiang, Q.H.; Peel, S.; Wang, X.X.; He, F.M. Effects of magnesium-substituted nanohydroxyapatite coating on implant osseointegration. *Clin. Oral Implant. Res.* **2013**, *24* (Suppl. S100), 34–41. [CrossRef]
173. Vu, A.A.; Robertson, S.F.; Ke, D.; Bandyopadhyay, A.; Bose, S. Mechanical and biological properties of ZnO, SiO2, and Ag2O doped plasma sprayed hydroxyapatite coating for orthopaedic and dental applications. *Acta Biomater.* **2019**, *92*, 325–335. [CrossRef]
174. Levengood, S.L.; Zhang, M. Chitosan-based scaffolds for bone tissue engineering. *J. Mater. Chem B* **2014**, *2*, 3161–3184. [CrossRef]
175. Balagangadharan, K.; Dhivya, S.; Selvamurugan, N. Chitosan based nanofibers in bone tissue engineering. *Int. J. Biol. Macromol.* **2017**, *104*, 1372–1382. [CrossRef]
176. Talebian, S.; Mehrali, M.; Mohan, S.; Raghavendran, H.R.B.; Mehrali, M.; Khanlou, H.M.; Kamarul, T.; Afifi, A.M.; Abass, A.A. Chitosan (PEO)/bioactive glass hybrid nanofibers for bone tissue engineering. *RSC Adv.* **2014**, *4*, 49144–49152. [CrossRef]
177. Padovani, G.C.; Feitosa, V.P.; Sauro, S.; Tay, F.R.; Duran, G.; Paula, A.J.; Duran, N. Advances in Dental Materials through Nanotechnology: Facts, Perspectives and Toxicological Aspects. *Trends Biotechnol.* **2015**, *33*, 621–636. [CrossRef]
178. Mishra, S.K.; Ferreira, J.M.; Kannan, S. Mechanically stable antimicrobial chitosan-PVA-silver nanocomposite coatings deposited on titanium implants. *Carbohydr. Polym.* **2015**, *121*, 37–48. [CrossRef]
179. Palla-Rubio, B.; Araujo-Gomes, N.; Fernandez-Gutierrez, M.; Rojo, L.; Suay, J.; Gurruchaga, M.; Goni, I. Synthesis and characterization of silica-chitosan hybrid materials as antibacterial coatings for titanium implants. *Carbohydr. Polym.* **2019**, *203*, 331–341. [CrossRef]
180. Valverde, A.; Perez-Alvarez, L.; Ruiz-Rubio, L.; Pacha Olivenza, M.A.; Garcia Blanco, M.B.; Diaz-Fuentes, M.; Vilas-Vilela, J.L. Antibacterial hyaluronic acid/chitosan multilayers onto smooth and micropatterned titanium surfaces. *Carbohydr. Polym.* **2019**, *207*, 824–833. [CrossRef]
181. Feng, W.; Geng, Z.; Li, Z.; Cui, Z.; Zhu, S.; Liang, Y.; Liu, Y.; Wang, R.; Yang, X. Controlled release behaviour and antibacterial effects of antibiotic-loaded titania nanotubes. *Mater. Sci. Eng. C Mater. Biol. Appl.* **2016**, *62*, 105–112. [CrossRef]

182. Lai, M.; Jin, Z.; Yanga, X.; Wang, H.; Xu, K. The controlled release of simvastatin from TiO2 nanotubes to promote osteoblast differentiation and inhibit osteoclast resorption. *Appl. Surf. Sci.* **2017**, *396*, 1741–1751. [CrossRef]

183. Rahnamaee, S.Y.; Bagheri, R.; Heidarpour, H.; Vossoughi, M.; Samadikuchaksaraei, A. Nanofibrillated chitosan coated highly ordered titania nanotubes array/graphene nanocomposite with improved biological characters. *Carbohydr. Polym.* **2020**, *254*, 117465. [CrossRef] [PubMed]

184. Song, Y.; Ren, S.; Xu, X.; Wen, H.; Fang, K.; Song, W.; Wang, L.; Jia, S. Immobilization of chitosan film containing semaphorin 3A onto a microarc oxidized titanium implant surface via silane reaction to improve MG63 osteogenic differentiation. *Int. J. Nanomed.* **2014**, *9*, 4649–4657. [CrossRef]

185. Mattioli-Belmonte, M.; Ferretti, C.; Cometa, S. Characterization and cytocompatibility of an antibiotic/chitosan/cyclodextrins nanocoating on titanium implants. *Carbohydr. Polym. Sci. Technol. Asp. Ind. Important Polysacch.* **2014**, *110*, 173–182. [CrossRef] [PubMed]

186. Ma, K.; Cai, X.; Zhou, Y.; Zhang, Z.; Jiang, T.; Wang, Y. Osteogenetic property of a biodegradable three-dimensional macroporous hydrogel coating on titanium implants fabricated via EPD. *Biomed. Mater.* **2014**, *9*, 015008. [CrossRef] [PubMed]

187. Rizeq, B.R.; Younes, N.N.; Rasool, K.; Nasrallah, G.K. Synthesis, Bioapplications, and Toxicity Evaluation of Chitosan-Based Nanoparticles. *Int. J. Mol. Sci.* **2019**, *20*, 5776. [CrossRef]

188. Hu, Y.L.; Wang, Q.; Feng, H.; Shao, J.Z.; Gao, J.Q. Toxicity evaluation of biodegradable chitosan nanoparticles using a zebrafish embryo model. *Int. J. Nanomed.* **2011**, *6*, 3351–3359.

189. Wang, Y.; Zhou, J.; Liu, L.; Huang, C.; Zhou, D.; Fu, L. Characterization and toxicology evaluation of chitosan nanoparticles on the embryonic development of zebrafish, Danio rerio. *Carbohydr. Polym.* **2016**, *141*, 204–210. [CrossRef]

190. Abou-Saleh, H.; Younes, N.; Rasool, K.; Younis, M.H.; Prieto, R.M.; Yassine, H.M.; Mahmoud, K.A.; Pintus, G.; Nasrallah, G.K. Impaired Liver Size and Compromised Neurobehavioral Activity are Elicited by Chitosan Nanoparticles in the Zebrafish Embryo Model. *Nanomaterials* **2019**, *9*, 122. [CrossRef]

191. Zhou, R.; Gao, H. Cytotoxicity of graphene: Recent advances and future perspective. *Wiley Interdiscip Rev. Nanomed Nanobiotechnol.* **2014**, *6*, 452–474. [CrossRef]

192. Zhao, Y.; Chen, P.; Bai, Y.; Li, Y.; Xia, T.; Wang, D.; Wu, Q. Response of microRNAs to in vitro treatment with graphene oxide. *ACS Nano* **2014**, *8*, 2100–2110.

193. Liao, K.H.; Lin, Y.S.; Macosko, C.W.; Haynes, C.L. Cytotoxicity of graphene oxide and graphene in human erythrocytes and skin fibroblasts. *Acs Appl. Mater. Interfaces* **2011**, *3*, 2607–2615. [CrossRef]

194. Chang, Y.; Yang, S.T.; Liu, J.H.; Dong, E.; Wang, Y.; Cao, A.; Liu, Y.; Wang, H. In vitro toxicity evaluation of graphene oxide on A549 cells. *Toxicol. Lett.* **2011**, *200*, 201–210. [CrossRef]

195. Kiew, S.F.; Kiew, L.V.; Lee, H.B.; Imae, T.; Chung, L.Y. Assessing biocompatibility of graphene oxide-based nanocarriers: A review. *J. Control. Release* **2016**, *226*, 217–228. [CrossRef]

196. Jelínek, M.; Smetana, K.; Kocourek, T.; Dvořánková, B.; Zemek, J.; Remsa, J.; Luxbacher, T. Biocompatibility and sp3/sp2 ratio of laser created DLC films. *Mater. Ence Eng. B* **2010**, *169*, 89–93. [CrossRef]

197. Pisarik, P.; Jelinek, M.; Smetana, K., jr.; Dvorankova, B.; Kocourek, T.; Zemek, J.; Chvostova, D. Study of optical properties and biocompatibility of DLC films characterized by sp3 bonds. *Appl. Phys. A* **2013**, *112*, 143–148. [CrossRef]

198. Wang, L.; Qiu, J.; Guo, J.; Wang, D.; Qian, S.; Cao, H.; Liu, X. Regulating the Behavior of Human Gingival Fibroblasts by sp(2) Domains in Reduced Graphene Oxide. *ACS Biomater. Sci. Eng.* **2019**, *5*, 6414–6424. [CrossRef]

199. Ming, G.; Lv, L.; Feng, D.; Niu, T.; Tong, C.; Xia, D.; Wang, S.; Zhao, X.; Liu, J.; Liu, Y. Effects of thermal treatment on the adhesion strength and osteoinductive activity of single-layer graphene sheets on titanium substrates. *Sci. Rep.* **2018**, *8*, 8141.

200. Hvard, B.; Haugen, J.; Shi, J.; Li, Z.X.; Wei, S. Citation: Graphene-Reinforced Titanium Enhances Soft Tissue Seal. *Front. Bioeng. Biotechnol.* **2021**, *9*, 294.

201. Gurunathan, S.; Kim, J.H. Synthesis, toxicity, biocompatibility, and biomedical applications of graphene and graphene-related materials. *Int. J. Nanomed.* **2016**, *11*, 1927–1945. [CrossRef]

202. Shin, S.R.; Li, Y.C.; Jang, H.L.; Khoshakhlagh, P.; Akbari, M.; Nasajpour, A.; Zhang, Y.S.; Tamayol, A.; Khademhosseini, A. Graphene-based materials for tissue engineering. *Adv. Drug Deliv. Rev.* **2016**, *105*, 255–274. [CrossRef]

203. Wang, C.; Hu, H.; Li, Z.; Shen, Y.; Zhang, Y. Enhanced Osseointegration of Titanium Alloy Implants with Laser Microgrooved Surfaces and Graphene Oxide Coating. *ACS Appl. Mater. Interfaces* **2019**, *11*, 39470–39483. [CrossRef] [PubMed]

204. Li, Q.; Wang, Z. Involvement of FAK/P38 Signaling Pathways in Mediating the Enhanced Osteogenesis Induced by Nano-Graphene Oxide Modification on Titanium Implant Surface. *Int. J. Nanomed.* **2020**, *15*, 4659–4676. [CrossRef] [PubMed]

205. Ji, H.; Sun, H.; Qu, X. Antibacterial applications of graphene-based nanomaterials: Recent achievements and challenges. *Adv. Drug Deliv. Rev.* **2016**, *105*, 176–189. [CrossRef] [PubMed]

206. Kang, M.S.; Jeong, S.J.; Lee, S.H.; Kim, B.; Hong, S.W.; Lee, J.H.; Han, D.W. Reduced graphene oxide coating enhances osteogenic differentiation of human mesenchymal stem cells on Ti surfaces. *Biomater. Res.* **2021**, *25*, 4. [CrossRef]

207. Zuo, Z.; Li, Y. Emerging Electrochemical Energy Applications of Graphdiyne. *Joule* **2019**, *3*, 899–903. [CrossRef]

208. Hu, X.; Wang, Y.; Tan, Y.; Wang, J.; Liu, H.; Wang, Y.; Yang, S.; Shi, M.; Zhao, S.; Zhang, Y.; et al. A Difunctional Regeneration Scaffold for Knee Repair based on Aptamer-Directed Cell Recruitment. *Adv. Mater.* **2017**, *29*, 1605235. [CrossRef]

209. Liu, J.; Chen, C.; Zhao, Y. Progress and Prospects of Graphdiyne-Based Materials in Biomedical Applications. *Adv. Mater.* **2019**, *31*, e1804386. [CrossRef]

210. Wang, R.; Shi, M.; Xu, F.; Qiu, Y.; Zhang, Y. Graphdiyne-modified TiO2 nanofibers with osteoinductive and enhanced photocatalytic antibacterial activities to prevent implant infection. *Nat. Commun.* **2020**, *11*, 4465. [CrossRef]
211. Martinez-Marquez, D.; Gulati, K.; Carty, C.P.; Stewart, R.A.; Ivanovski, S. Determining the relative importance of titania nanotubes characteristics on bone implant surface performance: A quality by design study with a fuzzy approach. *Mater. Sci. Eng. C* **2020**, *114*, 110995. [CrossRef]
212. Rahman, S.; Gulati, K.; Kogawa, M.; Atkins, G.J.; Pivonka, P.; Findlay, D.M.; Losic, D. Drug diffusion, integration, and stability of nanoengineered drug-releasing implants in bone ex-vivo. *J. Biomed. Mater. Res. Part A* **2016**, *104*, 714–725. [CrossRef]
213. Daish, C.; Blanchard, R.; Gulati, K.; Losic, D.; Findlay, D.; Harvie, D.J.E.; Pivonka, P. Estimation of anisotropic permeability in trabecular bone based on microCT imaging and pore-scale fluid dynamics simulations. *Bone Rep.* **2017**, *6*, 129–139. [CrossRef]
214. Aw, M.S.; Khalid, K.A.; Gulati, K.; Atkins, G.J.; Pivonka, P.; Findlay, D.M.; Losic, D. Characterization of drug-release kinetics in trabecular bone from titania nanotube implants. *Int. J. Nanomed.* **2012**, *7*, 4883–4892. [CrossRef]
215. Lopez-Heredia, M.A.; Goyenvalle, E.; Aguado, E.; Pilet, P.; Leroux, C.; Dorget, M.; Weiss, P.; Layrolle, P. Bone growth in rapid prototyped porous titanium implants. *J. Biomed. Mater. Research. Part A* **2008**, *85*, 664–673. [CrossRef]
216. Dong, J.; Zhang, S.; Liu, H.; Li, X.; Liu, Y.; Du, Y. Novel alternative therapy for spinal tuberculosis during surgery: Reconstructing with anti-tuberculosis bioactivity implants. *Expert. Opin. Drug. Deliv.* **2014**, *11*, 299–305. [CrossRef]
217. Garg, B.; Mehta, N. Current status of 3D printing in spine surgery. *J. Clin. Orthop. Trauma.* **2018**, *9*, 218–225. [CrossRef]
218. Wong, K.C. 3D-printed patient-specific applications in orthopedics. *Orthop. Res. Rev.* **2016**, *8*, 57–66. [CrossRef]
219. Losic, D. Advancing of titanium medical implants by surface engineering: Recent progress and challenges. *Expert. Opin. Drug. Deliv.* **2021**, 1–24. [CrossRef]

*nanomaterials*

MDPI

*Review*

# Microbial Decontamination and Antibacterial Activity of Nanostructured Titanium Dental Implants: A Narrative Review

Sepanta Hosseinpour, Ashwin Nanda, Laurence J. Walsh * and Chun Xu *

School of Dentistry, The University of Queensland, Herston, QLD 4006, Australia;
sp.hosseinpour@uqconnect.edu.au (S.H.); a.nanda@uq.edu.au (A.N.)
* Correspondence: l.walsh@uq.edu.au (L.J.W.); chun.xu@uq.edu.au (C.X.)

**Abstract:** Peri-implantitis is the major cause of the failure of dental implants. Since dental implants have become one of the main therapies for teeth loss, the number of patients with peri-implant diseases has been rising. Like the periodontal diseases that affect the supporting tissues of the teeth, peri-implant diseases are also associated with the formation of dental plaque biofilm, and resulting inflammation and destruction of the gingival tissues and bone. Treatments for peri-implantitis are focused on reducing the bacterial load in the pocket around the implant, and in decontaminating surfaces once bacteria have been detached. Recently, nanoengineered titanium dental implants have been introduced to improve osteointegration and provide an osteoconductive surface; however, the increased surface roughness raises issues of biofilm formation and more challenging decontamination of the implant surface. This paper reviews treatment modalities that are carried out to eliminate bacterial biofilms and slow their regrowth in terms of their advantages and disadvantages when used on titanium dental implant surfaces with nanoscale features. Such decontamination methods include physical debridement, chemo-mechanical treatments, laser ablation and photodynamic therapy, and electrochemical processes. There is a consensus that the efficient removal of the biofilm supplemented by chemical debridement and full access to the pocket is essential for treating peri-implantitis in clinical settings. Moreover, there is the potential to create ideal nano-modified titanium implants which exert antimicrobial actions and inhibit biofilm formation. Methods to achieve this include structural and surface changes via chemical and physical processes that alter the surface morphology and confer antibacterial properties. These have shown promise in preclinical investigations.

**Keywords:** decontamination; antibacterial agents; nano-modified dental implant; nanostructured titanium; dental implant

**Citation:** Hosseinpour, S.; Nanda, A.; Walsh, L.J.; Xu, C. Microbial Decontamination and Antibacterial Activity of Nanostructured Titanium Dental Implants: A Narrative Review. *Nanomaterials* **2021**, *11*, 2336. https://doi.org/10.3390/nano11092336

Academic Editor: May Lei Mei

Received: 14 August 2021
Accepted: 5 September 2021
Published: 8 September 2021

**Publisher's Note:** MDPI stays neutral with regard to jurisdictional claims in published maps and institutional affiliations.

## 1. Introduction

Today titanium implants have an essential place in dental procedures involving the bones of the jaws, ranging from supporting crowns, bridges and dentures to serving as anchorage points for various orthodontic devices. Titanium shows excellent biocompatibility with the surrounding hard and soft tissues. It has high mechanical strength and rigidity, and its surface can be modified. An increase in surface roughness boosts the anchorage of titanium dental implants with the surrounding bone, hence surface modification of implants has become commonplace [1].

In the 1970s and 1980s, implant surface modifications focused on the macro topography of the implant, including threads, serrations and hollow internal portions [1]. This trend then shifted to microtopographic surface modifications, including sandblasting, etching, abrasion, and laser machining, keeping where the implant form cylindrical with a tapering lower 1/3 pard, to mimic the root structure of a tooth [2]. Today, implant companies are moving to nano topographic modifications of the implant surface to gain superior integration with the bone compared to surfaces that have been sandblasted and acid-etched [3].

Current nano-topographic modifications comprise nanotubes, nanofilaments (fibers), nanodots, and nanocrystalline deposits on the implant surface. All of these can improve osseointegration through a greater surface area [4]. As the dental implant is placed in the prepared space in the jawbones, osteogenic (bone-forming) cells in the blood are attracted to the surface of the implant, and later differentiate into osteoblastic cells that lay down a layer of osteoid matrix. This matures to form bone on the implant. As surface area increases with nano modifications, osteoid deposition also increases, followed by the formation of bone, leading to a stable integration with the surrounding bone [5].

According to Thakral et al., there are four different ways to modify an implant surface at the nanoscale to enhance osseointegration (Table 1) [6]. In physical methods, the modification is carried out directly on the surface. These methods include self-assembly monolayers (where functional groups are attached to the surface to initiate bone formation), compaction of nanoparticles (where functionalized nanoparticles are attached to the surface), and ion beam deposition (where the beam creates nano irregularities) [7].

Chemical methods that generate nanoscale surface changes include treatments by chemicals alone or in concert with electrical changes, such as using electrochemistry. Such methods include etching with multiple acids, peroxidation (by the application of strong peroxides), treatment with strong alkalis (such as using NaOH to produce a Na-titania gel that allows deposition of hydroxyapatite particles), and anodization (where electrochemical techniques create nanotubules). In nanoparticle deposition, the nanoparticles can be bound to the surface [8], such as by using the sol-gel method or direct crystal deposition [9]. The current 3-dimensional (3D) printing method can also be used for surface modifications. Lithography and contact printing method are also used [10]. Anodization of the implant surface appears to provide the most predictable results, and hence it is the most used method for nano modification of implant surfaces [11].

**Table 1.** Summary of the methods for nano modification of the implant surface in order to enhance osteointegration [7].

| Methods of Modification | Types of Modification | Description | Reference |
|---|---|---|---|
| Physical method | Self-assembly monolayer | Functional group attachment for nano enhancement | [12] |
| | Compaction of nanoparticles | The attached nanoparticles increase bone integration | [13] |
| | Ion beam deposition | The laser beam causes nano modification | [13] |
| Chemical method | Acid-etching | Sandblasted and acid-etched treatment with acids | [14] |
| | Peroxidation | Peroxides causing gel for nano modification | [13] |
| | Alkaline treatment | NaOH forming gel to adhere bio-ceramics | [13] |
| | Anodization | Electrochemical nanotube formation | [15] |
| Nano deposition | Sol-gel | Gel formation to enhance nanoparticle adhesion | [16] |
| | Direct crystal deposition | Nanoparticle superimposed on the altered surface | [14] |
| 3D printed modification | Lithography | Nano printing outside the implant and later adhered to the surface | [13] |
| | Contact printing | Nano printing on the implant surface | [17] |

The nano-modified implant surface gains the advantage of an increased surface area for cell adhesion, although this same surface also develops complexities, such as allowing the adhesion and growth of other cells as well as microbial pathogens. If osteogenic cells dominate the surface, then bone formation will occur and there will be a firm bone to implant integration. Conversely, if bacterial growth dominates, the implant will fail to integrate and loss is likely [10].

Nowadays nano modification is directed to the creation of a surface that facilitates the attachment of osteogenic cells, rather than bacteria, which are repelled. Nanospike-like structures are one such example of a bioinspired surface [18]. If bacteria adhere to the surface, these spikes penetrate the cell, and cause rupture of their cell membrane, resulting in their death. A concern with this concept is that impaled cells on the nano spikes could allow other bacteria to attach. Hence, various methods of applying antibacterial medicaments were advocated, to maintain a bacterial-free layer on the nano-modified surface [19].

An ideal surface will support osseointegration and prevent bacterial adhesion and growth. This narrative review outlines treatment modalities that are carried out to eliminate bacterial biofilm and suppress its growth over the nano-modified implant surface. Figure 1 summarizes the current decontamination approaches for nano-modified titanium dental implants.

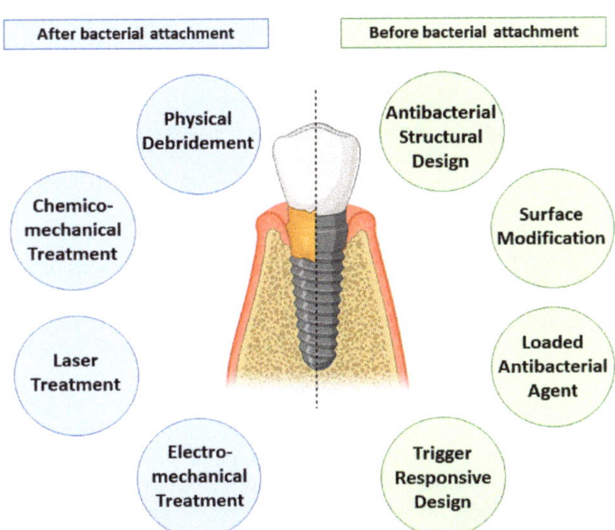


**Figure 1.** Summary of the current bacterial decontamination approaches before and after bacterial adhesion to the nano-modified titanium dental implants including various debridement techniques and inherent self-cleaning strategies.

For this narrative review, a comprehensive search of the MEDLINE, Scopus, EMBASE, Web of Science, and Google Scholar online databases was undertaken, for publications on microbial decontamination and antibacterial features of nano-modified titanium dental implants from all available years, to July 2021. All in vitro, in vivo, and clinical studies which investigated and discussed the topic were included. In addition, conference papers, systematic reviews, meta-analyses, narrative reviews, letters to the editor, book chapters, technical notes and theses were included, to retrieve all existing evidence. All record items had to be in the final published or "in the press" stage to be included in the review.

## 2. Surface Cleaning Techniques

The surface of a dental implant is much more complex than that of a natural tooth. As well, the supporting apparatus of the tooth, the periodontal ligament, contains a rich microvascular bed that allows immune cells to exit at any point around the surface of the root that sits within its socket. On the other hand, the peri-implant site originates by drilling the jawbone, so there is microscopic trauma and injury, followed by inflammation, which must resolve before bone will begin to form around the implant [20]. At any stage, bacteria from the saliva can adhere to exposed surfaces of the implant and begin to form a multispecies bacterial biofilm.

Until the late 1970s, peri-implant diseases were considered to be similar to periodontal diseases around natural teeth and were treated in a similar manner [21]. It is now known that the microbiota around an implant suite with bone loss (i.e., peri-implantitis) and a natural tooth with moderate to severe periodontitis is similar, but the former has more Gram-negative bacteria, with dominating clusters of spirochetes, as well as yeasts [22]. The etiology of peri-implantitis is similar to periodontitis, as both are caused by poor oral hygiene, with the mature biofilm extending into the gingival crevice and driving an inflammatory response. Likewise, the mild reversible forms of the disease, namely peri-implant mucositis and gingivitis around teeth, are similar in their etiology and management. When peri-implant mucositis develops, the circular gingival fibers surrounding the implant collar are broken down, which allows bacterial contamination to extend apically from the coronal portion of the implant. Nevertheless, the inflammation is confined to the soft tissue, and there is no loss of bone [23].

In peri-implantitis, the aggressive form of the disease, bone surrounding the implant is destroyed [24]. If left untreated, peri-implantitis could lead to the movement of implants or even implants failure. The treatment modalities for an implant affected by peri-implantitis begin with mechanical debridement methods adapted from the clinical treatment of periodontitis cases [25] and then extend to more complex methods, including surgical treatments. The treatment of peri-implantitis, which is focused on microbial decontamination of the dental implant surface, can be grouped into the following categories:

### 2.1. Physical Debridement

The physical removal of bacterial biofilms from titanium implant surfaces is the simplest and oldest form of treatment. Initially, hand-operated scalers and then powered (ultrasonic) scalers were used, and more recently particle beams were deployed [26]. A concern with all physical debridement methods is the extent of surface damage they cause. To reduce this, tips can be made of softer materials than stainless steel, such as plastic or carbon fiber. All physical debridement methods are most effective on the smooth parts of abutments and other components joined to implants and least effective on the aspects which have a macro or micro-roughness [27].

Deleterious changes to the surface include deposition of fragments of soft instrument tips over the implant surfaces, scratching and grooving of smooth areas, and flattening of projections on rough areas, thus disrupting the features of the implant surface. Such problems were noted with sandblasted and acid-etched (SLA) surfaces where the surface was modified to create a micro-roughness [28]. Hence, such methods would be contraindicated for nano-modified implant surfaces, because of the risk of distorting the nanoengineered surface features.

In particle beams, also known as air abrasion, suspended particles (such as sodium bicarbonate, calcium carbonate, glycine, erythritol, or hydroxyapatite) in a compressed airstrike or an air–water stream impact onto the implant surface. This detaches some parts of the biofilm but is not effective in areas that are protected from or inaccessible to the particle beam (such as parts of the threads) [29]. Although the particle beam method is superior to mechanical debridement using hand-operated or powered scalers, it has several drawbacks. Particles can be embedded into the implant surface, which can change its physical and chemical characteristics. The abrasive particles also degrade the surface microscopic features through fracture-based mechanisms. The compressed air also poses a risk of air emphysema around the implant [30].

With a nano-modified surface on the titanium implant, the particles may impact the surface, degrading nano projections and potentially leaving residues trapped between projections that may not be readily removed by the flow of water in the stream. Hence a particle beam method would be contraindicated for a nano-modified titanium implant surface [31].

### 2.2. Chemo-Mechanical Treatment

In chemo-mechanical treatment, chemicals are used combination with physical treatment. For instance, mineralized biofilms (e.g., dental calculus) are first removed with an ultrasonic scaler, and then the pharmacologically active substance is applied with a specialized brush made of plastic or titanium bristles. Chemical agents include antibiotics (such as tetracyclines), biocides (chlorhexidine, hydrogen peroxide), or weak acids (citric acid, reviewed in [32]). The brush is attached to a low-speed rotary handpiece, and the implant surface is cleaned using a rotary motion [33]. Concerns with this method are surface scratching and degradation, and entrapment of fragments. As well, to gain access to the implant threads, surgical access to the site may be needed [34]. All of these considerations argue against using a chemo-mechanical approach on a nano-modified implant surface. There are also concerns that any applied antibiotic agent will readily rinse away from the implant surface, through the action of saliva or blood, hence if antibiotics are desired, systemic administration would be preferred [35].

There is potential to incorporate biocompatible materials with low abrasive in this method. One material of interest is chitosan, a marine biopolymer that is based on chitin derived from the shells of marine crustaceans. It is approved for use in surgical bandages as a hemostatic agent, and it is safe when ingested as a dietary supplement. A split-mouth randomized clinical trial and case series studies using chitosan on an oscillating brush reported it to be effective in the treatment of mild peri-implantitis, with a rapid reduction in inflammation [36,37]. A further advantage is that if any residues remain, chitosan is non-allergenic and may exert anti-inflammatory actions [38].

### 2.3. Laser Ablation and Photodynamic Therapy

Infrared lasers when used with high peak powers can exert photothermal actions which will denature the cell walls of bacteria. Commonly used lasers include Er: YAG (Erbium: Yttrium Aluminum Garnet), Nd: YAG (Neodymium doped Yttrium Aluminum Garnet), and CO2 lasers, and GaAlAs (Gallium Aluminum Arnside) diode lasers [39]. Due to reflection, adverse actions on the surface are less for the longer wavelengths, particularly the Er: YAG laser. Use of the Nd: YAG laser is discouraged, as the wavelength is absorbed strongly by titanium, and surface melting and hot plasma effects can occur which would degrade the surface characteristics [40].

Photodynamic therapy is a non-thermal process that is based on the use of a low-power laser with an appropriate wavelength to absorb by a photosensitizer dye. The resulting oxygen radicals produced will kill bacteria to which the dye has bound [41]. A range of photosensitizers was used, including toluidine blue O (tolonium chloride), which has a strong safety profile [41]. Photodynamic therapy in canine animal models has shown a reduction in bacterial counts of Prevotella intermedia/nigrescens, Fusobacterium spp., and beta-hemolytic Streptococcus species [42]. This destruction of bacteria occurs without any damage to the underlying titanium surface [43]. The lack of surface effects makes this method attractive for use on nano-modified implant surfaces. On the other hand, the use of lasers with high peak powers on nano-modified surfaces could disrupt the integrity of the implant surface at the nanoscale. Additionally, such a laser system would be expensive and would require a specially trained and skillful operator, to minimize injury to adjacent tissues [43].

### 2.4. Electromechanical Treatment

This method for reducing or eliminating bacterial biofilms on titanium relies on electrical current flow and the generation of various chemical species that can disrupt biofilms or kill bacteria. Typically, the titanium implant is the anode, and the current flow is through an electrolyte that is specially designed to maximize biofilm disruption. A low voltage and a low current flow are used. High currents are avoided as these would potentially cause some microscopic surface loss from the implant. Conversely, if the current is too low, the decontaminant is process is not very effective [44].

The basic principles were laid down in 1992, and supporting evidence began to build from 2011 in preclinical models [45]. In 2021, the technology was deployed into clinical practice, moving from the preclinical phase, and animal studies have continued [46–49]. Schlee et al. (2019) documented the application of electrochemical decontamination of dental implants in the patient for the first time. The Galvo Surge GS-1000 device was used, with a sodium formate solution as the electrolyte. Effective disinfection was observed on the titanium implant surface, as hydrogen gas bubbles disrupted the biofilms and lifted them away. At 6 months follow up, the treated implants showed re-integration with the surrounding bone [50].

The electrochemical method of decontamination appears promising, but the current commercial system of this type requires surgical access to the affected site. Other methods that do not require surgery are thus attractive. These use lower voltages and currents, are not very technique sensitive, and cause almost no changes to the surface of the implant [51]. There is a need to explore further what effects the electrochemical method for bacterial biofilm elimination may have on the integrity of adjacent normal cells and tissues [47].

There is as yet no evidence of the use of electrochemical methods on nano-modified implant surfaces, as past work has focused on surfaces modified at the micro rather than at the nanoscale. It is hopeful that using low voltages, effective decontamination can be achieved. The method could also extend the process of anodization of the titanium surface whereby nanotubes are fabricated on the implant surface. With controlled parameters, ideally, biofilm elimination could be accompanied by an optimized nano-modified titanium implant surface.

## 3. Structural Enhancements and Experimental Designs

The above-mentioned methods focus on decontaminating the implant surface. Moving beyond that, it would be desirable to stop bacterial biofilms reforming on the treated surface. Thus, recent research has explored nano-scale modifications to the implant to functionalize the surface, endowing it with passive and/or active antibacterial activities [51]. According to Liu et al. [51], approaches for nano modification of implant surfaces to counter bacterial growth can be classified as based on the dimensions of the change, i.e., zero-dimensional (nanoparticles), one-dimensional (nanowires), two-dimensional (nanofilms) and three-dimensional (nano-blocks). Based on structure, these could be classified as antibacterial nanoparticles, antibacterial nano solids and antibacterial nano-assembled structures. Alternatively, based on the nature of the antibacterial active ingredient, they can be classified as using metallic ions or oxide photocatalysts [52].

A nano-modified surface with antibacterial features can be inspired by nature, such as where the nano protrusions on the surface mimic the wings of a dragonfly or cicada [53]. Sharp projections created on the surface of the metal can cause stress and deform the microbial cell membrane, leading to its rupture, and hence causing bacteriolysis. Such biomimicry thus not only kills the bacteria on the surface of the implant but also prevents future bacterial growth. If this was achieved, it would not be necessary to use chemical agents to eliminate the microorganisms [54]. The argument could be made that nano-scale surface projections could lower the mechanical strength of the implant by a trivial amount, while the antimicrobial property will enhance the integration of the implant with adjacent bone, thus boosting the overall success of the implant system. Table 2 summarizes various nano-structural modifications that can provide antibacterial properties.

**Table 2.** Nano-structural modifications of dental implants that enhance antibacterial properties.

| Nanostructures | Fabrication | Wettability | Surface Roughness ($R_a$ in Nanometers) | Antibacterial Effect/Rate |
| --- | --- | --- | --- | --- |
| (1) Nanoflowers of pure Titanium | Chemical etching → Hydrothermal oxidation | Hydrophilic surface | 829 | *S. aureus*: 43.12%/24 h Methicillin-resistant *S. aureus*: 73.15%/24 h |

**Table 2.** *Cont.*

| Nanostructures | Fabrication | Wettability | Surface Roughness ($R_a$ in Nanometers) | Antibacterial Effect/Rate |
|---|---|---|---|---|
| (2) Nanowires of grade V titanium alloy | Hydrothermal synthesis | Hydrophilic surface | – | *S. aureus*: 74%/18 h |
| (3) Regular nanotubes of grade V titanium alloy | Acid etching → anodic oxidation | Hydrophilic surface | 120 | *E. coli*: 72.6%/2 h<br>*S. aureus*: 68.2%/2 h |
| (4) Irregular nanotubes of grade V titanium alloy | Electrochemical anodization | Super Hydrophilic surface | 360 | *E. coli*: 48.7%/2 h<br>*S. aureus*: 50.8%/2 h |
| (5) Nanotubes of pure Titanium | Electrochemical anodization | Hydrophilic surface | 45.60 rms [Root-mean square] | *S. aureus*: 36.78%/16 h |

**Table 2.** *Cont.*

| Nanostructures | Fabrication | Wettability | Surface Roughness ($R_a$ in Nanometers) | Antibacterial Effect/Rate |
|---|---|---|---|---|
| (6) Nanoripples of pure Titanium<br>5000 nm | Femtosecond laser direct writing | Super hydrophilic surface | 274.6 | *E. coli*: 56%/24 h |
| (7) Nanoparticles of Aluminum-Titanium alloy<br>5000 nm | Aerosol flame synthesis | Super hydrophilic surface | – | *S. aureus*: 80% |
| (8) Nanopillars of pure Titanium<br>1000 nm | Plasma etching | Hydrophobic surface | – | *P. aeruginosa*: 87 ± 2%/24 h *S. aureus*: 72.5 ± 13%/24 h |

Among the various nanostructures, nanopillars have the strongest bactericidal action. Importantly, none of the nanostructures reduce the cellular activity of normal human cells, rather, the surface increases the metabolic activity of the cells that attach to it [55]. Antibacterial nanostructures can exert a modest bactericidal effect, but a limitation is that patterning the surface to mimic bioinspired features is more difficult for titanium than for materials such as polymers or silicones [51].

To create nanostructures on titanium surfaces, consideration must be given to the precise height, width and dimensions that are desired, as these play a key role. Until the present, there are only two methods by which nanostructures are fabricated on a titanium dental implant surface, namely two-photon polymerization and electron beam-induced deposition [56]. In the two-photon polymerization method, the Computer-Aided Design And Manufacturing (CAD/CAM) method of fabrication plays a vital role, as the accuracy of the nanostructures can be corrected and controlled to a precision of around 100 nm. This method has great versatility in terms of the types of nanostructures that can be created on the titanium surface [57].

The electron beam-induced deposition method, on the other hand, is more popular as a method of fabricating nanostructure. It employs vertical deposition of new material onto the surface, at a rate of approximately 1 nm/s. The rate of deposition can be altered by varying the deposition method, optimizing the gas injection system and/or changing the temperature [58,59].

Further work is needed to explain the mechanism of how nanostructure features cause the lysis of bacteria. There is evidence that the effect involves more than mechanical interlocking with nano-protrusion of projections into the bacterial cells. There may also be effects that are mediated by disruption of the extracellular polymeric substances that anchor bacteria to the surface [24]. As cells move, the projections may tear through the cell membrane, causing bacteriolysis [60]. A combination of these effects may drive severe plastic deformation of the bacterial cells, hence making the surface antibacterial in nature [61].

### 3.1. Topography and Chemistry of Titanium Nanotube and Their Antibacterial Activity

Nanotechnology holds considerable promise as a method to functionalize surfaces, endowing them with specific properties such as being "self-cleaning" [62,63]. Various topographical and chemical characteristics of titanium oxide nanotubes (TNTs) can influence bacterial attachment to nano-modified titanium implants. Titanium oxide-coated titanium structures can reduce bacterial adhesion and exert direct antibacterial properties, because of surface roughness at the nanoscale and higher surface energy [64,65]. TNTs have thus been introduced as an antibacterial coating candidate for dental implants [66–68].

Other features of TNTs are relevant to dental implants, including their highly ordered structure, high surface area and roughness, and capability of being loaded with therapeutic agents. These features make TNTs attractive for enhancing osseointegration and bone regeneration [63]. TNTs promote the adhesion, proliferation and differentiation of osteoprogenitor cells, especially for nanotubes with smaller diameters (<30 nm) compared to larger diameters (70 nm) [69]. Surface topography also influences this effect [69].

Bacteria cultured on surfaces with TNTs (40 to 60 nm diameters) have shown the greatest level of reduction in number when compared to smoother surfaces [70]. This could be due to the stress response of bacteria to TNTs which cause rupture of their cellular membrane [71]. Further work is needed to examine how inhibition of bacterial adhesion and proliferation may modulate drug resistance in bacteria [72], as this effect is not well understood [73].

There are contradictory reports regarding how the hydrophilicity of the surface with TNTs influences bacteria [73–75]. Greater hydrophilicity of the surface may enhance bacterial proliferation and adhesion. An increase in the diameter of the nanotubes may enhance bacterial adhesion [74,75]. One report has described how the number of bacteria first reduces and then rises, depending on the diameter of the nanotubes [73]. Overall, this process is complex and requires further investigation.

Shi et al. have shown excellent antibacterial properties of TNTs compared to smooth sheets of titanium [73]. They believed that this performance difference may be due to the impact of sterilization at the nanoscale. Ultraviolet C irradiation when used for sterilization creates highly oxidative holes on the surface which can react with oxygen and moisture, and produce free radicals [76]. These free radicals on the surface of the titanium implant may then rupture bacterial cell membranes via lipid peroxidation, and cause death [77].

The geometrical characteristics of TNTs influence their antibacterial properties [78,79], especially their diameter and surface area [78], with the greatest reduction in bacterial survival rate occurring for nanotubes with a diameter from 40 to 60 nm [73]. TNTs with a diameter of 60 nm have thin walls and greater photocatalytic actions compared with smaller diameters. Antibacterial actions fall away at a diameter of 100 nm.

### 3.2. Surface Modifications

There has been an interest in the application of antifouling polymers on the surface of dental implants, to inhibit bacterial attachment. These surface coatings consist of hydrophilic and zwitterionic polymers that reduce the adhesion of bacteria but do not kill bacteria outright [80]. The polymers create a hydrated layer on the surface that reduces protein adsorption. The effect varies according to the length of the polymer chains, and the uniformity and density of the polymer [81].

Polyethylene glycol (PEG) is a hydrophilic polymer that has known antifouling actions for dental implants because its hydrophilic chain prevents protein adsorption [82,83]. As summarized in Figure 2, Skovdal et al. showed that a coating of ultra-dense PEG on the surface of a titanium implant can reduce the adhesion of Staphylococcus aureus by 89–93%. In this manner, ultra-dense PEG coatings improved the treatment outcome for implant-associated infections in mice after 5 days [84]. However, this same adhesion blocking the action of PEG also hampers the adhesion of human cells, which would compromise osseointegration. Immobilizing specific bioactive molecules, such as integrin-binding peptide sequences, onto the implant surface may be a means to overcome this disadvantage [85]. Thus, an ideal therapeutic approach should use a surface modification approach where the antibacterial activity does not affect the adhesion, proliferation, and differentiation of the human cells that are needed for successful dental implant treatment.

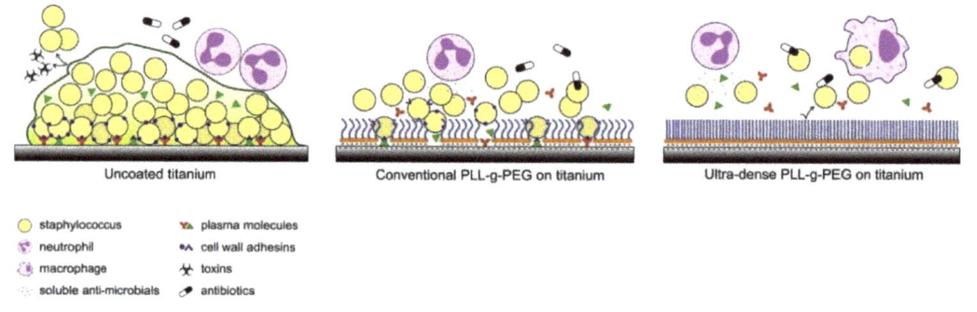


**Figure 2.** An ultra-dense PEG coating resists the binding of Staphylococcus epidermidis, which remains loosely adherent to the surface (Reprinted with permission from ref. [84]. Copyright 2018 Acta Biomater).

Polyphenols are another potential coating material of interest for titanium dental implants with nano-scale features. Polyphenols of plant origin have attracted much attention due to purported benefits for human health [86]. Tannic acid was reported to prevent surface colonization [87]. Other polyphenolic molecules of interest derived from natural sources include catechins and pyrogallol [88]. Polyphenol functionalization of titanium dental implants can give antibacterial effects as well as enhanced osteointegration and osteoinduction [89,90].

### 3.3. Loading Nanotubes with Drugs and Antibacterial Nanoparticles

Due to their structure, TNTs can be loaded with substantial amounts of various materials as cargo, including antibiotics, anti-inflammatory agents, nanoparticles, and ions [91–93]. Loading with antibiotics could greatly enhance antibacterial actions. TNTs loaded with vancomycin and antimicrobial peptides have shown enhanced antibacterial actions against Staphylococcus epidermidis and methicillin-resistant Staphylococcus aureus, and reduced adhesion of bacteria to the implant surface [92,94,95].

TNTs can also be loaded with nanoparticles and with various ions [Ag, Au, Cu] that have antibacterial actions, using techniques such as spin coating, sputtering, chemical reduction and drop-casting [93]. Jia et al. have presented a novel strategy for hierarchical $TiO_2/Ag$ coating which was able to reduce bacterial adhesion and lower their viability [96]. As shown in Figure 3, their proposed "trap-killing" principle involves multiple elements.

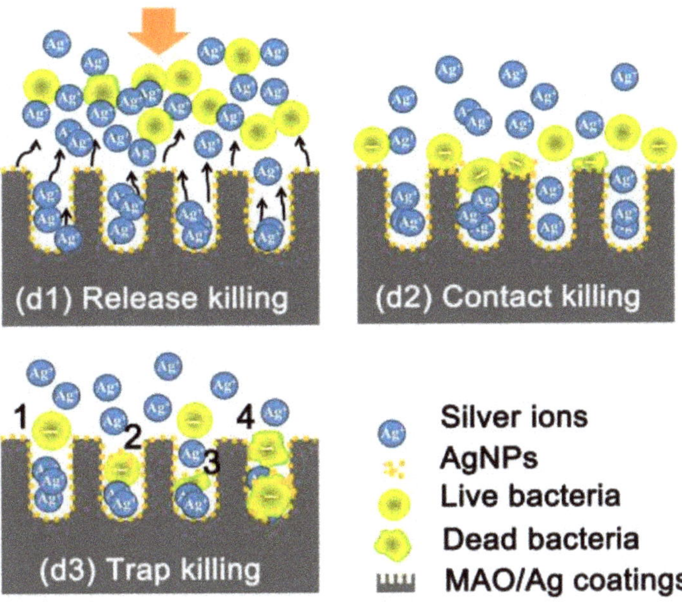


**Figure 3.** Schematic illustration of the possible antibacterial mechanisms involved during adhesion of bacteria to nanotubes: (**d1**) The majority of planktonic bacteria are repulsed from the surface by releasing Ag+ ions; (**d2**) Some of the landed bacteria are disrupted via contact with Ag nanoparticles on the surface; (**d3**) Surviving bacteria with a negative membrane charge are attracted into micropores (positively charged by interior silver) and killed (Reprinted with permission from ref. [96]. Copyright 2015 Biomaterials).

The use of ion/nanoparticle functionalization of TNTs holds great promise for applications on dental implants with nanoscale features. Issues with cytotoxicity from the released ions or nanoparticles need to be explored further, as these need to be balanced with their therapeutic efficiency.

### 3.4. Trigger-Responsive Therapy

A key concept is using a coating that can release the drug(s) only when needed, to give enhanced antibacterial activity, using trigger responsive release systems [84]. Such coatings would be responsive to changes in the local microenvironment, such as specific biomolecules whose concentrations would rise, to initiate the release of the cargo. As an example, an infection would lower the pH and raise the temperature. Hence, pH-responsive and temperature-sensitive materials could work as a release trigger for antibacterial agents. As proof of this concept, Dong et al. utilized a pH-responsive acetal linker and loaded silver nanoparticles into titanium nanotubes [97]. This coating maintained the silver nanoparticles at a pH around 7 (physiological pH), but then rapidly released the nanoparticles when the pH fell to approximately 5.5. In addition, Li et al. described a thermosensitive coating with a layer of hydrogel, which gave a highly efficient antibacterial action [98]. This was used to give a heat-triggered release of glycerin in an animal infection model Figure 4.

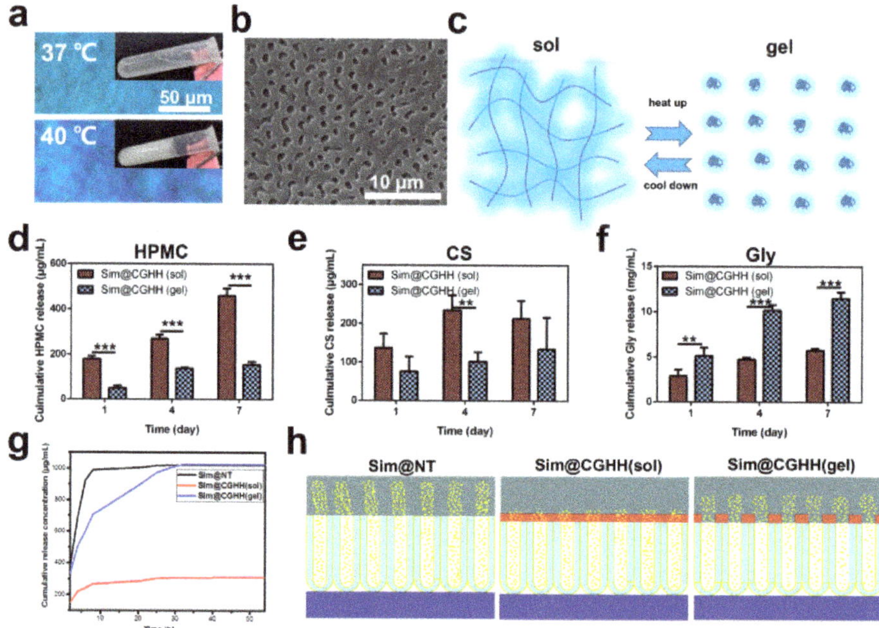

**Figure 4.** (**a**): Optical images of a chitosan-glycerin-hydroxypropyl methylcellulose hydrogel (CGHH) at 37 and 40 °C; (**b,c**): Thermal transition of CGHH between the sol and gel states; (**d–g**): the release rates for HPMC, CS, Gly, and Sim; (**h**): Simvastatin release from Sim@Nanotube (NT), Sim@CGHH and Sim@CGHH ** $p < 0.01$, *** $p < 0.001$ (Reprinted with permission from ref. [98]. Copyright 2021 Materials Sci. Eng. C.).

During bacterial infections, certain enzymes secreted by the bacteria could also act as a trigger. Vancomycin-loaded TNTs with specific enzyme responsive coatings were developed as a successful application of this concept [99]. A catechol-functionalized hyaluronic acid and chitosan coating was utilized as a multilayer coating of the TNTs. This coating degraded due to exposure to bacterial hyaluronidase in an infection model, which released the loaded vancomycin and ultimately killed the bacteria.

Another approach for titanium dental implants with nano topography is photo-catalytical processes linked to bactericidal coatings to give site- and time-specific antibacterial activity. Titanium dioxide has well-documented photocatalytic activity and antibacterial properties due to the production of reactive oxygen species (ROS), such as superoxide anions and hydroxyl radicals, following exposure to light [100]. These ROS degrade the cell membranes of bacteria. The photocatalytic effects of titanium oxide are mediated by the crystalline form [101], such as rutile and anatase. The latter has superior photoinduced antibacterial activity than the former [102]. Photo-catalytical antibacterial activity can be triggered by visible light, and especially by shorter wavelengths of light in the ultraviolet range. Further work is needed to determine how best to deliver short wavelengths of light in subgingival sites. The safety issues with ultraviolet (UV) light used for activation need to be addressed [102,103].

## 4. Conclusions

Although many studies have evaluated treatments for peri-implantitis, few have addressed the specific situation of titanium implants with nano-modified surfaces. Efficient removal of the biofilm remains paramount, supplemented by chemical treatments. Given the heterogeneity of studies and the combination of various methods, it is not yet possible to identify a single standard protocol for bacterial decontamination of nano-modified titanium dental implants. Despite this, there is a range of promising methods that will not influence nano-scale surface features. Further randomized clinical trials are required to establish the most cost-effective approaches such as photodynamic therapy with lasers and trigger-responsive therapies based on photo-catalytic actions. Such methods seem ideally suited to use with nano-modified antimicrobial titanium implants. Local delivery of ions or antibiotics to inhibit bacterial adhesion also appears very promising. Further preclinical investigations and randomized clinical trials are required to verify the existing preliminary findings, and to guide translation of these concepts into clinical practice.

**Author Contributions:** Conceptualization, L.J.W. and C.X. and S.H.; methodology, S.H. and A.N.; investigation, L.J.W. and C.X. and S.H. and A.N.; writing—original draft preparation, S.H. and A.N.; writing—review and editing, L.J.W. and C.X.; visualization, S.H. and A.N.; supervision, L.J.W. and C.X. All authors have read and agreed to the published version of the manuscript.

**Funding:** This research received no external funding.

**Acknowledgments:** S.H. acknowledges support from a University of Queensland post-graduate scholarship.

**Conflicts of Interest:** The authors declare no conflict of interest.

# References

1. John, A.A.; Jaganathan, S.K.; Supriyanto, E.; Manikandan, A. Surface modification of titanium and its alloys for the enhancement of osseointegration in orthopaedics. *Curr. Sci.* **2016**, *111*, 1003. [CrossRef]
2. Liu, Y.; Rath, B.; Tingart, M.; Eschweiler, J. Role of implants surface modification in osseointegration: A systematic review. *J. Biomed. Mater. Res. Part A* **2020**, *108*, 470–484. [CrossRef] [PubMed]
3. Souza, J.C.; Sordi, M.B.; Kanazawa, M.; Ravindran, S.; Henriques, B.; Silva, F.; Aparicio, C.; Cooper, L.F. Nano-scale modification of titanium implant surfaces to enhance osseointegration. *Acta Biomater.* **2019**, *94*, 112–131. [CrossRef] [PubMed]
4. Wang, N.; Li, H.; Lü, W.; Li, J.; Wang, J.; Zhang, Z.; Liu, Y. Effects of $TiO_2$ nanotubes with different diameters on gene expression and osseointegration of implants in minipigs. *Biomaterials* **2011**, *32*, 6900–6911. [CrossRef] [PubMed]
5. Rani, V.D.; Vinoth-Kumar, L.; Anitha, V.; Manzoor, K.; Deepthy, M.; Shantikumar, V.N. Osteointegration of titanium implant is sensitive to specific nanostructure morphology. *Acta Biomater.* **2012**, *8*, 1976–1989. [CrossRef] [PubMed]
6. Thakral, G.; Thakral, R.; Sharma, N.; Seth, J.; Vashisht, P. Nanosurface–the future of implants. *J. Clin. Diag. Res. JCDR* **2014**, *8*, ZE07. [CrossRef]
7. Subramani, K.; Mathew, R.T.; Pachauri, P. Titanium surface modification techniques for dental implants—From microscale to nanoscale. *Emerg. Nanotechnol. Dent.* **2018**, 99–124. [CrossRef]
8. Kulkarni, M.; Mazare, A.; Schmuki, P.; Iglič, A. Biomaterial surface modification of titanium and titanium alloys for medical ap-plications. *Nanomedicine* **2014**, *111*, 111.
9. Jaafar, A.; Hecker, C.; Árki, P.; Joseph, Y. SOL-gel derived hydroxyapatite coatings for titanium implants: A review. *Bioengineering* **2020**, *7*, 127. [CrossRef]
10. Prodanov, L.; Lamers, E.; Domanski, M.; Luttge, R.; Jansen, J.A.; Walboomers, X.F. The effect of nanometric surface texture on bone contact to titanium implants in rabbit tibia. *Biomaterials* **2013**, *34*, 2920–2927. [CrossRef] [PubMed]
11. Yao, C.; Webster, T.J. Anodization: A promising nano-modification technique of titanium implants for orthopedic applications. *J. Nanosci. Nanotechnol.* **2006**, *6*, 2682–2692. [CrossRef] [PubMed]
12. Christenson, E.M.; Anseth, K.S.; van den Beucken, J.J.; Chan, C.K.; Ercan, B.; Jansen, J.A.; Laurencin, C.T.; Wan-Ju, L.; Ramalingam, M.; Lakshmi, S.; et al. Nanobiomaterial applications in or-thopedics. *J. Orthopaedic Res.* **2007**, *25*, 11–22. [CrossRef]
13. Webster, T.J.; Ejiofor, J.U. Increased osteoblast adhesion on nanophase metals: Ti, Ti6Al4V, and CoCrMo. *Biomaterials* **2004**, *25*, 4731–4739. [CrossRef]
14. Klabunde, K.J.; Stark, J.; Koper, O.; Mohs, C.; Park, D.G.; Decker, S.; Jiang, Y.; Lagadic, I.; Zhang, D. Nanocrystals as stoichiometric reagents with unique surface chemistry. *J. Phys. Chem.* **1996**, *100*, 12142–12153. [CrossRef]
15. Zhao, G.; Zinger, O.; Schwartz, Z.; Wieland, M.; Landolt, D.; Boyan, B.D. Osteoblast-like cells are sensitive to submicron-scale surface structure. *Clin. Oral Implants Res.* **2006**, *17*, 258–264. [CrossRef]
16. Gutwein, L.G.; Webster, T.J. Increased viable osteoblast density in the presence of nanophase compared to conventional alumina and titania particles. *Biomaterials* **2004**, *25*, 4175–4183. [CrossRef] [PubMed]
17. Qin, J.; Yang, D.; Maher, S.; Lima-Marques, L.; Zhou, Y.; Chen, Y.; Atkins, G.; Losic, D. Micro- and nano-structured 3D printed titanium implants with a hydroxyapatite coating for improved osseointegration. *J. Mater. Chem. B* **2018**, *6*, 3136–3144. [CrossRef]
18. Patil, D.; Wasson, M.; Perumal, V.; Aravindan, S.; Rao, P. *Bactericidal Nanostructured Titanium Surface through Thermal Annealing Advances in Micro and Nano Manufacturing and Surface Engineering*; Springer: Amsterdam, The Netherlands, 2019; pp. 83–92.
19. Sjöström, T.; Nobbs, A.; Su, B. Bactericidal nanospike surfaces via thermal oxidation of Ti alloy substrates. *Mater. Lett.* **2016**, *167*, 22–26. [CrossRef]

20. Emecen-Huja, P.; Eubank, T.D.; Shapiro, V.; Yildiz, V.; Tatakis, D.N.; Leblebicioglu, B. Peri-implant versus periodontal wound healing. *J. Clin. Periodontol.* **2013**, *40*, 816–824. [CrossRef] [PubMed]
21. Koyanagi, T.; Sakamoto, M.; Takeuchi, Y.; Maruyama, N.; Ohkuma, M.; Izumi, Y. Comprehensive microbiological findings in peri-implantitis and periodontitis. *J. Clin. Periodontol.* **2013**, *40*, 218–226. [CrossRef]
22. Persson, G.R.; Renvert, S. Cluster of bacteria associated with peri-implantitis. *Clin. Implant Dent. Relat. Res.* **2014**, *16*, 783–793. [CrossRef] [PubMed]
23. Figuero, E.; Graziani, F.; Sanz, I.; Herrera, D.; Sanz, M. Management of peri-implant mucositis and peri-implantitis. *Periodontology 2000* **2014**, *66*, 255–273. [CrossRef] [PubMed]
24. Berglundh, T.; Jepsen, S.; Stadlinger, B.; Terheyden, H. Peri-implantitis and its prevention. *Clin. Oral Implants Res.* **2019**, *30*, 150–155. [CrossRef] [PubMed]
25. Schwarz, F.; Derks, J.; Monje, A.; Wang, H.L. Peri-implantitis. *J. Clin. Periodontol.* **2018**, *45*, S246–S266. [CrossRef]
26. Tran, C.; Walsh, L.J. Novel models to manage biofilms on microtextured dental implant surfaces. In *Microbial Biofilms—Importance and Applications*; Dhanasekaran, D., Thajuddin, N., Eds.; InTechOpen: London, UK, 2016; pp. 463–486.
27. Khammissa, R.A.G.; Feller, L.; Meyerov, R.; Lemmer, J. Peri-implant mucositis and peri-implantitis: Bacterial infection. *S. Afr. Dent. J.* **2012**, *67*, 70–74.
28. Tang, Z.; Cao, C.; Sha, Y.; Lin, Y.; Wang, X. Effects of non-surgical treatment modalities on peri-implantitis. *Zhonghua Kou Qiang Yi Xue Za Zhi/Zhonghua Kouqiang Yixue Zazhi/Chin. J. Stomatol.* **2002**, *37*, 173–175.
29. John, G.; Sahm, N.; Becker, J.; Schwarz, F. Nonsurgical treatment of peri-implantitis using an air-abrasive device or mechanical debridement and local application of chlorhexidine. Twelve-month follow-up of a prospective, randomized, controlled clinical study. *Clin. Oral Investig.* **2015**, *19*, 1807–1814. [CrossRef]
30. Renvert, S.; Lindahl, C.; Jansåker, A.-M.R.; Persson, G.R. Treatment of peri-implantitis using an Er:YAG laser or an air-abrasive device: A randomized clinical trial. *J. Clin. Periodontol.* **2011**, *38*, 65–73. [CrossRef]
31. Yan, S.; Li, M.; Komasa, S.; Agariguchi, A.; Yang, Y.; Zeng, Y.; Takao, S.; Zhang, H.; Tashiro, Y.; Kusumoto, T.; et al. Decontamination of titanium surface using different methods: An in vitro study. *Materials* **2020**, *13*, 2287. [CrossRef]
32. Mellado-Valero, A.; Buitrago-Vera, P.; Solá-Ruiz, M.F.; Ferrer-García, J.C. Decontamination of dental implant surface in pe-ri-implantitis treatment: A literature review. *Med. Oral. Patol. Oral. Cir. Bucal.* **2013**, *18*, e869–e876. [CrossRef]
33. Subramani, K.; Wismeijer, D. Decontamination of titanium implant surface and re-osseointegration to treat peri-implantitis: A literature review. *Int. J. Oral Maxillofac. Implants* **2012**, *27*, 1043–1054.
34. Mombelli, A. Microbiology and antimicrobial therapy of peri-implantitis. *Periodontology 2000* **2002**, *28*, 177–189. [CrossRef]
35. Leonhardt, Å.; Dahlén, G.; Renvert, S. Five-year clinical, microbiological, and radiological outcome following treatment of peri-implantitis in man. *J. Periodont.* **2003**, *74*, 1415–1422. [CrossRef] [PubMed]
36. Wohlfahrt, J.C.; Aass, A.M.; Koldsland, O.C. Treatment of peri-implant mucositis with a chitosan brush—A pilot randomized clinical trial. *Int. J. Dent. Hyg.* **2019**, *17*, 170–176. [CrossRef] [PubMed]
37. Wohlfahrt, J.C.; Evensen, B.J.; Zeza, B.; Jansson, H.; Pilloni, A.; Roos-Jansåker, A.M.; Di Tanna, G.L.; Aass, A.M.; Klepp, M.; Koldsland, O.C. A novel non-surgical method for mild pe-ri-implantitis- a multicenter consecutive case series. *Int. J. Implant Dent.* **2017**, *3*, 38. [CrossRef]
38. Guan, G.; Azad, M.A.K.; Lin, Y.; Kim, S.W.; Tian, Y.; Liu, G.; Wang, H. Biological effects and applications of chitosan and chi-to-oligosaccharides. *Front. Physiol.* **2019**, *10*, 516. [CrossRef] [PubMed]
39. Kreisler, M.; Kohnen, W.; Marinello, C.; Götz, H.; Duschner, H.; Jansen, B.; D'Hoedt, B. Bactericidal effect of the ER:Yag laser on dental implant surfaces: An in vitro study. *J. Periodontol.* **2002**, *73*, 1292–1298. [CrossRef] [PubMed]
40. Stübinger, S.; Etter, C.; Miskiewicz, M.; Homann, F.; Saldamli, B.; Wieland, M.; Sader, R. Surface alterations of polished and sandblasted and acid-etched titanium implants after Er: YAG, carbon dioxide, and diode laser irradiation. *Int. J. Oral Maxillofac. Impl.* **2010**, *25*, 104.
41. Dörtbudak, O.; Haas, R.; Bernhart, T.; Mailath-Pokorny, G. Lethal photosensitization for decontamination of implant surfaces in the treatment of peri-implantitis. *Clin. Oral Implants Res.* **2001**, *12*, 104–108. [CrossRef]
42. Shibli, J.A.; Martins, M.C.; Theodoro, L.H.; Lotufo, R.F.M.; Garcia, V.G.; Marcantonio, E., Jr. Lethal photosensitization in microbiological treatment of ligature-induced peri-implantitis: A preliminary study in dogs. *J. Oral Sci.* **2003**, *45*, 17–23. [CrossRef]
43. Aoki, A.; Mizutani, K.; Schwarz, F.; Sculean, A.; Yukna, R.A.; Takasaki, A.A.; Georgios, E.R.; Taniguchi, Y.; Sasaki, K.M.; Zeredoet, J.L.; et al. Periodontal and peri-implant wound healing following laser therapy. *Periodontology 2000* **2015**, *68*, 217–269. [CrossRef]
44. Al-Hashedi, A.A.; Laurenti, M.; Abdallah, M.-N.; Albuquerque, R.F., Jr.; Tamimi, F. Electrochemical treatment of contaminated titanium surfaces in vitro: An approach for implant surface decontamination. *ACS Biomater. Sci. Eng.* **2016**, *2*, 1504–1518. [CrossRef] [PubMed]
45. Mohn, D.; Zehnder, M.; Stark, W.J.; Imfeld, T. Electrochemical disinfection of dental implants-A proof of concept. *PLoS ONE* **2011**, *6*, e16157. [CrossRef] [PubMed]
46. Schneider, S.; Rudolph, M.; Bause, V.; Terfort, A. Electrochemical removal of biofilms from titanium dental implant surfaces. *Bioelectrochemistry* **2018**, *121*, 84–94. [CrossRef]
47. Kaiser, F.; Scharnweber, D.; Bierbaum, S.; Wolf-Brandstetter, C. Success and side effects of different treatment options in the low current attack of bacterial biofilms on titanium implants. *Bioelectrochemistry* **2020**, *133*, 107485. [CrossRef] [PubMed]

48. Schlee, M.; Rathe, F.; Brodbeck, U.; Ratka, C.; Weigl, P.; Zipprich, H. Treatment of peri-implantitis—Electrolytic cleaning versus mechanical and electrolytic cleaning—a randomized controlled clinical trial—Six-month results. *J. Clin. Med.* **2019**, *8*, 1909. [CrossRef]

49. Sahrmann, P.; Zehnder, M.; Mohn, D.; Meier, A.; Imfeld, T.; Thurnheer, T. Effect of low direct current on anaerobic multispecies biofilm adhering to a titanium implant surface. *Clin. Implant Dent. Related Res.* **2014**, *16*, 552–556. [CrossRef] [PubMed]

50. Heitz-Mayfield, L.; Salvi, G.; Mombelli, A.; Faddy, M.; Lang, N.; Group ICR. Anti-infective surgical therapy of peri-implantitis. A 12-month prospective clinical study. *Clin. Oral Implants Res.* **2012**, *23*, 205–210. [CrossRef]

51. Liu, J.; Liu, J.; Attarilar, S.; Wang, C.; Tamaddon, M.; Yang, C.; Xie, K.; Yao, J.; Wang, L.; Liu, C.; et al. Nano-modified titanium implant materials: A way toward im-proved antibacterial properties. *Front. Bioeng. Biotechnol.* **2020**, *8*. [CrossRef]

52. Wei, W.; Liu, Y.; Yao, X.; Hang, R. Na-Ti-O nanostructured film anodically grown on titanium surface have the potential to im-prove osteogenesis. *Surf. Coat. Technol.* **2020**, *397*, 125907. [CrossRef]

53. Kelleher, S.M.; Habimana, O.; Lawler, J.; Reilly, B.O.; Daniels, S.; Casey, E.; Cowley, A. Cicada wing surface topography: An investigation into the bactericidal properties of nanostructural features. *ACS Appl. Mater. Interfaces* **2016**, *8*, 14966–14974. [CrossRef]

54. Ivanova, E.P.; Hasan, J.; Webb, H.; Truong, V.K.; Watson, G.; Watson, J.; Baulin, V.; Pogodin, S.; Wang, J.; Tobin, M.; et al. Natural bactericidal surfaces: Mechanical rupture of pseudomonas aeruginosa cells by cicada wings. *Small* **2012**, *8*, 2489–2494. [CrossRef]

55. Linklater, D.P.; De Volder, M.; Baulin, V.A.; Werner, M.; Jessl, S.; Golozar, M.; Maggini, L.; Rubanov, S.; Hanssen, E.; Juodkazis, S.; et al. High aspect ratio nanostructures kill bacteria via storage and release of mechanical energy. *ACS Nano* **2018**, *12*, 6657–6667. [CrossRef] [PubMed]

56. Ovsianikov, A.; Chichkov, B.N. Three-dimensional microfabrication by two-photon polymerization technique. In *Computer-Aided Tissue Engineering*; Springer: Amsterdam, The Netherlands, 2012; pp. 311–325.

57. Qiao, H.; Li, Z.; Huang, Z.; Ren, X.; Kang, J.; Qiu, M.; Liua, Y.; Qia, X.; Zhonga, J.; Zhangb, H. Self-powered photodetectors based on 0D/2D mixed dimensional het-erojunction with black phosphorus quantum dots as hole accepters. *Appl. Mater. Today* **2020**, *20*, 100765. [CrossRef]

58. Friedli, V.; Utke, I. Optimized molecule supply from nozzle-based gas injection systems for focused electron- and ion-beam induced deposition and etching: Simulation and experiment. *J. Phys. D Appl. Phys.* **2009**, *42*, 125305. [CrossRef]

59. Skoric, L.; Sanz-Hernández, D.; Meng, F.; Donnelly, C.; Merino-Aceituno, S.; Fernández-Pacheco, A. Layer-by-layer growth of complex-shaped three-dimensional nanostructures with focused electron beams. *Nano Lett.* **2019**, *20*, 184–191. [CrossRef]

60. Limoli, D.H.; Jones, C.J.; Wozniak, D.J. Bacterial extracellular polysaccharides in biofilm formation and function. *Microbiol. Spectr.* **2015**, *3*. [CrossRef]

61. Bandara, C.D.; Singh, S.; Afara, I.O.; Wolff, A.; Tesfamichael, T.; Ostrikov, K.; Oloyede, A. Bactericidal effects of natural nanotopography of dragonfly wing on escherichia coli. *ACS Appl. Mater. Interfaces* **2017**, *9*, 6746–6760. [CrossRef]

62. Kim, T.N.; Balakrishnan, A.; Lee, B.C.; Kim, W.S.; Dvořánková, B.; Smetana, K.; Park, J.K.; Panigrahi, B.B.; Smetana, J.K. In vitro fibroblast response to ultra fine grained titanium produced by a severe plastic deformation process. *J. Mater. Sci. Mater. Electron.* **2007**, *19*, 553–557. [CrossRef]

63. Tian, A.; Qin, X.; Wu, A.; Zhang, H.; Xu, Q.; Xing, D.; Yang, H.; Qiu, B.; Xue, X.; Zhang, N.; et al. Nanoscale $TiO_2$ nanotubes govern the biological behavior of human glioma and osteosarcoma cells. *Int. J. Nanomed.* **2015**, *10*, 2423–2439. [CrossRef]

64. Puckett, S.D.; Taylor, E.; Raimondo, T.; Webster, T.J. The relationship between the nanostructure of titanium surfaces and bacterial attachment. *Biomaterials* **2010**, *31*, 706–713. [CrossRef] [PubMed]

65. Colon, G.; Ward, B.C.; Webster, T.J. Increased osteoblast and decreased Staphylococcus epidermidis functions on nanophase ZnO and $TiO_2$. *J. Biomed. Mater. Res. Part A* **2006**, *78*, 595–604. [CrossRef]

66. Kummer, K.M.; Taylor, E.N.; Durmas, N.G.; Tarquinio, K.M.; Ercan, B.; Webster, T.J. Effects of different sterilization techniques and varying anodized $TiO_2$ nanotube dimensions on bacteria growth. *J. Biomed. Mater. Res. Part B Appl. Biomater.* **2013**, *101*, 677–688. [CrossRef] [PubMed]

67. Roguska, A.; Belcarz, A.; Piersiak, T.; Pisarek, M.; Ginalska, G.; Lewandowska, M. Evaluation of the antibacterial activity of Ag-loaded $TiO_2$ nanotubes. *Eur. J. Inorg. Chem.* **2012**, *2012*, 5199–5206. [CrossRef]

68. Zhao, L.; Wang, H.; Huo, K.; Cui, L.; Zhang, W.; Ni, H.; Zhang, Y.; Wu, Z.; Chu, P.K. Antibacterial nano-structured titania coating incorporated with silver nanoparticles. *Biomaterials* **2011**, *32*, 5706–5716. [CrossRef]

69. Oh, S.; Daraio, C.; Chen, L.H.; Pisanic, T.R.; Finones, R.R.; Jin, S. Significantly accelerated osteoblast cell growth on aligned $TiO_2$ nanotubes. *J. Biomed. Mater. Re. Part A* **2006**, *78*, 97–103. [CrossRef]

70. Ercan, B.; Taylor, E.; Alpaslan, E.; Webster, T.J. Diameter of titanium nanotubes influences anti-bacterial efficacy. *Nanotechnology* **2011**, *22*, 295102. [CrossRef]

71. Kang, S.; Herzberg, M.; Rodrigues, D.F.; Elimelech, M. Antibacterial effects of carbon nanotubes: Size does matter! *Langmuir* **2008**, *24*, 6409–6413. [CrossRef]

72. Qiu, Z.; Yu, Y.; Chen, Z.; Jin, M.; Yang, D.; Zhao, Z.; Wang, J.; Shen, Z.; Wang, X.; Qian, D.; et al. Nanoalumina promotes the horizontal transfer of multiresistance genes mediated by plasmids across genera. *Proc. Natl. Acad. Sci. USA* **2012**, *109*, 4944–4949. [CrossRef]

73. Shi, X.; Sun, H.; Dong, C.; Xu, Q.; Tian, A.; Xue, X.; Yang, H. Antibacterial activities of $TiO_2$ nanotubes on Porphyromonas gingivalis. *RSC Adv.* **2015**, *5*, 34237–34242. [CrossRef]

74. Ji, J.; Zhang, W. Bacterial behaviors on polymer surfaces with organic and inorganic antimicrobial compounds. *J. Biomed. Mater. Res. Part A* **2009**, *88*, 448–453. [CrossRef]

75. Lee, J.H.; Lee, S.J.; Khang, G.; Lee, H.B. Interaction of fibroblasts on polycarbonate membrane surfaces with different micropore sizes and hydrophilicity. *J. Biomater. Sci. Polym. Ed.* **1999**, *10*, 283–294. [CrossRef] [PubMed]

76. Storz, G.; Imlayt, J.A. Oxidative stress. *Curr. Opin. Microbiol.* **1999**, *2*, 188–194. [CrossRef]

77. Davies, K.; Delsignore, M.; Lin, S. Protein damage and degradation by oxygen radicals. II. Modification of amino acids. *J. Biol. Chem.* **1987**, *262*, 9902–9907. [CrossRef]

78. Cui, Q.; Feng, B.; Chen, W.; Wang, J.-X.; Lu, X.; Weng, J. Effects of morphology of anatase $TiO_2$ nanotube films on photo-catalytic activity. *J. Inorg. Mater.* **2010**, *25*, 916–920. [CrossRef]

79. Demetrescu, I.; Pirvu, C.; Mitran, V. Effect of nano-topographical features of Ti/$TiO_2$ electrode surface on cell response and electrochemical stability in artificial saliva. *Bioelectrochemistry* **2010**, *79*, 122–129. [CrossRef]

80. Liu, L.; Li, W.; Liu, Q. Recent development of antifouling polymers: Structure, evaluation, and biomedical applications in nano/micro-structures. *Wiley Interdiscip. Rev. Nanomed. Nanobiotechnol.* **2014**, *6*, 599–614. [CrossRef]

81. Chen, Z.; Wang, Z.; Qiu, W.; Fang, F. Overview of antibacterial strategies of dental implant materials for the prevention of PE-RI-implantitis. *Bioconjug. Chem.* **2021**, *32*, 627–638. [CrossRef]

82. Mas-Moruno, C.; Su, B.; Dalby, M.J. Multifunctional coatings and nanotopographies: Toward cell instructive and antibacterial implants. *Adv. Health Mater.* **2018**, *8*, e1801103. [CrossRef]

83. Buwalda, S.; Rotman, S.; Eglin, D.; Moriarty, F.; Bethry, A.; Garric, X.; Guillaume, O.; Nottelet, B. Synergistic anti-fouling and bactericidal poly(ether ether ketone) surfaces via a one-step photomodification. *Mater. Sci. Eng. C* **2020**, *111*, 110811. [CrossRef]

84. Skovdal, S.M.; Jørgensen, N.P.; Petersen, E.; Jensen-Fangel, S.; Ogaki, R.; Zeng, G.; Johansen, M.I.; Wang, M.; Rohde, H.; Meyer, R.L. Ultra-dense polymer brush coating reduces Staphylococcus epidermidis biofilms on medical implants and improves antibiotic treatment outcome. *Acta Biomater.* **2018**, *76*, 46–55. [CrossRef] [PubMed]

85. Buxadera-Palomero, J.; Calvo, C.; Torrent-Camarero, S.; Gil, F.J.; Mas-Moruno, C.; Canal, C.; Rodríguez, D. Biofunctional polyethylene glycol coatings on titanium: An in vitro -based comparison of functionalization methods. *Colloids Surf. B Biointerfaces* **2017**, *152*, 367–375. [CrossRef] [PubMed]

86. Daglia, M. Polyphenols as antimicrobial agents. *Curr. Opin. Biotechnol.* **2012**, *23*, 174–181. [CrossRef] [PubMed]

87. Hancock, V.; Dahl, M.; Vejborg, R.M.; Klemm, P. Dietary plant components ellagic acid and tannic acid inhibit Escherichia coli biofilm formation. *J. Med. Microbiol.* **2010**, *59*, 496–498. [CrossRef] [PubMed]

88. Khan, N.; Mukhtar, H. Tea polyphenols in promotion of human health. *Nutrients* **2018**, *11*, 39. [CrossRef] [PubMed]

89. Steffi, C.; Shi, Z.; Kong, C.H.; Chong, S.W.; Wang, D.; Wang, W. Use of polyphenol tannic acid to functionalize titanium with strontium for enhancement of osteoblast differentiation and reduction of osteoclast activity. *Polymers* **2019**, *11*, 1256. [CrossRef]

90. Lee, S.; Chang, Y.-Y.; Lee, J.; Perikamana, S.K.M.; Kim, E.M.; Jung, Y.-H.; Yun, J.-H.; Shin, H. Surface engineering of titanium alloy using metal-polyphenol network coating with magnesium ions for improved osseointegration. *Biomater. Sci.* **2020**, *8*, 3404–3417. [CrossRef] [PubMed]

91. Popat, K.C.; Eltgroth, M.; LaTempa, T.J.; Grimes, C.A.; Desai, T.A. Decreased Staphylococcus epidermis adhesion and increased os-teoblast functionality on antibiotic-loaded titania nanotubes. *Biomaterials* **2007**, *28*, 4880–4888. [CrossRef]

92. Popat, K.C.; Eltgroth, M.; LaTempa, T.J.; Grimes, C.A.; Desai, T.A. Titania nanotubes: A novel platform for drug-eluting coatings for medical implants? *Small* **2007**, *3*, 1878–1881. [CrossRef]

93. Chopra, D.; Gulati, K.; Ivanovski, S. Understanding and optimizing the antibacterial functions of anodized nano-engineered titanium implants. *Acta Biomater.* **2021**, *127*, 80–101. [CrossRef]

94. Zhang, H.; Sun, Y.; Tian, A.; Xue, X.X.; Wang, L.; Alquhali, A.; Bai, X. Improved antibacterial activity and biocompatibility on van-comycin-loaded $TiO_2$ nanotubes: In vivo and in vitro studies. *Int. J. Nanomed.* **2013**, *8*, 4379. [CrossRef] [PubMed]

95. Ma, M.; Kazemzadeh-Narbat, M.; Hui, Y.; Lu, S.; Ding, C.; Chen, D.; Hancock, R.; Wang, R. Local delivery of antimicrobial peptides using self-organized $TiO_2$ nanotube arrays for peri-implant infections. *J. Biomed. Mater. Res. Part A* **2012**, *100*, 278–285. [CrossRef]

96. Jia, Z.; Xiu, P.; Li, M.; Xu, X.; Shi, Y.; Cheng, Y.; Wei, S.; Zheng, Y.; Xi, T.; Cai, H.; et al. Bioinspired anchoring AgNPs onto micro-nanoporous $TiO_2$ orthopedic coatings: Trap-killing of bacteria, surface-regulated osteoblast functions and host responses. *Biomaterials* **2015**, *75*, 203–222. [CrossRef]

97. Dong, Y.; Ye, H.; Liu, Y.; Xu, L.; Wu, Z.; Hu, X.; Maa, J.; Pathake, J.L.; Liua, J.; Wub, G. pH dependent silver nanoparticles releasing titanium implant: A novel ther-apeutic approach to control peri-implant infection. *Colloids Surf. B Biointerfaces* **2017**, *158*, 127–136. [CrossRef] [PubMed]

98. Li, B.; Zhang, L.; Wang, D.; Peng, F.; Zhao, X.; Liang, C.; Lia, H.; Wanga, H. Thermosensitive-hydrogel-coated titania nanotubes with controlled drug release and immunoregulatory characteristics for orthopedic applications. *Mater. Sci. Eng. C* **2021**, *122*, 111878. [CrossRef]

99. Yuan, Z.; Huang, S.; Lan, S.; Xiong, H.; Tao, B.; Ding, Y.; Liu, Y.; Liu, P.; Cai, K. Surface engineering of titanium implants with enzyme-triggered an-tibacterial properties and enhanced osseointegration in vivo. *J. Mater. Chem. B* **2018**, *6*, 8090–8104. [CrossRef] [PubMed]

100. Suketa, N.; Sawase, T.; Kitaura, H.; Naito, M.; Baba, K.; Nakayama, K.; Wennerberg, A.; Atsuta, M. An antibacterial surface on dental implants, based on the photocatalytic bactericidal effect. *Clin. Implant Dent. Relat. Res.* **2005**, *7*, 105–111. [CrossRef]

101. Jain, S.; Williamson, R.S.; Marquart, M.; Janorkar, A.V.; Griggs, J.A.; Roach, M.D. Photofunctionalization of anodized titanium surfaces using UVA or UVC light and its effects againstStreptococcus sanguinis. *J. Biomed. Mater. Res. Part B Appl. Biomater.* **2017**, *106*, 2284–2294. [CrossRef] [PubMed]
102. Pantaroto, H.N.; Ricomini-Filho, A.P.; Bertolini, M.M.; da Silva, J.H.D.; Neto, N.F.A.; Sukotjo, C.; Rangel, E.C.; Barão, V.A. Antibacterial photocatalytic activity of different crystalline $TiO_2$ phases in oral multispecies biofilm. *Dent. Mater.* **2018**, *34*, e182–e195. [CrossRef]
103. Hatoko, M.; Komasa, S.; Zhang, H.; Sekino, T.; Okazaki, J. UV Treatment improves the biocompatibility and antibacterial prop-erties of crystallized nanostructured titanium surface. *Int. J. Mol. Sci.* **2019**, *20*, 5991. [CrossRef]

 *nanomaterials*

*Review*

# Periodontal and Dental Pulp Cell-Derived Small Extracellular Vesicles: A Review of the Current Status

Shu Hua [1], Peter Mark Bartold [2], Karan Gulati [2], Corey Stephen Moran [2], Sašo Ivanovski [1,2,*] and Pingping Han [1,2,*]

[1] Epigenetics Nanodiagnostic and Therapeutic Group, Center for Orofacial Regeneration, Rehabilitation and Reconstruction (COR3), School of Dentistry, Faculty of Health and Behavioural Sciences, The University of Queensland, Brisbane, QLD 4006, Australia; s.hua@uq.net.au

[2] School of Dentistry, The University of Queensland, Brisbane, QLD 4006, Australia; mark.bartold@adelaide.edu.au (P.M.B.); k.gulati@uq.edu.au (K.G.); corey.moran@uq.edu.au (C.S.M.)

* Correspondence: s.ivanovski@uq.edu.au (S.I.); p.han@uq.edu.au (P.H.)

**Abstract:** Extracellular vesicles (EVs) are membrane-bound lipid particles that are secreted by all cell types and function as cell-to-cell communicators through their cargos of protein, nucleic acid, lipids, and metabolites, which are derived from their parent cells. There is limited information on the isolation and the emerging therapeutic role of periodontal and dental pulp cell-derived small EVs (sEVs, <200 nm, or exosome). In this review, we discuss the biogenesis of three EV subtypes (sEVs, microvesicles and apoptotic bodies) and the emerging role of sEVs from periodontal ligament (stem) cells, gingival fibroblasts (or gingival mesenchymal stem cells) and dental pulp cells, and their therapeutic potential *in vitro* and *in vivo*. A review of the relevant methodology found that precipitation-based kits and ultracentrifugation are the two most common methods to isolate periodontal (dental pulp) cell sEVs. Periodontal (and pulp) cell sEVs range in size, from 40 nm to 2 μm, due to a lack of standardized isolation protocols. Nevertheless, our review found that these EVs possess anti-inflammatory, osteo/odontogenic, angiogenic and immunomodulatory functions *in vitro* and *in vivo*, via reported EV cargos of EV–miRNAs, EV–circRNAs, EV–mRNAs and EV–lncRNAs. This review highlights the considerable therapeutic potential of periodontal and dental pulp cell-derived sEVs in various regenerative applications.

**Keywords:** extracellular vesicles; exosomes; nanomedicine; regeneration; cell-free therapy

**Citation:** Hua, S.; Bartold, P.M.; Gulati, K.; Moran, C.S.; Ivanovski, S.; Han, P. Periodontal and Dental Pulp Cell-Derived Small Extracellular Vesicles: A Review of the Current Status. *Nanomaterials* **2021**, *11*, 1858. https://doi.org/10.3390/nano11071858

Academic Editor: Alicia Rodríguez-Gascón

Received: 23 June 2021
Accepted: 15 July 2021
Published: 19 July 2021

**Publisher's Note:** MDPI stays neutral with regard to jurisdictional claims in published maps and institutional affiliations.



## 1. Introduction

Extracellular vesicles (EVs) are membrane-bound bilayered lipid particles that are secreted from both prokaryotic and eukaryotic cells, carrying a cargo of biological molecules (i.e., protein, nucleic acid, lipids and metabolites) from their parent cells [1]. Initially, EVs were considered 'cellular dust', generated by cellular metabolism, until their biological role in the mineralization of bone was recognized [2,3]. A principal role of EVs is as an intercellular communicator of biological information into a recipient cell. This interaction can trigger signaling cascades and modulate cell behavior [4]. The biological function of EVs is defined by the parent cells from which they originate. EVs are involved in almost all cellular interactions, especially tumor metastasis, tissue homeostasis, and inflammatory regulation [4,5]. Due to their constituent biological molecules, EVs hold great promise as a therapeutic delivery system in regenerative medicine.

The definition, terminology and subtypes of EVs are still being debated. The International Society of Extracellular Vesicles (ISEV) recommends a division of EV subtypes based on their size: medium/large EV (>150 nm) and small EV (<150 nm) [6]. However, considering the discrepancies in the published literature, for simplification purposes, this review will define EV subtypes based on both their size and biogenesis (Figure 1a): small extracellular vesicles (also known as exosomes) (sEVs, <200 nm), microvesicles (MVs,

50–1000 nm) and apoptotic bodies (ApoBDs, 50–2000 nm). Furthermore, it is noteworthy that all EVs have various membrane proteins (e.g., tetraspanin, MHC, and HSP) and components (e.g., dsDNA, RNA, microRNA, circular RNA [7], and proteins) (Figure 1b).

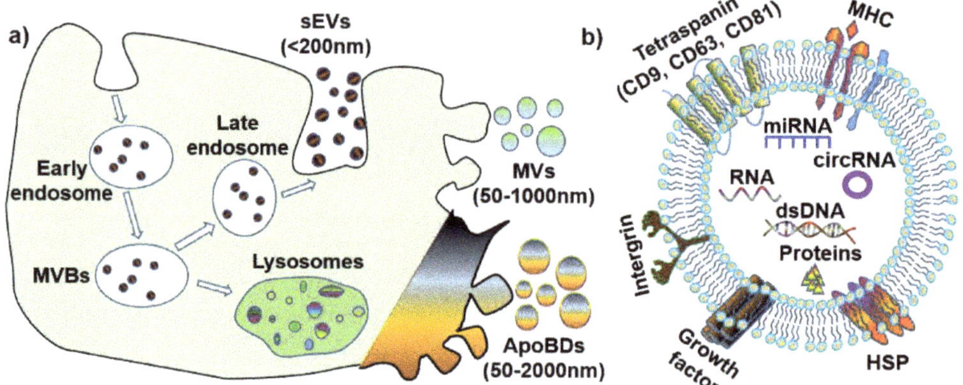


**Figure 1.** The biogenesis and contents of extracellular vesicles (EVs). (**a**) Biogenesis and size of three EV subtypes. (**b**) Common surface markers and cargos of EVs. sEVs: small extracellular vesicles; MVs: microvesicles; ApoBDs: apoptotic bodies; MVBs: multi-vesicular bodies; MHC: major histocompatibility complex; HSP: heat-shock protein; dsDNA: double-stranded DNA; miRNA: microRNA; circRNA: circular RNA.

### 1.1. Small EV (Exosomes)

Small EVs (sEV), or exosomes, originate from endosomes, and are biological nanoparticles that are smaller than 200 nm [5]. Further, sEVs are produced through an endocytic pathway, and their particle size is partially overlaid with that of microvesicles and apoptotic bodies. The biogenesis process for sEV is unique, whereby the endosomal network is the source of sEV that produce, classify, distribute and define the proper destination of the secreted sEV [2,8]. Endosome production can be categorized into the following three subtypes, according to each stage of development: early endosomes, late endosomes, and recycling endosomes. Early endosomes are formed by inward budding of the cell membrane, before a second inward budding of the endosomal membrane that results in the formation of late endosomes—intraluminal vesicles (ILVs). Late endosomes containing IVLs are named multi-vesicular bodies (MVBs), and the MVBs either fuse with lysosomes to degrade or follow the endocytic pathway for sEV generation. Once fusion with the plasma membrane is completed, the small membrane-enclosed vesicles are released into the extracellular matrix.

The biogenesis of sEV is affected by the following two main pathways that can induce multi-vesicular bodies (MVBs) generation: the endosomal sorting complex, required for the transport (ESCRT)-dependent pathway and ESCRT-independent pathway [9]. For the ESCRT-dependent pathway, ESCRT I and ESCRT II mediate the invagination of the late endosomal membrane, and ESCRT III will be recruited to the invaginated membrane sites. The cargo proteins are then deubiquitinated, and this stimulates the departure of the vesicle and the formation of MVBs. In the ESCRT-independent pathway, the neutral sphigomylinase2 (nSMase2) takes sphingolipids as substrates and converts sphingolipids to ceramide at the endosomal membrane. Following this, the microdomain is prepared for merging into a larger structure, which accelerates the endosomal budding and biogenesis of MVBs [10]. Moreover, sEVs that are produced by these different pathways possess different biomarkers, except CD63, which is the most common biomarker for all sEVs [11,12]. With respect to the ESCRT-dependent pathway, if endocytosis is mediated by Ras-related protein 27A/B (RAB27A/B), TSG101 is a biomarker of sEV. If the endocytosis is mediated by phospholipase D2 (PLD2) and RAB7, through the ESCRT-independent pathway, the biomarkers of sEV are alix, syntenin, and syndecan. As for the RAB11/35-mediated ESCRT-independent pathway, CD81, Wnt and proteolipid protein (PLP) are the preferred biomarkers.

The function of sEV in intercellular communication is determined by the interconnection between sEV surface proteins and receptors on the recipient cells that subsequently activates a variety of signaling pathways [5]. Further, sEVs arising from different cell types have different cargos that dictate and direct different biological effects. The sEVs are highly abundant in biofluids [13–15], and they have been demonstrated to be associated with immune response, viral pathogenicity, osteogenesis, odontogenesis, neuroprotection, angiogenesis, and anti-tumor functions [16]. For example, oral cancer cell-derived sEVs create a mechanism that can promote tumor progression by modifying vesicular contents and establishing a distant premetastatic niche with molecules that favor cancer cell proliferation, migration, invasion, metastasis, angiogenesis, and even drug resistance [17]. Evidence that sEVs play an important role in cell differentiation suggests that sEVs may have a potential role in tissue regeneration.

### 1.2. Microvesicles

Microvesicles (MVs) are membrane vesicles of different sizes, surrounded by a lipid layer of membrane, and they range in size from 50 nm to 1 μm. Microvesicles are generated by the outward budding of the plasma membrane, and are abundant in tissues/cells and biofluids [18]. The contents of MVs are similar to that of sEV. The MV components of note include CD40, selectins, integrins, cytoskeletal proteins, and cholesterol [19].

The biogenesis of MV involves the contraction of cytoskeletal proteins and phospholipid redistribution, contributing to a dynamic interplay in the plasma membrane and the resultant formation of microvesicles. Within the plasma membrane, the aminophospholipid translocase regulates phospholipid distribution, transferring phospholipids from one leaflet to another. Once phosphatidylserine (PS) is translocated to the leaflet of the outer membrane, the outward blebbing of the membrane and microvesicle formation is initiated. The interaction between actin and myosin causes the cytoskeletal structure contraction, which mediates membrane budding [20].

MVs have been reported to maintain tissue homeostasis during tissue regeneration, angiogenesis, anti-tumor effects, and in pathologies such as tumorigenesis, chronic inflammation, and atherosclerosis [19]. MVs that are produced by blood cells (e.g., neutrophils, macrophages, and platelets) are involved in the pro-coagulatory response [21]. MVs can be both pro-inflammatory and anti-inflammatory; this is determined by the induction or stimulation that is received by their parent cells. MVs that are produced by tumor cells enhance invasiveness and accelerate cancer progression, as well as strengthen the drug resistance of tumor cells [22]. This indicates that MVs are potential therapeutic agents for tissue regeneration; however, the function of MVs in periodontal tissue healing and regeneration requires further investigation.

### 1.3. Apoptotic Bodies

Apoptotic bodies (ApoBDs) are produced by cells undergoing apoptosis, and vary in size from 50 nm to 2 μm [23,24]. ApoBDs result from the formation of subcellular fragments when an apoptotic cell disassembles. They are comprised of molecular components from living cells and provide a rich molecular pool for recipient cells. However, ApoBDs are engulfed by macrophages and digested by phagolysosomes shortly after they are released [25]. ApoBDs and apoptosis are not related to an inflammatory reaction, the constituents in dying cells and ApoBDs are not released automatically to the environment, and anti-inflammatory cytokines are not generated during engulfing. ApoBDs have phosphatidylserine (PS) on their surfaces, to attract engulfing cells, and are considered to be specific biomarkers for ApoBDs [26]. Autoimmune diseases may be associated with defects in the clearance of ApoBDs. ApoBDs may stimulate the formation of thrombus and improve anti-cancer immunity.

Increasing evidence suggests that ApoBDs have important immune regulatory roles, in autoimmunity, cancer, and infection [24], as well as promoting osteogenesis [27]. For example, ApoBDs that are derived from mature osteoclasts can induce osteoblast differentiation by activating the protein kinase B/phosphoinositide 3-kinases (PI3K/AKT) pathway [27]. However, knowledge of their function and role is still limited and more studies are required in this field.

## 2. The Source and Characteristics of Periodontal (Dental Pulp) Cells

Dental tissue-derived (or stem) cells have remarkable characteristics for therapeutic application, being easily accessible and a rich source of stem cells with a well-known regenerative capacity. A great variety of multipotent adult or postnatal stem cells can be retrieved from dental tissues, especially from periodontal tissue and dental pulp from extracted permanent teeth (dental pulp stem cells—DPSCs) and exfoliated deciduous teeth (SHED). A healthy periodontium consists of soft (periodontal ligament-PDL and gingiva) and hard (alveolar bone and cementum) tissue, and cells residing within the healthy periodontal tissues include periodontal ligament (stem) cells (PDLSCs), PDL and gingival fibroblasts (PDLF, GFs), or gingival stem cells (GMSCs), osteoblasts (OBs), osteoclasts (OCs), and various immune cells (Figure 2) [28,29]. Moreover, stem cells can be obtained from dental apical papilla tissues (SCAP) and dental follicles (DFSCs, or DFCs) of the developing tooth [28,29]. Importantly, EV that is derived from these cells can be detected within periodontal tissues and biofluid (i.e., gingival crevicular fluid) (Figure 2).

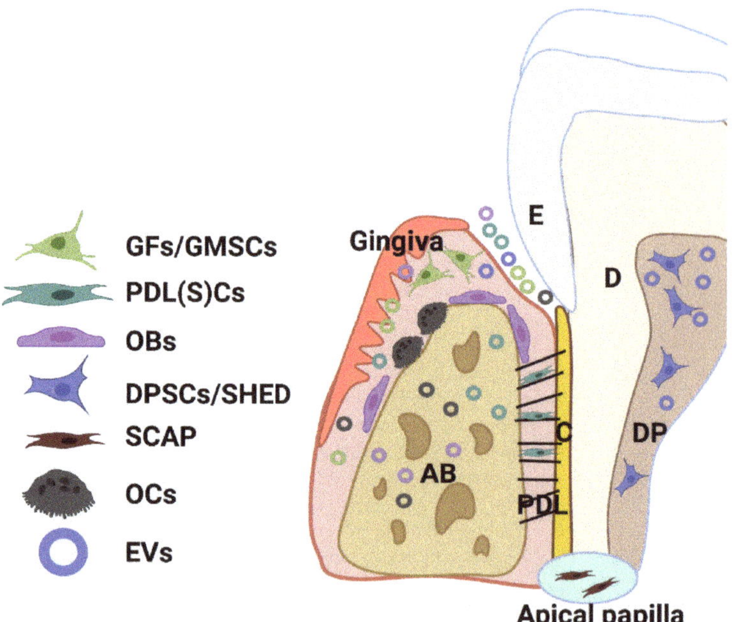


**Figure 2.** Schematic showing the main cell population and cell products (EVs) within a healthy periodontium. Various cells reside in the periodontium, such as periodontal ligament (stem) cells (PDLSCs), fibroblasts (GFs) and stem cells (GMSCs) from the gingiva, osteoblasts (OBs), osteoclasts (OCs), and various immune cells. AB: alveolar bone; C: cementum; D: dentin; DP: dental pulp; PDL: periodontal ligament; SHED: dental pulp cells from human exfoliated deciduous teeth (SHED); SCAP: cells from periodontal apical papilla tissues (SCAP).

Dental mesenchymal stem cells originate from the neural crest ectomesenchyme and reside in stromal niches (perivasculature and peripheral nerve-associated glia cells). The current consensus holds that both perivascular cells [30] and glia cells [31] are responsible for dental MSCs origin, as revealed in mouse experiments [31]. Much like bone-marrow-derived MSCs that originate from mesoderm [32], dental stem cells express MSCs markers and exhibit multipotent linage regeneration (i.e., osteogenic, chondrogenic, neurogenic) and immunomodulatory capabilities. These properties make these cells suitable candidates for therapeutic application (reviewed by Chalisserry et al. in [33]) in neurological disorders, angiogenesis, dentin-pulp regeneration and periodontal regeneration. PDLSCs, GFs, DPSCs, SHED, and DFSCs have been demonstrated to promote multiple-tissue regeneration, both *in vitro* and *in vivo* [34–40]. However, cell therapy has several challenges, including high cost, insufficient cell number, and associated regulatory barriers. On the other hand, a cell-free approach, centered around cell products (i.e., EVs derived from these cells), has been proposed, and there is an emerging focus on cell-derived EVs as potential therapeutic agents to promote periodontal regeneration. The utilization of sEVs for dental tissue regeneration is emerging as a viable cell-free treatment option, with 'proof of concept' studies reported using bone marrow or adipose MSC-derived sEVs (reviewed in [41–43]); yet, periodontal or dental pulp cell sources are likely to uniquely reflect the functional complexity of the periodontium and oral cavity.

The following sections will summarize the current methods for cell-derived sEV isolation and characterization, with particular emphasis on sEVs from periodontal and dental pulp cells.

## 3. Cell-Derived sEV Isolation Methods

### 3.1. General Concepts

Although sEVs have been studied for decades, there is still no standardized protocol for their isolation. Despite the presence of recommended guidelines for EV isolation and characterization, such as the Minimal Information for Studies of Extracellular Vesicles 2014 (MISEV2014) and MISEV 2018, these guidelines are not always followed.

Prior to the isolation of sEV, sequential centrifugation is commonly used to remove cell debris and large EVs, as follows:

i      Step 1: the cell conditional media (CM) is harvested and centrifuged at $300–400\times g$ to remove cells, and the supernatant (SN) is collected;

ii      Step 2: the SN collected in step 1 is centrifuged at $2000–3000\times g$ to remove cells debris and apoptotic bodies. The SN is collected from this step;

iii      Step 3: SN from step 2 is centrifuged at $10,000–20,000\times g$ to remove the aggregates of biopolymers, microvesicles, and the other structures with a buoyant density higher than sEVs. The SN is collected from this step;

iv      Step 4: then, the following isolation methods are used to enrich the sEVs: ultracentrifugation, sucrose gradient centrifugation, size exclusion chromatography, precipitation-based isolation, immunoaffinity chromatography, and ultrafiltration.

Given the growing interest in EVs, technical standardization is critical, as many different methodologies have been utilized for isolation and analysis. The influence of these various techniques on the downstream composition and functionality of EV cargos remains unclear; accordingly, the ISEV position papers [6,44] have raised the need to define 'good practices' and ultimately archive standardization. However, many researchers are not following these four steps, due to a lack of standardized protocols. Here, our review briefly introduces each isolation method, and discusses its merits and disadvantages (listed in Table 1).

**Table 1.** Representative advantages and disadvantages of various EV isolation methods.

| Method | Time | Advantages | Disadvantage |
|---|---|---|---|
| Ultracentrifuge ($100,000\times$–$200,000\times g$ for 1–2 h) | 1.5 h to 10 h | • Well-known 'gold-standard' method<br>• Easy to access<br>• Straightforward methodology | • Low recovery rate of sEV<br>• Time consuming (normally will need 2 steps of UC)<br>• Impure sEV with non-EV contamination and aggregates |
| Floatation-related methods (sucrose gradient centrifugation) | 250 min to 1 day | • Pure EV population<br>• No protein contamination | • Fails to separate large vesicles with similar sedimentation rates |
| Size exclusion chromatography (SEC) | ~30 min (including column washing) | • Time-efficient<br>• Pure EV product | • sEV and microvesicles cannot be separated |
| Precipitation based isolation (sodium acetate, PEG, protamine) | Overnight incubation | • Low-speed centrifuge ($1500\ g$) to retrieve the sEV sample<br>• Straightforward method<br>• Many samples can be processed | • Low EV recovery<br>• Co-precipitation of protein and other molecules<br>• Further purification step is required |
| Immunoaffinity chromatography | ~240 min | • Very pure EV subpopulation (i.e., CD9+ EV) | • Low EV yield<br>• Low scalability |
| Membrane filtration/Ultrafiltration | ~130 min | • Small sample volume<br>• Simple procedure<br>• Higher yield than UC method | • High contamination of non-EV protein |

### 3.2. Ultracentrifuge

Ultracentrifugation is the gold-standard method for isolating sEV, as the equipment is relatively easy to access and the methodology is technically straightforward. The method involves an ultracentrifugation step at $100,000\times$–$200,000\times g$ to pellet sEV [45]. However, ultracentrifugation has disadvantages, in that it leads to a low recovery rate of sEV, it is time consuming (1.5–10 h), contains non-vesicular macromolecule contamination, and results in EV aggregation.

### 3.3. Floatation-Related Methods (Sucrose Gradient Centrifugation)

Floatation-related methods distribute molecules based on the buoyant density, and the protein aggregates and sEV can be sufficiently separated. Differential gradient centrifugation (usually takes 250 min—1 day) takes advantage of buoyant density to fractionate EVs using sucrose or idoxanol gradients [45]. The sEVs can be separated by the discontinuous gradient sucrose solution, with each layer containing the desired size of EV. Other chemical reagents (i.e., iodixanol) can also be utilized instead of sucrose, for continuous EV harvest with no layers. Non-vesicular protein contaminants are distributed at a reduced level within this method, resulting in less protein contamination. However, sucrose gradient centrifugation cannot separate large particles that have a similar sedimentation rate.

### 3.4. Size Exclusion Chromatography (SEC)

SEC can be used to isolate small sEV, based on the size of the molecules, where large particles pass through the gel earlier than the small-sized molecules. The small-sized particles are trapped in the tiny pores on the surface of the gel, while the larger molecules can bypass the gel or receive less interference from the gel [46]. This technique has been well established with commercialized SEC columns, including qEV (iZON Science), Exo-spin™ SEC columns (Cell Guidance Systems Ltd.) and Pure-EVs SEC columns (HandaBioMed, Lonza). SEC has been proposed as an effective alternative method for pure sEV isolation, with a key advantage being its time efficiency (~30 min, including 10 mL of column washing with PBS). However, the similarly sized sEV and microvesicles cannot be separated by SEC.

*3.5. Precipitation-Based Isolation (Sodium Acetate, PEG, Protamine)*

Precipitation-based isolation has the following two mechanisms: polymeric precipitation and neutralizing charges [47]. In polymeric precipitation, a soluble polymer, usually polyethylene glycol (PEG), is mixed with EV samples and the mixture is incubated overnight, and EVs are sedimented by low-speed centrifugation at 1500 *g*. PEG precipitation enables a simple process for a large number of samples. Commercial kits, such as ExoQuick (System Biosciences), total exosome isolation reagent (Invitrogen), EXO-Prep (HansaBioMed), exosome purification kit (Norgen Biotek), and miRCURY exosome isolation kit (Exiqon), are based on this principle. For the other precipitation method, all EVs possess negative charges, so positively charged molecules (i.e., sodium acetate and protamine) are chosen for the precipitation. This method is popular due to its straightforward protocol; however, these precipitation methods lead to low sEV purity due to co-precipitation of the components from CM or biofluids, such as protein, DNA and RNA, and hence further purification is required.

*3.6. Immunoaffinity Chromatography*

The monoclonal antibodies (mAbs) against specific sEV surface proteins (i.e., CD 9) are fixed on the column, to capture a specific sEV population [48]. Once the CM passes through the column, the EVs, which express certain exosomal markers on their membrane, will be captured by the mAbs. This method leads to a very pure EV population, but low yield and scalability. This is attributed to the fact that this step needs to be repeated several times to ensure the mAbs can capture sufficient EVs (~240 min).

*3.7. Ultrafiltration*

Semi-permeable membranes (ranging from 3 kDa to 100 kDa) are adapted for sEV fractionation within filtration-based isolation; the membrane function is determined by its pore size. However, sEVs cannot be fractionated according to their biogenesis or biomarkers, but it is normally used to concentrate sEVs. It is still an efficient way to eliminate the minimal sample volume (~130 min) with a simple procedure, and has been proven to yield higher recovery of sEVs than ultracentrifugation [49]. However, ultrafiltration might lead to low EV protein, but a rather higher concentration of non-EV proteins (i.e., albumin).

*3.8. Current Isolation Challenge*

As mentioned previously, the current challenges of sEV isolation include time-consuming procedures, impurities, insufficient EV yield, and low scalability [50]. Although many researchers have investigated combinations of these isolation methods, an urgent demand has arisen to investigate high-yielding and time-effective isolation protocols. Currently, there is no optimal sEV isolation method; however, a combination of ultracentrifugation, SEC, and ultrafiltration has been used for the pure sEV population, which is a critical factor for downstream therapeutic applications.

## 4. sEV Isolation and Characterization Methods for Periodontal (and Dental Pulp) Cells

To date, there are no standardized protocols for sEV isolation and characterization. From the 33 studies that are reported in this review, we have summarized periodontal (dental pulp) cell-derived sEV isolation and characterisation methods [51–83]. Various isolation methods have been used for periodontal (dental pulp) cells sEVs, including ultracentrifugation (UC), precipitation-based methods, and ultrafiltration (Figure 3a). Regarding EV characterization, the latest MISEV 2018 guidelines [6] suggest that all EV researchers should characterize sEV from at least three different aspects, such as EV particle numbers, EV morphology, and EV-enriched protein markers. However, most of the current studies did not follow the MISEV guidelines, and this requires additional attention for all EV research.

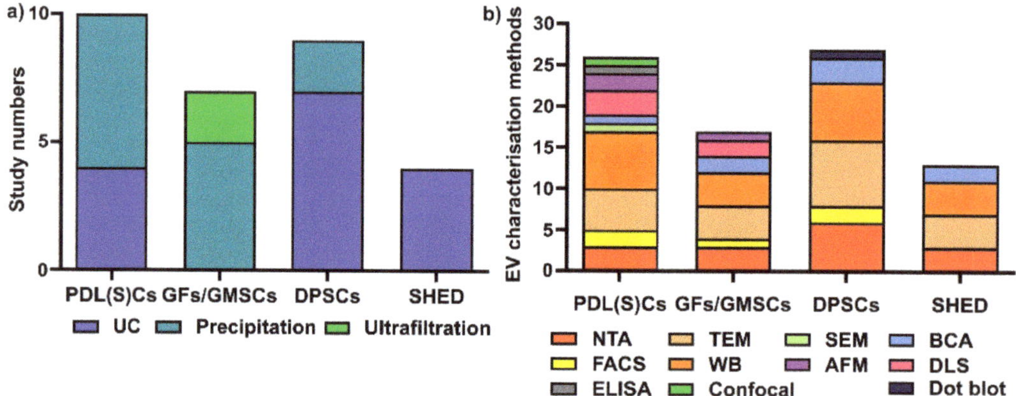

**Figure 3.** Various sEV isolation methods (**a**) and characterization methods (**b**) are used for periodontal (dental pulp) cells. UC: ultracentrifugation; NTA: nanoparticle tracking analysis; TEM: transmission electron microscopy; WB: Western blot; SEM: scanning electron microscopy; AFM: atomic force microscopy; BCA: bicinchoninic acid assay; DLS: dynamic light scattering; ELISA: enzyme-linked immunosorbent assay; confocal: confocal microscopy.

Different sEV isolation methods have been utilized for various cells (Figure 3a), with precipitation and ultracentrifugation methods being the two most commonly used techniques. In PDL(S)C-derived sEV isolation (10 studies), the precipitation-based method (i.e., a commercial ExoQuick kit) is the most commonly used ($n = 6$, 60%), followed by ultracentrifugation ($n = 4$, 40%). Among six studies in GFs/GMSC-derived sEVs, the precipitation-based method ($n = 4$, 66.7%) and ultrafiltration ($n = 2$, 33.3%) were used for GFs/GMMSCs–sEV isolation. Regarding DPSC-derived sEV, most researchers selected ultracentrifugation ($n = 7$, 77.8%), with one study using the precipitation-based method ($n = 2$, 22.2%). For SHED–sEVs, all of the studies ($n = 4$, 100%) used the ultracentrifugation method to isolate sEVs from SHED.

Concerning EV characterisation [6], NTA and DLS are common methods to quantify EV particle number, size, and distribution; TEM, SEM, and AFM can be used for EV morphology and size; BCA is for EV protein quantification; and WB is to determine EV-enriched protein markers. We have summarized the various EV characterisation methods for periodontal cell-derived sEVs (Figure 3b). For PDL(S)Cs—sEV (10 studies included in this review), WB is the most commonly used characterization method ($n = 7$ studies), followed by TEM ($n = 5$), NTA ($n = 3$), AFM ($n = 2$), flow cytometry ($n = 2$), SEM ($n = 1$), BCA assay ($n = 1$), ELISA ($n = 1$), and confocal microscopy ($n = 1$). In GFs/GMMSCs–sEVs, WB ($n = 4$) was utilized to detect CD9, CD63, and TSG101, as well as TEM ($n = 4$), NTA ($n = 3$), DLS ($n = 2$), BCA assay ($n = 2$), AFM ($n = 1$), and FACS ($n = 1$). The characterisation of DPSC–EVs are mostly performed using WB ($n = 7$), TEM ($n = 8$), NTA ($n = 6$), BCA assay ($n = 3$), FACS ($n = 2$), and dot blot ($n = 1$). For the characterization of SHED–sEVs (4 studies), TEM and WB ($n = 4$) are most commonly applied; NTA ($n = 3$) and BCA ($n = 2$) were also used for OBs–sEVs.

In summary, ultracentrifugation and precipitation-based methods are the two most common methods used for periodontal (dental pulp) cells sEV isolation. WB, TEM and NTA are the most common methods for periodontal (dental pulp) cell-derived sEVs characterisation.

## 5. The Function of sEVs Derived from Periodontal (Dental Pulp) Cells

Current studies mainly focus on small EV biogenesis and function in the periodontal regeneration field; thus, this review summarizes 33 studies [51–83] on periodontal cell-, gingival cell- and dental pulp (DPSCs and SHED) cell-derived sEV isolation, characterization, and their therapeutic role in tissue regeneration. Most of this research has focused on the function of sEVs in cell differentiation, and 11 studies investigated the cargos of sEVs (i.e., miRNA [52,53,61,63,64,72,75,81], circRNA [51,71], lncRNA [51], and EV-mRNA [67,72]) during this process.

### 5.1. Periodontal Ligament Fibroblasts or Stem Cells (hPDL(S)Cs)–sEV

A total of 10 studies investigated sEVs derived from human PDL (stem) cells or fibroblasts [51–60], and are summarized in Table 2. Eight studies isolated sEV from hPDLSCs [51–53,55,56,58–60], with one study each using sEVs from a human PDL fibroblast (hPDLFs) cell line [54] and hPDLCs [57]. Three of these studies investigated hPDLCs sEV function *in vivo*, using animal models [56,59,60].

According to the latest MISEV 2018 guidelines [6], it is critical to consider several factors influencing the collection of EV, including characteristics of primary cell source (donor health status, age, gender), primary cell passage number, confluence at harvest, culture volume, media change frequency, CM harvesting conditions, as well as all culture media composition and preparation details. Thus, our review includes detailed donor information for primary cells (if mentioned), the cell culture condition prior to CM collection, and detailed sEV isolation protocols. This will allow future EV researchers to select appropriate protocols for CM harvesting and sEV isolation.

Donor age was disclosed in only two studies (18–30 [51] and 18–21 years old [55]), while there is no clear information in the other studies [52–54,56–60]. The cells at passages 2–3 were used in five studies [51–53,56,59], passage 3–7 in one study [54], with no passage information provided in the remaining studies [55,57,58,60]. Since fetal calf serum (FBS or FCS) contains a large amount of EV, it is crucial to state how cells are cultured before CM harvesting. Currently, either EV-depleted FBS or FBS starvation is used before CM harvest for PDLCs–sEV isolation; from the 10 studies that were reviewed, 5 did not state how the cells were cultured before CM collection [53,56,58–60], 4 studies used EV-depleted FBS [52,54,55,57], and one study used FBS starvation [51]. While it is of considerable importance to clearly articulate the cell source, passage number, and CM harvest condition, this is something that is currently under-reported in many studies.

The following three aspects of hDPL(S)Cs–sEV analysis were evaluated in the 10 studies that have been included in this review: (1) EV size, (2) EV content (protein, RNA, etc.), and (3) EV function in cell differentiation *in vitro* and *in vivo*. Regarding the size of hPDL(S)Cs–sEV, three studies did not characterize the sEV size [52,53,60]. There is a large deviation for the reported EV size: <200 nm in five studies [51,54,55,57,58], two populations (90 ± 20 nm and 1200 ± 400 nm) in one study [59], and 100–710 nm in one study [56]. It is noted that two studies engineered the hPDLCs–EV using polyethyleneimine (PEI, yielding PEI–EV) [56,59]. The following two factors may contribute to this deviation: the EV isolation method (UC or precipitation methods), and the EV size characterization methods (TEM, or DLS, or NTA). We will define EV size <200 nm as sEV, and unclear EV size as EV.

**Table 2.** The isolation, characterization, and function of PDL(S)C-derived EVs.

| Reference | Cell Source | EV Isolation | EV Characterization | Key Findings |
|---|---|---|---|---|
| Xie et al., 2021 [51] | • hPDLSCs<br>• Donor: Healthy patients aged 18–30 years old, with no system diseases, underwent impacted third molar or orthodontic extractions<br>• Passage 3 | • FBS starvation for 48 h<br>• 3000 $g$ for 20 min; 16,500 $g$ for 20 min; 0.2 μm filter; ultracentrifuge at 100,000× $g$ for 70 min<br>• Centrifuge temperature unclear<br>• Exosomes from hPDLSCs before (EX0) and after osteogenic induction for 5 (EX5) and 7 (EX7) days | • NTA<br>• Flow cytometry<br>• TEM<br>• WB (CD63 and CD81) | • The size of sEV ranged from 20 to 100 nm.<br>• 69–557 sEV–circRNAs and 2907–11,581 exosomal lncRNAs were found in EX0, EX5 and EX7 by RNA sequencing.<br>• Within hPDLSCs—exosomes, compared with EX0, 3 circRNAs and 2 lncRNAs were upregulated and 39 circRNAs and 5 lncRNAs downregulated in EX5 and EX7.<br>• Exosomal circRNAs may function as competing endogenous RNAs through a circRNA–miRNA–mRNA network via TGF-β pathway, MAPK pathway, mTOR pathway and FoxO signaling pathways during hPDLSCs osteogenic differentiation *in vitro*. |
| Zhang et al., 2020 [52] | • hPDLSCs<br>• Donors: Periodontal ligaments from premolars of healthy and periodontitis patients<br>• Age unclear<br>• Passage 2–5 | • hPDLSCs were cultured with a vesicle-free medium.<br>• Centrifuge for 10 min at 500× $g$, 30 min at 16,000× $g$; ultracentrifugation for 70 min at 150,000× $g$ | • Flow cytometry<br>• TEM<br>• WB (TSG101 and CD63)<br>• SEM | • EV size was not mentioned.<br>• When co-cultured with TNF-alpha, the angiogenesis of HUVECs was promoted by hPDLSCs exosomes.<br>• The angiogenesis of HUVECs was downregulated when the secretion of exosomes was blocked.<br>• Inflammation influenced pro- angiogenesis of hPDLSCs via regulating the exosome-mediated transfer of VEGFA targeted by miR-17-5p. |
| Chiricosta et al., 2020 [53] | • hPDLSCs<br>• Donor: Five healthy patients from tooth removal for orthodontic purposes<br>• Passage 2 | • Conditioned medium of hPDLSCs at (CM; 10 mL) after 48 h of incubation.<br>• 15 min at 3000× $g$; mixed with ExoQuick TC reagent at 4 °C overnight 1500× $g$ for 30 min. | • No EV characterization performed before RNA sequencing | • EVs size was not mentioned<br>• The EV derived from hPDLSCs contain several non-coding RNAs, especially the following five miRNAs: miR24-2, miR142, miR296, miR335, and miR490. The target genes of these miRNAs are involved in Ras protein signal transduction and actin/microtubule cytoskeleton organization. |

**Table 2.** *Cont.*

| Reference | Cell Source | EV Isolation | EV Characterization | Key Findings |
|---|---|---|---|---|
| Zhao et al., 2019 [54] | • Human periodontal ligament fibroblasts (hPDLFs) cell line<br>• MG-63 osteoblast cells line<br>• Passages 3–7 | • *P. gingivalis* LPS-treated hPDLFs for 4 h and cultured with exosome-depleted FBS/1% PS for another 24 h.<br>• $1000 \times g$ for 10 min and $10,000 \times g$ for 15 min, 0.22 μm Millipore filter; concentrated CM (3–5 mL) using a 100 kDa ultracentrifuge filter at $5000 \times g$ in for 30 min; ExoQuick-TC reagent was added and incubated with CM overnight, centrifuge the mixture at $1500 \times g$ for 30 min | • BCA<br>• TEM<br>• Dynamic light scattering (DLS)<br>• WB (CD9, CD63, TSG101) | • Exosomes size: 70–100 nm, peaked at 84 nm.<br>• Exosome-enriched protein and total exosomal protein levels were higher in the LPS-treated hPDLFs than those in non-LPS-treated hPDLFs.<br>• Upregulated IL-6, TNF-α and inhibited expression ALP, collagen-I, RUNX2, and OPG were found in MG-63 OBs after uptaking the exosomes derived from LPS-treated hPDLFs. |
| Čebatariūn-ienė et al., 2019 [55] | • hPDLSCs grew in a bioreactor on gelatin-coated microcarriers<br>• Donor: healthy periodontal tissues of two Caucasian females (18 and 21 years old) using explant outgrowth method | • Supernatants (SN) from PDLSC in basal medium supplemented with 10% of EV-depleted FCS every 72 h<br>• All centrifugation steps were performed at 4 °C<br>• $300\,g$ for 10 min, $2000\,g$ for 10 min; $20,000\,g$ for 30 min; then ultracentrifuged at $100,000 \times g$ for 70 min.<br>• The pellets were washed in 40 mL PBS and ultracentrifuged again at $100,000 \times g$ for 70 min | • TEM<br>• NTA<br>• WB (CD63, MFG-E8) | • The size of sEV derived from microcarrier cell cultures of PDLSCs: 112–182 nm.<br>• hPDLSCs–sEV suppressed basal and LPS-induced activity of NF-kB in PDLSCs.<br>• Combined treatment with EV and anti-TLR4 antibody attenuated the inhibitory effect on the NF-kB activity<br>• Uptake of hPDLSCs–sEV in hPDLSC-activated phosphorylation of Akt and GSK3β (Ser 9) indicating that PI3K/Akt signaling pathway may act as a suppressor of NF-kB activity<br>• EVs did not significantly affect osteogenic mineralization of hPDLSCs cultures, but EV significantly increased expression of ALP, OPN, BSP, and CP23 gene expression, but downregulated BMP2 expression on the 10 days of osteogenic differentiation. |

**Table 2.** *Cont.*

| Reference | Cell Source | EV Isolation | EV Characterization | Key Findings |
|---|---|---|---|---|
| Pizzicannel-la et al., 2019 [56] | • hPDLSCs<br>• Donor: Five healthy participants undergoing teeth removal for orthodontic purpose; age unclear<br>• Passage 2 | • The CM from $15 \times 10^3/\text{cm}^2$ hPDLSCs was centrifuged at $3000\times g$ for 15 min; 2 mL of ExoQuick TC was added to 10 mL of CM; incubated overnight at 4 °C without rotation, centrifuge at $1500\times g$ for 30 min, resuspend with 200 µL PBS<br>• EVs were coated with branched polyethylenimine (PEI, yielding PEI-EV) | • DLS<br>• AFM | • hPDLSCs–EV: 100–710 nm; engineered PEI–EV: 1050–7700 nm by DLS.<br>• *In vitro*: increased expression of osteogenic markers (RUNX2, COL1A1, BMP2/4, VEGFA and VEGFR2) in hPDLSCs cultured in a collagen membrane with EVs and PEI–EVs, as well as increased protein levels of VEGF and VEGFR2.<br>• *in vivo*: EV and PEI–EV groups increased VEGF and VEGFR2 protein expression at 6 weeks post transplanting into a rat calvarial defect, as well as enhanced vascular and bone formation. While PEI–EV showed better bone and vascularization than that of the EV group. |
| Wang et al., 2019 [57] | • Human PDL cells (hPDLCs) were prepared from the PDL of fully erupted lower third molar teeth<br>• A mouse macrophage-like cell line (J774.1)<br>• Human monocyte-like cell line THP-1 | • Exosome-depleted FBS was prepared using the FBS exosome depletion kit (Norgen, Thorold, ON, Canada)<br>• The PureExo R exosome isolation kit (101Bio, Mountain View) was used for exosome isolation.<br>• SN was centrifuged at $3000\times g$ for 15 min; 2 mL of SN mixed with a solution of PureExo isolation kit and incubated at 4 °C for 30 min, centrifuged at $5000\times g$ for 3 min, air-dried, re-suspension in 100 µL PBS; centrifuge for 5 min at $5000\times g$;<br>• Processed on a PureExo column, centrifuged at $1000\times g$ for 5 min. | • TEM<br>• WB (CD9)<br>• ELISA using PS CaptureTM exosome ELISA kit for CD63 | • D: 30 and 100 nm (average 50 nm).<br>• Cyclic stretch (CS) exposed PDL cells generated 30 times more exosomes compared to non-CS treated normal PDL cells at 24 h<br>• CS hPDLC exosomes inhibited IL-1β production in LPS/nigericin-stimulated J774.1 macrophages and the nuclear translocation of NF-kB as well as NF-kB p65 DNA-binding activity in LPS-stimulated macrophages, suggesting that exosomes suppress IL-1β production by inhibiting the NF-kB signaling pathway. |

**Table 2.** *Cont.*

| Reference | Cell Source | EV Isolation | EV Characterization | Key Findings |
|---|---|---|---|---|
| Kang et al., 2018 [58] | • hPDLSCs and THP-1 cells lines<br>• The passage not mentioned for hPDLSCs | • hPDLSCs were treated with 1 µg/mL of LPS for 1 h and CM was collected after 24 h of incubation and filtered using a 0.22 µm filter.<br>• CM was centrifuged at 2000× $g$ for 10 min at 4 °C and filtration through a 0.22 µm filter; ultracentrifuged at 100,000× $g$ for 60 min at 4 °C. | • NTA<br>• WB (CD81 and CD63) | • The median sEV particle sizes were 151.3 nm and 146.9 nm for the EVs from control hPDLSCs and LPS-preconditioned hPDLSCs.<br>• sEV particle number was significantly decreased in the sEV from LPS-preconditioned PDLSCs compared to those from the control hPDLSCs.<br>• sEV from LPS-preconditioned hPDLSCs induced M1 polarization in THP-1 cells, with increased mRNA expression of IL-6 and TNF-$\alpha$, and TNF-$\alpha$ protein. The M1 polarization was abolished by DNase I treatment of sEV. |
| Diomede et al., 2018 [59] | • hPDLSCs culturing on collagen membranes (Evolution-Evo).<br>• Donor: Five participants, either patient for orthodontic purposes or healthy volunteers. Unclear age.<br>• Passage 2 | • After 48 h of incubation, the conditioned medium (CM; 10 mL) was collected from hPDLSCs.<br>• CM was centrifuged at 3000× $g$ for 15 min; 2 mL ExoQuick TC was added to 10 mL CM; incubated overnight at 4 °C without rotation, centrifuge at 1500× $g$ for 30 min<br>• EVs were engineered by noncovalently coating EVs with PEI | • DLS<br>• AFM | • DLS analysis showed that hPDLSCs–EV had two populations of vesicles, with an average diameter of 90 ± 20 nm and 1200 ± 400 nm. hPDLSCs–PEI-EV size: 250 ± 50 nm and 3600 ± 500 nm for two populations.<br>• Evo enriched with EV and PEI-EV showed high biocompatibility and osteogenic properties both *in vitro* and *in vivo*.<br>• PEI-EV promoted the expression of osteogenic genes, such as TGFβ1, MMP8, TUFT1, TFIP11, BMP2, and BMP4 after 6 weeks of *in vitro* osteogenic differentiation in hPDLSCs.<br>• PEI-EV group led to an *in vivo* organized extracellular matrix showing mineralization areas and blood-vessel formation, with upregulated BMP2/4 in collagen membrane enriched with PEI-EV and hPDLSCs after 6 weeks in a rat calvarial defect. |

**Table 2.** *Cont.*

| Reference | Cell Source | EV Isolation | EV Characterization | Key Findings |
|---|---|---|---|---|
| Rajan et al., 2016 [60] | • hPDLSCs<br>• Donor: Five human periodontal ligament biopsies from human premolar teeth of healthy and relapsing-remitting multiple sclerosis (RR-MS) patients. | • CM was centrifuged at $3000 \times g$ for 15 min; 2 mL ExoQuick TC was added to 10 mL CM recovered from hPDLSCs; incubated overnight at 4 °C without rotation, centrifuge at $1500 \times g$ for 30 min. | • Confocal image of CD 63 using fluorescent lipid probes<br>• No appropriate characterisation | • EV size was not mentioned.<br>• After intravenous administration of hPDLSCs–EV into EAE rats, pro-inflammatory cytokines IL-17, IFN-$\gamma$, IL-1$\beta$, IL-6, and TNF-$\alpha$ were reduced, while the anti-inflammatory cytokine, IL-10, was upregulated. Meanwhile, apoptosis-related STAT1, p53, Caspase 3, and Bax expressions were attenuated.<br>• hPDLSCs–EV from MS patients and healthy donors block experimental autoimmune encephalomyelitis (EAE), a mouse model of MS, by inducing anti-inflammatory and immunosuppressive effects in the spinal cord and spleen, and reverse disease progression by restoring tissue integrity via remyelination in the spinal cord. |

**Abbreviations:** TGF$\beta$, transforming growth factor $\beta$; MAPK, mitogen-activated protein kinase; mTOR, mechanistic target of rapamycin; FoxO, forkhead box protein O; VEGFA, vascular endothelial growth factor A; Ras, Ras GTPase; LPS, lipopolysaccharide; MFG-E8, milk fat globule-EGF factor 8 protein; TLR4, Toll-like receptor 4; PKB/Akt, protein kinase B; GSK-3$\beta$, glycogen synthase kinase 3; ALP, alkaline phosphatase; OPN, osteopontin; BSP, bone sialoprotein; CP23, cementum protein 23; BMP2, bone morphogenetic protein 2; RUNX2, runt-related transcription factor 2; COL1A1, alpha-1 type I collagen; VEGFR2, vascular endothelial growth factor receptor 2; IL-1$\beta$, interleukin-1 beta; IL-6, interleukin-6; TNF-$\alpha$, tumor necrosis factor-$\alpha$; MMP8, matrix metalloproteinase-8; TUFT1, Tuftelin 1; TFIP11, Tuftelin-interacting protein 11; IL-17, interleukin-17; IFN-$\gamma$, interferon-$\gamma$; IL-10, interleukin-10; STAT1, signal transducer and activator of transcription 1; Bax, Bcl-2-associated X protein.

Regarding the EV content, it seems that hPDLCs–EV contain miRNAs [52,53] and circular RNAs [51] that may alter the recipient cells functions. RNA sequencing of hPDLSCs–EV (where EV size was unclear) revealed that hPDLSCs–EV contains 955 non-coding transcripts, with five representative miRNAs, including MIR24-2, MIR142, MIR296, MIR335, and MIR490 [53]. The hDPLSCs–EV–miR-17-5p can regulate the angiogenesis of human umbilical vein endothelial cells (HUVECs) during inflammatory stimulation by TNF-α [52]. Furthermore, circular RNA and long non-coding RNAs (lncRNAs) were also found in the sEV from hPDLSCs, after five and seven days of osteogenic differentiation, with 69–557 circRNAs and 2907–11,581 lncRNAs detected by RNA sequencing. Compared with the sEV from hPDLSCs before osteoinduction, 3 sEV–circRNAs and 2 sEV–lncRNAs were upregulated, while 39 sEV–circRNAs and 5 sEV–lncRNAs were downregulated after 5 and 7 days of osteoinduction. RT-qPCR validation showed that three sEV–circRNAs (*hsa_circ_0087960*, *hsa_circ_0000437*, and *hsa_circ_0000448*) were upregulated after osteogenic differentiation, while one was downregulated (*hsa_circ_0000448*). However, three selected lncRNAs (*small nucleolar RNA host gene5—SNHG5, LOC100130992,* and *ATP6VB1-AS1*) showed no difference between the groups [51].

There is increasing evidence demonstrating that hPDL(S)Cs–EV can modulate *in vitro* angiogenesis (in HUVECs [52]), osteogenesis (in MG-63 OBs [54] and hPDLSCs [55,56,59]), anti-inflammation (in LPS-treated hPDLSCs [55,60] and J774.1 macrophages [57]), and immunoregulation (induced M1 polarization in THP-1 cells [58]) via modulating the TGF-beta pathway, MAPK pathway, mTOR pathway and FoxO signaling pathways [51], and PI3K/Akt signaling [55] and NF-kB signaling pathways [57]. The *in vivo* function of hPDL(S)Cs–EV was explored, either in rat calvaria defect [56,59] or intravenous administration in mouse multiple sclerosis disease [60] models. Pizzicannel-la et al. [56] created a calvarial defect, with a diameter of 4 mm and a height of 0.25 mm in male Wistar rats (300–350 g; *n* = 4 for each group). The hPDLSCs–EV and hPDLSCs–PEI–EV were loaded on collagen membranes and transplanted into the rat calvaria defect for 6 weeks, leading to enhanced bone and vascularization compared to the no-EV groups, with the PEI–EV group inducing better osteogenesis and vascularization compared to the EV group. Diomede et al. [59] revealed similar results, showing that hPDLSCs–PEI–EV leads to increased blood vessel formation after 6 weeks of the transplantation of hPDLSCs–EV- and hPDLSCs–PEI–EV-loaded collagen membranes into a rat calvarial defect. Rajan et al. [60] established a mouse model of MS disease, and intravenous administration of hPDLSCs–EVs decreased apoptosis and inflammation in the diseased mice.

In summary, the size of hPDL(S)Cs–EV ranges from 20 nm to 1600 nm when using different EV isolation methods, with under-reporting of sufficient detail about the cell source and cell culture conditions before CM collection. The hPDL(S)Cs–EV contains miRNAs, circRNAs, and lncRNAs, and they modulate the angiogenesis, osteogenesis, and inflammation of recipient cells, through TGF-β, MAPK, mTOR and FoxO pathways [51], and PI3K/Akt [55] and NF-kB signaling pathway [57]. However, none of the three *in vivo* studies [56,59,60] used either a periodontal defect or a periodontitis animal model.

*5.2. Human Gingival Fibroblasts (hGFs)–sEV*

Table 3 summarizes seven studies [61–67]) of EV from fibroblasts (hGFs [62,63] or MSCs (hGMSCs [61,64–67]) from human gingiva tissues. There are two studies that investigated the *in vivo* role of hGMSCs–EV using animal models [66,67]. The cells from either 20-to-40-year-old donors [66] or unclear age human donors [62–65,67] were used at passage 2 [64], passage 4–6 [62], <6 passage [66], or unclear [63,65,67]. EV-depleted FBS [61,63,66], FBS starvation [65], and unclear cell culture conditions [62,64,67] were applied in the studies before CM collection for EV isolation. The size of hGFs/hGMSCs–EV varied from different studies, as follows: <200 nm [61–63,66], 50–500 nm [65], unclear size [64], and a combination of two populations (93 ± 24 nm and 1200 ± 400 nm) [67]. Engineered hGMSCs–PEI–EV had the following two populations: 250 ± 50 nm and 3600 ± 500 nm [67].

**Table 3.** The isolation, characterization, and function of EV from hGFs or hGSMCs.

| Reference | Cell Source | EV Isolation | EV Characterization | Key Findings |
|---|---|---|---|---|
| Nakao et al., 2021 [61] | • Human gingival mesenchymal stem cells (hGMSCs)<br>• Donor details are unclear<br>• Passage 4–6 | • CM was collected after 48 h in FBS-free media and centrifuged at $10,000\times g$ for 30 min.<br>• hGMSCs exosomes were isolated using MagCapture TM exosome isolation kit PS (FUJIFILM Wako). | • TEM<br>• NTA<br>• WB (CD9, CD63 and CD81) | • Mode of sEVs: $109 \pm 3.1$ nm and $104 \pm 1.8$ nm for hGMSCs–sEVs and TNF-$\alpha$ pre-treated hGMSCs-derived sEVs.<br>• *In vitro*: TNF-$\alpha$ stimulation increased the number of hGMSCs–sEVs and exosomal CD73, as well as induced anti-inflammatory M2 macrophage polarization. The hGMSCs–sEVs–miR-1260b can target Wnt5a-mediated RANKL expression.<br>• *in vivo*: in a ligature-induced mice periodontitis model, a local injection of GMSC-derived exosomes significantly reduced periodontal bone resorption. |
| Yin et al., 2020 [62] | • hGFs<br>• Donor: Five normal gingival tissues (*n*) and 1 idiopathic gingival fibromatosis (IGF) gingival tissues<br>• Passage 4–6 | • ExoQuick TC (no detailed isolation protocol) | • BCA assay<br>• TEM<br>• FACS (CD63 and CD81) | • D: 50–200 nm<br>• IGF–GFs–Exo increased cell proliferation of normal hGFs at 24 h and 48 h by MTS assay.<br>• The expression of *Ki67*, *PCNA*, *Bcl-2*, and *Bax* were enhanced after 24 h treated with IGF–GFs–Exo. After 48 h, the level of *PCNA*, *Bcl-2* and *Bax* was significantly downregulated, while the expression of Ki67 was not varied significantly. |
| Zhuang et al., 2020 [63] | • hGFs<br>• Donor: Fresh human gingiva from donors with wisdom tooth extraction<br>• hBMSCs<br>• Donor: Fresh human bone marrow from the iliac bone of jaw cysts during the reconstruction of bone defects with hydroxyapatite powder and bone marrow after surgery curettage | • Exosome-depleted medium was obtained after 25,000 r.p.m. ultracentrifugation for 90 min. Then, 10 mL of CM was mixed with ExoQuick exosome precipitation solution and refrigerated overnight.<br>• CM was centrifuged at 1500 r.p.m. for 30 min at 4 °C and then at 3000 r.p.m. for 5 min. | • TEM<br>• WB (CD63, CD81 and tubulin) | • TEM of EV diameter: 50 to 200 nm.<br>• Irradiation-activated hGFs–exosome inhibited osteogenic differentiation of hBMSCs for 7 days, with reduced *ALP*, *COL1* and *RUNX2* gene expression via an exosomal miR-23a/CXCL12 axis. |

**Table 3.** *Cont.*

| Reference | Cell Source | EV Isolation | EV Characterization | Key Findings |
|---|---|---|---|---|
| Silvestro et al., 2020 [64] | • hGMSCs<br>• Donor: six healthy adult volunteers with no gingival inflammation during teeth removal for orthodontic purpose<br>• Passage 2 | • The conditioned medium (CM; 10 mL) after 48 h of incubation were collected from hGMSCs at passage 2.<br>• The CM was centrifuged at $3000 \times g$ for 15 min; 2 mL ExoQuick TC was added to 10 mL of CM and incubated overnight at 4 °C without rotation; one centrifugation step was performed at $1500 \times g$ for 30 min to sediment the EVs | • No EV characterization | • The size of EV was not mentioned.<br>• RNA sequencing analysis showed that 15,380 genes were identified in GMSCs–EVs. There were 1067, 886, 808, 768, 562, and 541 protein-coding genes for hydrolase, enzyme modulators, the transcription factor, transferase, the receptor and the transporter, respectively. There were 1155 non-coding RNA genes for anti-sense RNAs, lncRNAs and miRNAs (miR1302 family, miR451 family, miR24 family, miR219 family and miR194 family).<br>• hGMSCs–EV also contain mRNAs for proteins of the interleukins, TGF-, BMPs, GDFs, Wnt, VEGF, FGF, and neurotrophins are critical for basic or neuronal, bone or vascular development. |
| Coccè et al., 2019 [65] | • Human MSCs were isolated from gingival papilla (named GinPaMSCs)<br>• Donor details are unclear | • GinPaMSCs cultured with FBS-free media for 72 h cultures before EV collection.<br>• CM was collected at 24 and 48 h of incubation and centrifuged on a 100 kDa filter device at $5000 \times g$ for 15 min. The two fractions (i.e., EV: F > 100 kDa; free PTX: F < 100 kDa) | • DLS<br>• NTA<br>• TEM | • TEM of exosomes: 50 to 500 nm. DLS of EV: 200–300 nm. NTA detected the following 3 different EV populations: 135 nm, 200–300 nm, and 435 nm.<br>• PTX was presented in PTX-treated GinPaMSCs-secreted EVs.<br>• *GinPaMSCs*–EV/PTX have anti-cancer activity in human pancreatic carcinoma and squamous carcinoma cells. |

**Table 3.** *Cont.*

| Reference | Cell Source | EV Isolation | EV Characterization | Key Findings |
|---|---|---|---|---|
| Mao et al., 2019 [66] | • hGMSCs: gingival tissues were obtained from five healthy human subjects aged from 20-to-40 years, who underwent routine dental procedures<br>• A rat Schwann cell (SC) line RT4-D6P2T<br>• hGMSCs: passage < 6<br>• SCs passage < 4 | • hGMSCs cultured in media with 1% exosome-depleted FBS (System Biosciences, SBI) for 48 h to collect CM.<br>• The CM was centrifuged at $1000\times g$ for 30 min; 20 mL of culture media was filtered a Vivaspin 20 ultrafiltration device (100 kDa) and centrifuged at $3000\times g$ for 60 min to get 1 mL of the concentrated medium; mixed with 0.2 mL ExoQuick-TC exosome precipitation solution (5:1) and incubated overnight at 4 °C; centrifuged at $1500\times g$ for 30 min at 4 °C | • BCA assay<br>• NTA<br>• WB (CD 63 and CD9) | • GMSC-derived sEV had a mean size of 103.8 ± 2.1 nm by NTA.<br>• *in vivo* studies mimicking clinical nerve repair showed that hGMSCs-derived sEV promoted functional recovery, axonal repair and regeneration of crush-injured mice sciatic nerves.<br>• *in vitro*: GMSC-derived EVs promoted proliferation and migration of Schwann cells, with upregulated c-JUN, Notch1, GFAP (glial fibrillary acidic protein), and SRY (sex-determining region Y)-box 2 (SOX2). |
| Diomede et al., 2018 [67] | • hGMSCs seeded on 3D poly (lactide) (PLA) scaffolds<br>• Donor: gingival tissue biopsies were obtained from healthy adult volunteers with no gingival inflammation | • After 48 h of incubation, EVs were collected from hGMSCs at passage 2.<br>• centrifuged at $3000\times g$ for 15 min; 2 mL ExoQuick TC was added to 10 mL CM and incubated overnight at 4 °C; $1500\times g$ for 30 min to sediment the EVs<br>• EVs were engineered with PEI (MW 25,000). | • DLS<br>• AFM<br>• WB (CD9, CD63, CD81, and TSG101) | • DLS of hGMSCs–EVs: 93 ± 24 nm and 1200 ± 400 nm; engineered hGMSCs–PEI–EV: 250 ± 50 nm and 3600 ± 500 nm.<br>• EV and PEI–EV increased calcium deposits after 6 weeks of osteogenic differentiation in hGMSCs, with increased RUNX2 and BMP2/4 gene and protein expression.<br>• RNA sequencing of the transcriptome of hGMSCs, EV and PEI–EV revealed that 31 genes were differentially expressed between groups. GO analysis showed that these 31 genes involved in "regulation of ossification" and "ossification" were upregulated in the PEI–EV group compared to hGMSCs group through the TGF-β signaling.<br>• *in vivo* results showed that PEI–EV with/without hGMSCs in PLA scaffolds enhanced bone and blood vessels formation in a rat cortical calvaria defect by histology and microCT. |

**Abbreviation:** MTS, 3-(4,5-dimethylthiazol-2-yl)-5-(3-carboxymethoxyphenyl)-2-(4-sulfophenyl)-2H-tetrazolium; Ki-67, antigen KI-67/marker of proliferation Ki-67; PCNA, proliferating cell nuclear antigen; CXCL12, C-X-C motif chemokine 12/ stromal cell-derived factor 1; FGF, fibroblast growth factor; GDNF, glial cell-derived neurotrophic factor; PTX, paclitaxel; c-JUN, jun proto-oncogene/AP-1 transcription factor subunit; Notch1, Notch homolog 1(translocation-associated).

RNA sequencing data from Silvestro et al. showed that hGMSCs–EV comprises 15,380 genes (for interleukins, TGF-β, BMPs, GDFs, Wnt, VEGF, FGF, and neurotrophins), and 1155 non-coding RNA (lncRNAs and miRNAs—miR1302, miR451, miR24, miR219 and miR194) [64]. The miRNA microarray data from Nako et al. [61] showed that 655 universal differentially expressed miRNAs were found in Exo-TNF compared to Exo-Ctrl, particularly miR-1260b (ranked in the top three of the most highly upregulated miRNAs, by using TNF- α preconditioning). RNA sequencing from Diomede et al. [67] demonstrated that 31 ossification genes were enhanced in hGMSCs–PEI–EV compared to hGMSCs-EV through the TGF-β signaling pathway.

The *in vitro* functional assays showed that hGFs/hGMSCs–sEV facilitates cell proliferation (in hGFs [62] and Schwann cells line [66]), anti-osteoclastogenic [61] and osteogenic differentiation (in hBMSCs [63] and hGMSCs [67]), as well as an anti-carcinogenesis effect (in human pancreatic carcinoma and squamous carcinoma cells [65]). This may be mediated by an miR-1260b/Wnt 5A/RANKL pathway [61], miR-23a/CXCL12 axis [63], interleukins, TGF-β, BMPs, GDFs, Wnt, VEGF, FGF, and neurotrophins [64], and TGF-β signaling [67].

In their *in vivo* investigations, Nakao et al. [61] created a ligature-induced periodontitis mice model, and locally injected hGMSCs–sEV or TNF-α-preconditioned GMSC-derived exosomes (hGMSCs–sEV–TNF) into the palatal gingiva of the ligated second maxillary molar. One week post-injection, both the interventions significantly reduced periodontal bone loss compared to the PBS control group, while hGMSCs–sEV–TNF further reduced the distance from the cementoenamel junction to the alveolar bone crest (CEJ–ABC) and the number of tartrate-resistant acid phosphatase (TRAP)-positive osteoclasts, indicating an anti-osteoclastic property for hGMSCs–sEV [61]. Moreover, Mao et al. [66] transplanted hGMSCs–sEV-loaded gelfoam sheets into the crush-injury sites of sciatic nerves in C57BL/6J mice, and the EV group had comparable beneficial effects on the functional recovery of the injured sciatic nerves of mice compared to the hGMSCs group. Further, hGMSCs–sEV enhanced the expression of neuronal and Schwann cell markers (β-tubulin III and S100 calcium-binding protein B—S100B) at one-month post-injury, compared with hGMSCs controls, suggesting that hGMSCs–sEV can promote neuron regeneration *in vivo*. Diomede et al. [67] loaded hGMSCs–EV or hGMSCs–PEI–EV into 3D-printed PLA scaffolds with/without hGMSCs, and transplanted them into rat calvaria defects for 6 weeks. Both the hGMSCs–EV and hGMSCs–PEI–EV groups enhanced bone and blood vessel formation, yet hGMSCs–PEI–EV performed better than the EV group.

In summary, the EV diameter from hGFs/hGSMCs is different among studies, ranging from 50 nm to 1600 nm. EV–mRNAs and EV–miRNAs [61,64,67] may contribute to their *in vitro* and *in vivo* function in cell proliferation [62,66], and reduce bone resorption [61], osteogenic differentiation [63,67] and nerve regeneration [66]. More *in vivo* studies are required in order to explore the function of EV from gingival tissue-derived cells.

### 5.3. Human Dental Pulp Cells (hDPSCs)–sEV

Table 4 summarizes nine studies investigating EV from human primary DPSCs [68–76], with three of these including *in vivo* models [69,72,76]. Cells were isolated from donors who were 24–41 years old [69], 16–25 years old [70], 20 years old [71], 19–28 years old [73], 22–36 years old [75], or an unclear donor age [68,72,74,76]. The cells at passage 2 [71], passage 3–6 [70], passage <4 [76], passage 3–7 [75], passage 3–5 [73], or unclear passage number [68,69,72,74] were used in these studies. Prior to CM collection, the cells were cultured in EV-depleted FBS [68,70–73] or FBS starvation [69,74–76]. The mode size of hPDSCs–sEV was smaller than <200 nm in most studies [68–72,75], with one study reporting 50–400 nm [72], 80–400 nm [73], and 30–250 nm [74], and unclear EV size [76].

**Table 4.** The isolation, characterization, and function of hDPSCs–EV.

| Reference | Cell Source | EV Isolation | EV Characteri-zation | Key Findings |
| --- | --- | --- | --- | --- |
| Faruqu et al., 2020 [68] | • Human dental pulp pluripotent-like stem cells (hDPPSCs) from healthy human third molars extracted for orthodontic and prophylactic reasons<br>• Umbilical cord-derived mesenchymal stem cells<br>• The passage is not mentioned | • Exosome-depleted FBS was obtained after ultracentrifugation at $100,000\times g$ for 18 h at 4 °C.<br>• CM was filtered through 0.22 μm filter, mixed with sucrose solution with 25% $w/w$ in deuterium oxide (D2O) and ultracentrifugation at $100,000\times g$ for 1.5 h at 4 °C and another $100,000\times g$ for 1.5 h at 4 °C. | • NTA<br>• Dot blot (CD9, CD63, alix, TSG101, and calnexin) | • sEV from hDPPSCs spheroids culturing in KnockOut™ serum replacement (KO-medium) at both day 1–12 and day 13–24 samples were detected similar particle sizes (168.7 ± 7.2 nm and 156 ± 7.6 nm). |
| Zhou et al., 2020 [69] | • hDPSCs<br>• Donor: 5 healthy donors (male: 2; female: 3; age: 24~41 years) and periodontitis teeth (n = 6), named H-DPSCs and P-DPSCs<br>• Passage 3~5<br>• H-DPSCs and P-DPSCs-derived EVs are named H-EV and P-EV. | • At 90% confluence, hDPSCs were washed three times with PBS prior to culturing with serum-free media for 48 h.<br>• CM was centrifuged at $300\times g$ for 10 min; $2000\times g$ for 10 min; centrifuged at $10,000\times g$ for 30 min; ultracentrifuged at $100,000\times g$ for 70 min; washed with PBS at $100,000\times g$ for 70 min | • TEM<br>• NTA<br>• WB (alix, HSP70, CD9, CD63 and CD81)<br>• BCA assay | • Both H-EV and P-EV sizes ranged from 30 to 200 nm.<br>• hPDSCs–sEV from periodontitis patients promoted endothelial cells proliferation and angiogenesis with higher expression levels of VEGF and AngII genes/proteins compared with hPDSCs–sEV from healthy patients.<br>• *in vivo*: The vascularization of mouse skin defects and wound healing were promoted by both H-EV and P-EV, while the treatment with the latter brought a faster repairing process, as well as the formation of fresh vessels. |
| Ivica et al., 2020 [70] | • hDPSCs from healthy third molars (n = 3, 16 to 25 years old) were extracted<br>• Human bone marrow-derived mesenchymal stem cells (HBMMSCs). Passage 3–6 | • Exosome-free medium (Invitrogen) was used for hDPSCs culture for 48 h before CM collection.<br>• CM centrifuged at $300\times g$ for 10 min; $2000\times g$ for 30 min; total exosome isolation agent (Invitrogen) was added; the mixture was centrifuged at $10,000\times g$ for 30 min | • TEM<br>• WB (CD9, and Grp94) | • hDPSC–sEV ranged from 45 to 156 nm by TEM, with a mean size of 90 nm.<br>• hDPSC–sEV encapsulated in fibrin gel promoted hBMMSCs migration and proliferation. |

**Table 4.** *Cont.*

| Reference | Cell Source | EV Isolation | EV Characterization | Key Findings |
|---|---|---|---|---|
| Xie et al., 2020 [71] | • hDPSCs<br>• Donor: one healthy patient aged 20 years old, free of periodontal or endodontic problems. | • Exosomes secreted by hDPSCs during the starvation of 48 h without FBS, marked as EX0. Osteogenic-induced DPSC-derived exosomes (OI-DPSC-Ex) after culturing with osteogenic media with 15% exosome-free FBS. Exosomes secreted by these osteogenic-induced DPSCs at days 5 and 7 were extracted and marked as EX5 and EX7.<br>• CM was centrifugated at $3000 \times g$ for 20 min; centrifuged at $16,500 \times g$ for 20 min; filtered with a 0.2 micron filter to collect the filtrate; ultracentrifuged at $100,000 \times g$ for 70 min<br>• Passage 2 | • TEM<br>• NTA<br>• Flow cytometry (CD63 and CD81) | • TEM and NTA detected that exosomes size range from 20 to 120 nm.<br>• The OI-DPSC-Ex induced the osteogenic differentiation of recipient parent hDPSCs via exosomal circPAR1 binding with hsa-miR-31. |
| Shen et al., 2020 [72] | • hDPSCs<br>• Donor: from exfoliated teeth of healthy donors<br>• Passage not mentioned | • 80% confluent hDPSCs cultured with medium supplemented with 10% exosome-free FBS (centrifuged at $120,000 \times g$ for 18 h) for 3 days.<br>• CM was centrifuged at $300 \times g$ for 10 min; centrifuged at $16,500 \times g$ for 20 min; ultracentrifuged at $120,000 \times g$ for 2.5 h at 4 °C<br>• DPSC-Exo encapsulated in chitosan hydrogel (CS) was named DPSC-Exo/CS | • TEM<br>• NTA<br>• WB (CD9, HSP70 and TSG 101)<br>• Flow cytometry (CD 63 and CD 81) | • NTA of EV: 50 to 400 nm, peaked at 155.4 nm.<br>• In an *in vivo* periodontitis mice model, DPSC-Exo/CS promoted the regeneration of alveolar bone and periodontal epithelium in the periodontitis mice model after 10 days.<br>• DPSC-Exo/CS had anti-inflammatory effects and facilitated the immune response by switching macrophages from a pro-inflammatory phenotype to an anti-inflammatory phenotype at both *in vitro* and *in vivo* mice with periodontitis via hDPSCs-derived exosomal miR-1246. |
| Li et al., 2019 [73] | • hDPSCs<br>• Donor: healthy patients (aged 19–28 years old, n = 12) impacted third molars<br>• LPS pretreated overnight in hDPSCs. EV from LPS-treated cells were named LPS-exo<br>• Passage 3–5 | • hDPSCs at 70–80% confluence cultured for 48 h in media exosome-depleted FBS (Systembio).<br>• CM was centrifuged at $500 \times g$ for 10 min at 4 °C, centrifuged at $2000 \times g$ for 10 min; centrifuged at $10,000 \times g$ for 1 h at 4 °C and filtrated through a 0.22 μm filter; ultracentrifuged at $100,000 \times g$ for 2 h; ultracentrifuged at $100,000 \times g$ at 4 °C for 70 min | • BCA assay<br>• WB (alix, CD9, CD63 and GM130)<br>• TEM<br>• NTA | • EV: 80–400 nm, peaked at 116 nm.<br>• LPS-treated cells generated more exosomes particles.<br>• Both exo and LPS-exo facilitated the production of dentin sialoprotein and mineralization of Schwann cells (SCs).<br>• LPS-exo had a higher promoted proliferation, migration and odontoblast differentiation of Schwann cells (SCs) compared to Exo only, with increased DSPP, DMP1, OCN, and RUNX2 gene expression after 14 days. |

**Table 4.** *Cont.*

| Reference | Cell Source | EV Isolation | EV Characterization | Key Findings |
|---|---|---|---|---|
| Ji et al., 2019 [74] | • hPDSCs were isolated from healthy dental pulp tissues and (n = 8) from the caries-free teeth that need to be extracted due to orthodontics<br>• hBMMSCs were isolated from bone marrow aspirates of healthy people (n = 8). | • At 80% confluence, the medium was replaced with a serum-free medium for 48 h before CM collection.<br>• The CM was centrifuged at $300 \times g$ for 10 min, centrifuged at $16,500 \times g$ for 30 min at 4 °C; passed through a 0.2 μm; ultracentrifuged at 4 °C at $100,000 \times g$ for 70 min; ultracentrifuged again at 4 °C at $100,000 \times g$ for 70 min | • TEM<br>• NTA<br>• BCA<br>• WB (CD9 and CD63) | • NTA of EV: 30–250 nm, mean size of 135 nm.<br>• hDPSCs–EV suppressed the differentiation of CD4+ T cells into T helper 17 cells (Th17), but stimulated the polarization of CD4+ T cells into regulatory T cells (Treg).<br>• hDPSCs–EV and hBMMSCs–EV inhibited the secretions of pro-inflammatory factors (IL-17 and TNF-α) and increased anti-inflammatory factors (IL-10 and TGF-β) in CD4+ T cells after 72 h of treatment.<br>• Both hDPSCs–EV and hBMMSCs–EV inhibited proliferation, but increase the apoptosis of CD4+ T cells. |
| Hu et al., 2019 [75] | • hDPSCs<br>• Donor: healthy pulp tissues isolated from caries-free teeth of patients (5 females, age 24–35 years; 5 males, age 22–36 years) undergoing extraction of fully erupted third molars<br>• Passage 3–7 | • Exosomes from hDPSCs in either growth (UN-Exo) or odontogenic differentiation media (OD-Exo) for 10 days<br>• Cells were washed in serum-free PBS and cultured for 48 h in serum-free media.<br>• The Exo-spin (Cell Guidance) exosome isolation reagent (no detailed description) | • WB (CD9 and CD63)<br>• TEM | • UN-Exo and OD-Exo range from 30 to 150 nm in diameter by TEM.<br>• OD-Exo promoted the odontogenic differentiation in hDPSCs with increased *RUNX2, DMP-1, DSP* and *ALP* gene expression.<br>• RNA sequencing analysis showed that 28 microRNAs significantly changed in OD-Exo isolated under odontogenic conditions, of which 7 miRNAs increased and 21 miRNAs decreased. The qRT-PCR analysis showed that miR-5100 and miR-1260a levels in OD-Exo increased, while miR-210-3p and miR-10b-5p decreased, which were consistent with the miRNA sequencing. GO analysis showed that the differentially expressed miRNAs are associated with the TGFβ1 pathway.<br>• OD-exo activated the TGFβ1 pathway by upregulating TGFβ1, TGFR1, p-Smad2/3, and Smad4 in DPSCs, compared to the control group and UN-Exotreated group.<br>• Compared with UN-Exo, miR-27a-5p was expressed 11 times higher in OD-Exo.<br>• The luciferase reporter assay demonstrated that miR-27a-5p can target the 3′-UTR of LTBP1 directly.<br>• This suggests that exosomal miRNAs promoted odontogenic differentiation via the TGFβ1/smads signaling pathway by downregulating LTBP1. |

Table 4. *Cont.*

| Reference | Cell Source | EV Isolation | EV Characterization | Key Findings |
|---|---|---|---|---|
| Huang et al., 2016 [76] | • hDPSCs and primary hMSCs are not clearly detailed.<br>• Both hDPSCs and hMSCs used were passage <4 | • Exosomes were isolated from hDPSCs in either growth (DPSC-Exo) or odontogenic differentiation media (DPSC-OD-Exo) for 4 weeks. Cells were washed in serum-free media and cultured for 24 h in serum-free media before CM collection.<br>• ExoQuick-TC exosome isolation reagent (non-detailed description) | • WB (CD63 and CD9)<br>• TEM | • Size of EV was not demonstrated.<br>• The endocytosis of exosomes was dose-dependent and manner-saturable for both hMSCs and hDPSCs *in vitro*, which initiated the MAPK pathway through the caveolar endocytic mechanism.<br>• hDPSCs–EV can bind to fibronectin and type I collagen via an exosomal integrin-mediated process.<br>• DPSC-Exo and DPSC-OD-Exo induce increased expression of odontogenic marker genes in DPSCs in 3D culture within type I collagen hydrogels.<br>• DPSCs-Exo and DPSC-OD-Exo in a collagen membrane on a filled root canal spaces of human tooth root slices were implanted subcutaneously in athymic nude mice for 2 weeks. DPSC-Exo and DPSC-OD-Exo triggered increased expression of odontogenic differentiation marker proteins DMP1 and DPP, and only the DPSC-OD-Exo group improved active blood vessels formation and endothelial cell marker von Willebrand factor (vWF).<br>• DPSC-Exo and DPSC-OD-Exo in hMSCs after 48 h of *in vitro* treatment increased in the expression levels of several growth factors and ECM proteins along with the transcription factor Runx2. |

**Abbreviations**: circPAR1, circular Prader–Willi/Angelman region-1; DMP-1, dentin matrix acidic phosphoprotein 1; DSPP, dentin phosphophoryn; TGFR1, transforming growth factor beta receptor I; p-Smad2/3, mothers against decapentaplegic homolog 2/3 or SMAD family member 2/3; Smad4, SMAD family member 4/mothers against decapentaplegic homolog 4; UTR, untranslated region; LTBP1, latent TGF-beta binding protein.

The hPDSCs–EV modulates angiogenesis in endothelial cells [69], migration/proliferation (in hBMMSCs [70], Schwann cells (SCs) [73], and CD4+ T cells [74]), osteogenic differentiation in hDPSCs [71], anti-inflammation (in DPSCs [72] and CD4+ T cells [74]), and odontogenic differentiation of Schwann cells (SCs) [73], hDPSCs [75,76], and hMSCs [76]. This may be regulated through hDPSCs–sEV–circPAR1 binding with hsa-miR-31 [71], hDPSCs–sEV–miR-1246 [72] and hDPSCs–sEV–miR-27a-5p [75]. RNA sequencing data from Hu et al. [75] demonstrated that 7 increased sEV–miRNAs and 21 decreased sEV–miRNAs were found in odontogenic differentiated hDPSCs–sEV, and these miRNAs are associated with the TGFβ1/smads signaling pathway. The authors concluded that sEV–miR-27a-5p can modulate odontogenic differentiation via the TGFβ1/smads signaling pathway, by downregulating latent-transforming growth factor beta-binding protein 1 (LTBP1).

With respect to *in vivo* studies, Zhou et al. [69] created a full-thickness excisional skin wound-healing model in male C57BL/6 mice (8 weeks old), and then subcutaneously injected hDPSCs–sEV (200 µg in 100 µL) from healthy or periodontitis patients derived hDPSCs–sEV (200 µg in 100 µL PBS) for 4, 9, and 14 days. Both the sEV groups promoted the wound healing process and vascularization compared to the PBS control group, while hDPSCs–sEV from the periodontitis patients increased the wound closure rate and the number of newly formed microvessels, with more CD31- and VEGF-positive cells compared to the sEV from a healthy patient. Shen et al. [72] established a ligation-induced periodontitis model in 6–8-week-old male C57BL/6J mice, and a chitosan hydrogel (CS) loaded with 50 µg of hDPSCs–sEV (hPSDCs–sEV–CS group) was locally injected after ligature removal, with a local injection of PBS or hPSDCs–sEV used as the controls. The results showed that the hDPSCs–sEV–CS group led to increased bone formation, a thick layer of epithelial layers, less inflammatory cells, and a lower amount of TRAP-positive osteoclasts, at 10 days post-treatment. Furthermore, hPSDCs–sEV–CS treatment significantly reduced pro-inflammatory cytokines (IL-23, IL-1α, TNF-α, IL-12, IL-1β, IL-27, and IL-17), and NF-κB p65 and p38 MAPK signaling, in periodontal tissues compared with other groups. RNA sequencing analysis of the periodontium showed that 7351 differentially expressed genes (DEGs) were found between the hDPSCs–sEV–CS and CS groups. GO term enrichment analysis of the top 200 DEGs demonstrated that they are associated with chemotaxis pathways and the immune response, which were downregulated in the hDPSCs–sEV–CS group. Most importantly, hDPSCs–sEV–CS induced macrophages converting from a proinflammatory phenotype to an anti-inflammatory phenotype in the periodontium of periodontitis mice, with more CD206+ anti-inflammatory macrophages and significantly decreased CD86+ in pro-inflammatory macrophages [72]. This indicates that hDPSCs–sEV can promote bone formation, epithelium re-growth, and reduce inflammation in a periodontitis mice model. Furthermore, Huang et al. [76] loaded hDPSCs–EV into clinical-grade type I collagen membranes, and then placed them on a human tooth root slice (3–4 mm in thickness), before subcutaneously transplanting into athymic nude mice for 2 weeks. They resulted in enhanced dental-pulp-like tissues, with increased odontogenic proteins (dentin matrix acidic phosphoprotein 1—DMP1, and dentin phosphophoryn—DSPP) and endothelial cell marker protein (von Willebrand factor—vWF).

To summarize, hDPSCs–EV, ranging from 30 nm to 400 nm among nine studies, and containing circRNA [71], miRNAs [72,75], and mRNAs [72], may modulate angiogenesis [69], migration/proliferation [70,73,74], osteogenic differentiation [71], anti-inflammation [72,74] and odontogenic differentiation [73,75,76] in recipient cells. Among the three *in vivo* studies, the skin wound-healing model [69], periodontitis disease model [72], and subcutaneous transplantation [76] were employed, and the results showed that hPDSCs–EV can promote angiogenesis, osteogenesis, dentin-pulp regeneration, and reduce inflammation and osteoclastic activity. Further *in vivo* studies are required to validate the function of hDPSCs–EV.

*5.4. SHED/SCAP/DFCs–sEVs*

Table 5 summarizes seven investigations (five *in vivo* studies) examining sEVs from dental cells, including SCAP [77,78], SHED [79–82], and DFCs [83]. Cells were isolated from 5–8-year-old donors [81], 12–15 years old [77,78], 13–19 years old [83], or unknown age [79,80,82], at passage 3–4 [79], 4–7 [80], 4 [81], 3–6 [82], 5 [83], or unknown [77,78]. EV-depleted FBS [77,79,81,82] and FBS-starvation [78,80,83] were applied for CM collection.

The size of SHED/SCAP/DFCs–sEVs was smaller than 200 nm in all the studies; these sEVs promote angiogenesis in HUVECs [77,82], anti-inflammation in mBMSCs [80] and chondrocytes [81], osteogenesis in PDLCs [79], mBMSCs [80] and rBMSCs [82], and dentinogenesis in BMMSCs [78,83] *in vitro*, by the Cdc42 pathway [77], Wnt/β-catenin and BMP/Smad signaling pathways [79], miR-100–5p/mTOR pathway [81], and AMPK pathway [82].

The function of SCAP–sEVs was investigated *in vivo* on gingival soft tissue [77] and dentin-pulp regeneration [78]. Liu et al. [77] created full-thickness circular gingival wounds in C57BL/6J mice, using a biopsy punch (soft tissue defects with a diameter of 2.0 mm). Following this, 40 μg of SCAP–sEVs, SCAP–siCdc42–sEVs, or PBS, was injected submucosally into the palates of the wounds sites. Seven days post-injection, SCAP–sEVs promoted palatal gingival tissue regeneration by enhancing vascularization in the early phase [77]. Zhuang et al. [78] loaded 50 μg/mL SCAP–sEVs and $4 \times 10^5$ BMMSCs with gelatin sponge onto a dentin slice, before subcutaneously transplanting them into immunodeficient mice. Significant dentin-pulp regeneration was observed 12 weeks post-transplantation in the SCAP–sEVs group compared to the PBS control group.

The action of SHED–sEVs on periodontitis disease and periodontal defect *in vivo* has been investigated in a mouse [80] and rat model [82], respectively. Wei et al. [80] locally injected 20 μg of SHED–sEVs into buccal and lingual sides of the first molar once per week, over 2 weeks, in ligature-induced periodontitis mice. After 2 weeks, SHED–sEVs reduced bone loss, with a decreased CEJ–ABC distance compared to the controls. Moreover, Wu et al. [82] generated a periodontal defect ($4 \times 2 \times 1.5$ mm$^3$) in their rat model, at the buccal alveolar bone of the first-to-third mandibular molars. SHED–sEVs were loaded into a β-TCP scaffold before placing them into the periodontal defect for four weeks, resulting in enhanced neovascularization and new bone formation compared to the β-TCP/PBS scaffold.

In their study, Shi et al. [83] injected gelatin hydrogels (100 μL), loaded with LPS–DFCs–sEVs (sEVs derived from LPS-treated DFCs) or DFCs–sEVs, into the periodontal pocket of the right maxillary second molar in a ligature-induced periodontitis rat model. The intervention was once a week for up to 8 weeks, and resulted in significantly reduced alveolar bone loss and TRAP-positive osteoclasts, as well as enhanced well-oriented PDL fibers in the LPS–DFCs–sEVs group.

In summary, SHED/SCAP/DFCs–sEVs are smaller than 200 nm, and those containing miR-100–5p [81] may modulate angiogenesis [77,82], inflammation [80,81], osteogenesis [79,80,82], and dentinogenesis [78,83] *in vitro*. More importantly, five *in vivo* studies showed that SHED/SCAP/DFCs–sEVs can promote angiogenesis [77,82], dentin-pulp complex [78], alveolar bone [82], and well-organized PDL fiber formation [82]. It is noted that two studies utilized a ligature-induced periodontitis disease model [80,83] and one study used a periodontal defect model [82]. More studies are needed to further validate the *in vivo* functional role of SHED/SCAP/DFCs–sEVs.

**Table 5.** The isolation, characterization, and function of sEVs from SHED, SCAP and DFCs.

| Reference | Cell Source | EV Isolation | EV Characterization | Key Findings |
|---|---|---|---|---|
| Liu et al., 2021 [77] | • Stem cells from apical papilla (SCAP) <br>• Donor: healthy third molars with immature roots from healthy donors aged 12 to 15 years <br>• Passage number is unclear | • SCAP cells were cultured in exosome-free medium for 48 h and CM was centrifuged at 4 °C in an ultracentrifuge at the following three different speeds: $3000 \times g$ for 20 min, $20{,}000 \times g$ for 30 min, and $120{,}000 \times g$ for 2 h <br>• Ultracentrifugation method | • TEM <br>• NTA <br>• WB (CD9, CD63, and Alix) | • sEVs mode: 120.1 nm; mean: $139.2 \pm 62.5$ nm <br>• *in vivo*: SCAP–sEVs promoted vascularization to accelerate tissue regeneration of the palatal gingiva via Cdc42-mediated vascularization in a mouse gingival wound healing model. <br>• *in vitro*: SCAP–sEVs enhanced the cell migration and angiogenic capacity of HUVECs via Cdc42-mediated cytoskeletal reorganization. |
| Zhuang et al., 2021 [78] | • SCAP <br>• Donor: Human impacted third molar with immature roots were collected from healthy patient (12–15 years old) <br>• Passage number is unclear | • SCAP cells at 60–80% confluence were cultured with serum-free media for 48 h before CM collection. <br>• The CM was centrifuged sequentially at 4 °C: $3000 \times g$ for 20 min, $20{,}000 \times g$ for 30 min, and $120{,}000 \times g$ for 2 h. <br>• Ultracentrifugation | • TEM <br>• NTA <br>• BCA <br>• WB (CD9, and Alix) | • sEVs peaked at 120.6 nm. <br>• SCAP-Exo promoted mouse BMMSC-based dentine-pulp complex regeneration *in vivo* and *in vitro* dentinogenesis of BMMSCs. |
| Wang et al., 2020 [79] | • Stem cells from human-exfoliated deciduous teeth (SHED) <br>• hPDLCs <br>• Donor: age is unclear; pulp tissue from non-carious primary teeth extracted from children for orthodontic reasons <br>• Passage: 3–4 | • SHED and hPDLCs cells were cultured in 15% and 10% exosome-free media, respectively. <br>• The CM was centrifuged at $300 \times g$ for 10 min, $2000 \times g$ for 10 min, and $20{,}000 \times g$ for 30 min and $100{,}000 \times g$ for 70 min <br>• Ultracentrifugation | • TEM <br>• BCA <br>• NTA <br>• WB (CD9, CD63, TSG101 and Calnexin) | • DLS revealed that the SHED–sEVs diameter ranges from 40 to 140 nm. <br>• SHED–sEVs promote *in vitro* osteogenic differentiation in PDLCs via Wnt/β-catenin and BMP/Smad signaling pathways. |

**Table 5.** *Cont.*

| Reference | Cell Source | EV Isolation | EV Characteri-zation | Key Findings |
| --- | --- | --- | --- | --- |
| Wei et al., 2020 [80] | • SHED were purchased<br>• Mouse bone marrow stromal cells (mBMSCs) were isolated from femur and tibia bone marrow of CD-1 mice (9–10 months old).<br>• Passage: 4–7 for SHED–sEVs | • At 70% confluence, SHED cells were cultured in serum-free media for 24 h.<br>• The CM was collected and centrifuged at $300\times g$ for 10 min, $2000\times g$ for 10 min, $10,000\times g$ for 60 min before a 0.22 μm filter. $100,000\times g$ for 70 min twice to pellet sEVs.<br>• Ultracentrifugation | • TEM<br>• WB (CD63) | • Diameter: ~100nm<br>• *in vivo*: local injection of SHED–sEVs rescued ligature-induced periodontitis bone loss in mice.<br>• *in vitro*: SHED–sEVs promoted cell proliferation, osteogenesis and reduced adipogenesis and inflammatory cytokines secretion in mBMSCs. |
| Luo et al., 2019 [81] | • SHED was purchased from a cell bank and cells were from nine normal human deciduous incisors collected from 5- to 8-year-old individuals.<br>• Chondrocytes were isolated from cartilage tissues of five patients with condylar fracture<br>• Passage 4 for both SHED and chondrocytes | • SHED cells were confluent before culturing in exosome-free media for 48 h.<br>• CM was collected and centrifuged at 4 °C: $300\times g$ for 10 min, $2000\times g$ for 10 min, $20,000\times g$ for 30 min and $100,000\times g$ for 70 min<br>• Ultracentrifugation | • TEM<br>• NTA<br>• WB (CD9, CD63 and TSG101) | • The sizes of SHED–sEVs range from 30 to 100 nm.<br>• SHED–sEVs inhibited pro-inflammatory cytokines expression in chondrocytes *in vitro* via an exosomal miR-100-5p and mammalian target of rapamycin (mTOR) pathway. |
| Wu et al., 2019 [82] | • SHED and HUVECs cells: commercial cells<br>• rBMSCs: femoral bones of SD rats<br>• Passage: 3–6 (SHED) | • Exosome-depleted FBS was obtained after ultracentrifuging at $100,000\times g$ for 12 h.<br>• CM collection was not clear.<br>• The CM was centrifuged at $300\times g$ for 10 min, and $2000\times g$ for 15 min, $10,000\times g$ for 30 min and concentrated using ultrafiltration, followed by centrifuging at $100,000\times g$ for 1 h.<br>• Ultracentrifugation | • TEM<br>• NTA<br>• BCA<br>• WB (CD81, CD9 and TSG101) | • D: 50 to 200 nm, with two peaks, at 101 and 144 nm.<br>• *in vitro*: SHED—sEVs promoted proliferation, migration and angiogenesis in HUVECs and osteogenic differentiation in rBMSCs via adenosine monophosphate-activated protein kinase (AMPK) pathway.<br>• *in vivo*: SHED–sEVs promoted neovascularization and new bone formation in a periodontal bone defect rat model. |

**Table 5.** *Cont.*

| Reference | Cell Source | EV Isolation | EV Characterization | Key Findings |
|---|---|---|---|---|
| Shi et al., 2020 [83] | <ul><li>Dental follicle cells (DFCs)</li><li>Donor: dental follicle tissue obtained from the immature third molars was selected in young patients (age 13−19 years).</li><li>hPDLCs from chronic periodontitis (age 40−55 years).</li><li>Passage: 5 (for DFCs)</li></ul> | <ul><li>At 80% confluence, DFCs were treated with/without LPS for 24 h prior to culturing in serum-free media for 48 h.</li><li>The CM was centrifuged at $2000\times g$ for 30 min and filtered by a 0.22 μm filter. Then the supernatant was ultrafiltered using 100 KD ultrafiltration at $5000\times g$ or 30 min. Then total exosome isolation reagent was added to the concentrated solution and put into a 4 °C refrigerator overnight and centrifuged at $10,000\times g$ for 1 h.</li><li>Ultracentrifuge</li></ul> | <ul><li>TEM</li><li>NTA</li><li>WB (CD63 and TSG101)</li></ul> | <ul><li>The diameter of DFCs–sEVs was peaked at 120nm.</li><li>LPS precondition increased the secretion of sEV from DFCs.</li><li>*in vitro*: LPS–DFCs–sEVs promoted proliferation, migration and osteogenic differentiation in periodontitis derived hPDLCS.</li><li>*in vivo*: LPS–DFCs–sEVs enhanced orientated periodontal ligament formation, periodontal bone formation, as well as reduced TRAP-positive osteoclasts cells and RANKL/OPG expression.</li></ul> |

**Abbreviations:** Cdc42, cell division control protein 42 homolog; RANKL/OPG, receptor activator of nuclear factor kappa-B ligand/osteoprotegerin.

## 6. Summary and Discussion

Periodontal cells (PDLCs/SCAP and GFs/GMSCs) and dental pulp (DPSCs/SHED)-derived EVs can play an important role in augmenting the function of recipient cells, such as proliferation and osteo/odontogenic differentiation, as well as anti-inflammation and anti-cancer properties [51–83]. In particular, one study of GMSCs–sEVs [61] and DPSCs–sEVs [72], two studies of SHED–sEVs [80,82], and one study of DFCs–sEVs [83] can promote alveolar bone, vasculature and well-organized PDL fibers regeneration, and reduced inflammation in a periodontitis animal model or a periodontal defect model. As such, these EVs may serve as potential 'cell-free' therapeutics to facilitate periodontal regeneration; however, more *in vivo* studies are required to confirm this concept.

As stated in the latest MISEV 2018 guidelines [6], it is critical to clearly describe the primary cell source (i.e., donor age, health status, gender), primary cell passage number, cell culture conditions (using either EV-depleted FBS or FBS starvation before CM collection), and detailed EV isolation and characterization protocols. Among 33 studies in our review, only two studies used human or mouse cell lines [54,57], and 31 studies isolated EV from primary cells, with only 12 out of 31 studies stating a clear age range for the human or mouse donors [51,55,66,69–71,73,75,77,78,81,83], and 13 out of 31 studies were unclear about cell passage numbers [55,57,58,60,63,65,67–69,72,74,77,78]. Since FBS is largely EV contaminated, EV-depleted FBS or FBS starvation should be used for cell culture before CM collection. EV-depleted FBS was used in 12 studies [61,63,66,68,70–73,77,79,81,82], FBS starvation in 8 studies [65,69,74–76,78,80,83], and unclear cell culture conditions in 11 studies. Although all the studies used the two most common sEV (or exosome) isolation methods (precipitation and ultracentrifugation), the EV size in these studies (excluding studies with no EV characterization) is not consistent, with 22 studies generating <200 nm sEVs [51,54,55,57,58,61–63,66,68–72,75,77–83]. This may be attributed to the different CM collection, EV isolation and characterization methods among the studies. Thus, appropriate methods should be chosen to prepare CM, and isolate and characterize cell-derived EV according to the MISEV guidelines. Our review has defined <200nm EV as sEV (small EV) and unclear size or >200 nm as simply EV.

Among 33 studies, 12 studies performed *in vivo* research to investigate the EV function of hPDLCs–sEV [56,59,60], hGMSCs–EV [66,67], hDPSCs–EV [69,72,76], SCAP–sEVs [77,78], SHED–sEVs [80,82] and DFCs–sEVs [83]. Furthermore, three studies engineered the EV using polyethyleneimine (PEI), yielding PEI–EV [56,59,67], and all three studies reported that the PEI–EV group enhanced *in vivo* osteo/odontogenic and angiogenic properties compared to the EV group. Animal studies employed either defect or disease models, such as calvaria defects [56,59,67], nerve injury model [66,67], skin wound-healing model [69], subcutaneous transplantation [76], and multiple sclerosis [60], ligation-induced periodontitis [69,72,80,83] and a periodontal defect [82].

EVs were administrated either by loading into biomaterials, such as collagen membrane [56,59,76], gelfoam sheets [66], gelatin sponge [78] and 3D-printed PLA scaffold [67], or via intravenous administration [60], subcutaneous injection [69], local injection [61,72,80,83], or submucosal injection [77]. More pre-clinical models (i.e., periodontal defects or periodontitis disease models) and EV delivery systems need to be investigated to explore the potential of periodontal cell-derived EV in the regeneration of anatomically complex tissues, such as the periodontium.

All of the above factors are critical for a successful therapeutic outcome; thus, it is of great importance to follow the relevant guidelines and consider the above-discussed variables with more comparisons between different parameters.

## 7. Conclusions and Future Perspectives

This review demonstrates that sEV can be isolated from periodontal and pulp cells, with 11 studies investigated the EV cargos, including sEV–miRNA [52,53,61,63,64,72,75,81], EV–circRNA [51,71], EV–lncRNA [51] and EV–mRNA [67,72]. We summarize the common EV–miRNA and EV–circRNA within periodontal (or dental pulp) cells (Figure 4a,b).

From the included studies, except for one common EV–miRNA (miR-1260b) between DPSCs/SHED and GFs/GMSCs, there appears to be no common EV–miRNA detected between these cell types (shown in Venn diagram, Figure 4a). We also listed reported EV–miRNAs and EV–circRNAs from PDL(S)Cs–EV and hDPSCs—EV (Figure 4b). However, this needs further confirmation with more studies. Furthermore, 38 EV–miRNAs, 69–557 EV–circRNAs, 254–15,380 EV–mRNAs and 2907–11,581 EV–lncRNAs were reported for EV from periodontal (dental pulp) cells by RNA sequencing analysis. We have outlined that these EVs possess anti-inflammation, osteo/odontogenesis, anti-osteoclastogenesis, angiogenesis and immunomodulatory functions *in vitro* and *in vivo*. Thus, we propose that periodontal cell-derived EVs can modulate the cell function via EV cargos (Figure 4c). However, more studies for periodontal cell-derived EVs are required to further confirm this concept.

Given that cell source, CM collection, and EV isolation and characterization are critical in obtaining pure EV populations, future studies should take these factors into account and follow the latest MISEV guidelines. Researchers should consider adding EV purity (EV particles per μg protein), DNase/RNase/proteinase treatment and EV engineering before *in vivo* therapeutic research. Although current research has not yet standardized these factors, data from all 33 studies in this review suggest that periodontal (dental pulp) cell-derived EVs can function as potential therapeutics to promote periodontal regeneration and impart anti-inflammatory properties. However, investigating the effect of periodontal cell-derived EV on *in vivo* periodontal regeneration models is required to understand their potential therapeutic role in periodontal regeneration.

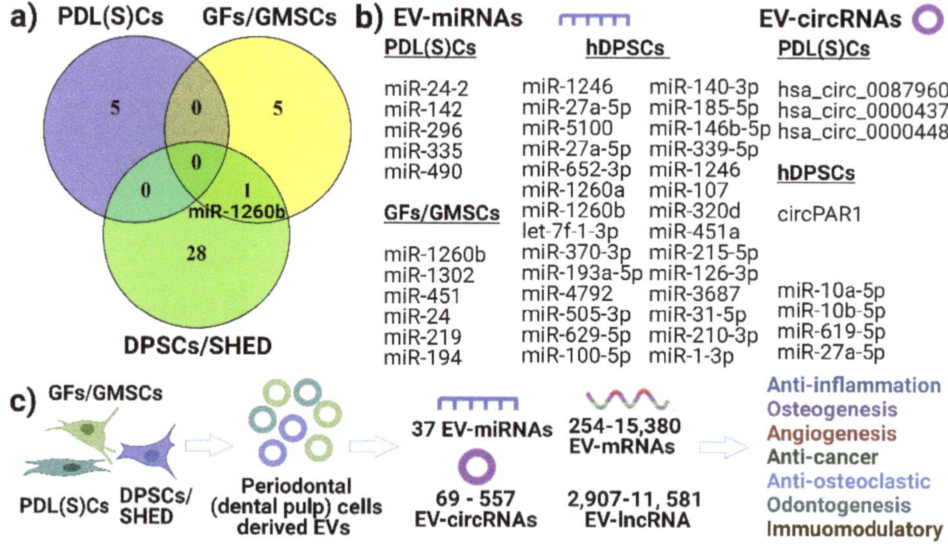

**Figure 4.** Summary of EV–miRNAs, circular RNAs (**a,b**) and proposed (**c**) function of periodontal cell-derived EV on recipient cells function. (**a**) Venn diagram showing no common EV–miRNAs found from PDL(S)Cs, GFs/GMSCs and DPSCs. (**b**) Listed EV–miRNAs and EV–circRNAs. (**c**) Proposed mechanism of how periodontal cell-derived EVs modulate inflammation, angiogenesis, osteo/odontogenesis via EV cargos, such as miRNA, mRNAs, lncRNAs, and circRNAs.

**Funding:** This work was supported by the Australian Dental Research Foundation (ADRF) grant number 534-2019. Karan Gulati is supported by the National Health and Medical Research Council (NHMRC) Early Career Fellowship (APP1140699).

**Institutional Review Board Statement:** Not applicable.

**Informed Consent Statement:** Not applicable.

**Data Availability Statement:** Not applicable.

**Conflicts of Interest:** The authors declare no conflict of interest.

# References


1. Raposo, G.; Stahl, P.D. Extracellular vesicles: A new communication paradigm? *Nat. Rev. Mol. Cell Biol.* **2019**, *20*, 509–510. [CrossRef]
2. Johnstone, R.M.; Adam, M.; Hammond, J.; Orr, L.; Turbide, C. Vesicle formation during reticulocyte maturation. Association of plasma membrane activities with released vesicles (exosomes). *J. Biol. Chem.* **1987**, *262*, 9412–9420. [CrossRef]
3. Hirsch, J.G.; Fedorko, M.E.; Cohn, Z.A. Vesicle fusion and formation at the surface of pinocytic vacuoles in macrophages. *J. Cell Biol.* **1968**, *38*, 629. [CrossRef]
4. Van Niel, G.; D'Angelo, G.; Raposo, G. Shedding light on the cell biology of extracellular vesicles. *Nat. Rev. Mol. Cell Biol.* **2018**, *19*, 213–228. [CrossRef] [PubMed]
5. Kalluri, R.; LeBleu, V.S. The biology, function, and biomedical applications of exosomes. *Science* **2020**, *367*. [CrossRef]
6. Théry, C.; Witwer, K.W.; Aikawa, E.; Alcaraz, M.J.; Anderson, J.D.; Andriantsitohaina, R.; Antoniou, A.; Arab, T.; Archer, F.; Atkin-Smith, G.K.; et al. Minimal information for studies of extracellular vesicles 2018 (MISEV2018): A position statement of the International Society for Extracellular Vesicles and update of the MISEV2014 guidelines. *J. Extracell. Vesicles* **2018**, *7*, 1535750. [CrossRef]
7. Jiao, K.; Walsh, L.J.; Ivanovski, S.; Han, P. The emerging regulatory role of circular RNAs in periodontal tissues and cells. *Int J. Mol. Sci.* **2021**, *22*, 4636. [CrossRef] [PubMed]
8. Pan, B.-T.; Teng, K.; Wu, C.; Adam, M.; Johnstone, R.M. Electron microscopic evidence for externalization of the transferrin receptor in vesicular form in sheep reticulocytes. *J. Cell Biol.* **1985**, *101*, 942–948. [CrossRef]
9. Boulbitch, A. Deflection of a cell membrane under application of a local force. *Phys. Rev. E* **1998**, *57*, 2123. [CrossRef]
10. Bratton, D.L.; Fadok, V.A.; Richter, D.A.; Kailey, J.M.; Guthrie, L.A.; Henson, P.M. Appearance of phosphatidylserine on apoptotic cells requires calcium-mediated nonspecific flip-flop and is enhanced by loss of the aminophospholipid translocase. *J. Biol. Chem.* **1997**, *272*, 26159–26165. [CrossRef]
11. Pols, M.S.; Klumperman, J. Trafficking and function of the tetraspanin CD63. *Exp. Cell Res.* **2009**, *315*, 1584–1592. [CrossRef] [PubMed]
12. Beinert, T.; Münzing, S.; Possinger, K.; Krombach, F. Increased expression of the tetraspanins CD53 and CD63 on apoptotic human neutrophils. *J. Leukoc. Biol.* **2000**, *67*, 369–373. [CrossRef] [PubMed]
13. Han, P.; Bartold, P.M.; Salomon, C.; Ivanovski, S. Salivary small extracellular vesicles associated miRNAs in periodontal status—A pilot study. *Int. J. Mol. Sci.* **2020**, *21*, 2809. [CrossRef]
14. Han, P.; Bartold, P.M.; Salomon, C.; Ivanovski, S. Salivary outer membrane vesicles and DNA methylation of small extracellular vesicles as biomarkers for periodontal status: A pilot study. *Int. J. Mol. Sci.* **2021**, *22*, 2423. [CrossRef]
15. Han, P.; Lai, A.; Salomon, C.; Ivanovski, S. Detection of salivary small extracellular vesicles associated inflammatory cytokines gene methylation in gingivitis. *Int. J. Mol. Sci.* **2020**, *21*, 5273. [CrossRef]
16. Shi, Q.; Huo, N.; Wang, X.; Yang, S.; Wang, J.; Zhang, T. Exosomes from oral tissue stem cells: Biological effects and applications. *Cell Biosci.* **2020**, *10*, 108. [CrossRef] [PubMed]
17. Zhan, C.; Yang, X.; Yin, X.; Hou, J. Exosomes and other extracellular vesicles in oral and salivary gland cancers. *Oral Dis.* **2020**, *26*, 865–875. [CrossRef] [PubMed]
18. Cocucci, E.; Racchetti, G.; Meldolesi, J. Shedding microvesicles: Artefacts no more. *Trends Cell Biol.* **2009**, *19*, 43–51. [CrossRef] [PubMed]
19. Ratajczak, M.Z.; Ratajczak, J. Extracellular microvesicles/exosomes: Discovery, disbelief, acceptance, and the future? *Leukemia* **2020**, *34*, 3126–3135. [CrossRef] [PubMed]
20. Muralidharan-Chari, V.; Clancy, J.W.; Sedgwick, A.; D'Souza-Schorey, C. Microvesicles: Mediators of extracellular communication during cancer progression. *J. Cell Sci.* **2010**, *123*, 1603–1611. [CrossRef] [PubMed]
21. Tricarico, C.; Clancy, J.; D'Souza-Schorey, C. Biology and biogenesis of shed microvesicles. *Small GTPases* **2017**, *8*, 220–232. [CrossRef]
22. Panfoli, I.; Santucci, L.; Bruschi, M.; Petretto, A.; Calzia, D.; Ramenghi, L.A.; Ghiggeri, G.; Candiano, G. Microvesicles as promising biological tools for diagnosis and therapy. *Exp. Rev. Proteom.* **2018**, *15*, 801–808. [CrossRef]
23. King, K.; Cidlowski, J. Cell cycle regulation and apoptosis. *Annu. Rev. Physiol.* **1998**, *60*, 601–617. [CrossRef]
24. Caruso, S.; Poon, I.K.H. Apoptotic cell-derived extracellular vesicles: More than just debris. *Front. Immunol.* **2018**, *9*. [CrossRef]
25. Kakarla, R.; Hur, J.; Kim, Y.J.; Kim, J.; Chwae, Y.-J. Apoptotic cell-derived exosomes: Messages from dying cells. *Exp. Mol. Med.* **2020**, *52*. [CrossRef] [PubMed]
26. Opferman, J.T.; Korsmeyer, S.J. Apoptosis in the development and maintenance of the immune system. *Nat. Immunol.* **2003**, *4*, 410–415. [CrossRef] [PubMed]
27. Ma, Q.; Liang, M.; Wu, Y.; Ding, N.; Duan, L.; Yu, T.; Bai, Y.; Kang, F.; Dong, S.; Xu, J. Mature osteoclast-derived apoptotic bodies promote osteogenic differentiation via RANKL-mediated reverse signaling. *J. Biol. Chem.* **2019**, *294*, 11240–11247. [CrossRef] [PubMed]
28. Ivanovski, S.; Gronthos, S.; Shi, S.; Bartold, P.M. Stem cells in the periodontal ligament. *Oral Dis.* **2006**, *12*, 358–363. [CrossRef]

29. Guo, T.; Gulati, K.; Arora, H.; Han, P.; Fournier, B.; Ivanovski, S. Orchestrating soft tissue integration at the transmucosal region of titanium implants. *Acta Biomater.* **2021**, *124*, 33–49. [CrossRef]

30. Shi, X.; Mao, J.; Liu, Y. Pulp stem cells derived from human permanent and deciduous teeth: Biological characteristics and therapeutic applications. *Stem Cells Transl. Med.* **2020**, *9*, 445–464. [CrossRef] [PubMed]

31. Kaukua, N.; Shahidi, M.K.; Konstantinidou, C.; Dyachuk, V.; Kaucka, M.; Furlan, A.; An, Z.; Wang, L.; Hultman, I.; Ährlund-Richter, L.; et al. Glial origin of mesenchymal stem cells in a tooth model system. *Nature* **2014**, *513*, 551–554. [CrossRef]

32. Komada, Y.; Yamane, T.; Kadota, D.; Isono, K.; Takakura, N.; Hayashi, S.; Yamazaki, H. Origins and properties of dental, thymic, and bone marrow mesenchymal cells and their stem cells. *PLoS ONE* **2012**, *7*, e46436. [CrossRef] [PubMed]

33. Chalisserry, E.P.; Nam, S.Y.; Park, S.H.; Anil, S. Therapeutic potential of dental stem cells. *J. Tissue Eng.* **2017**, *8*. [CrossRef]

34. Dan, H.X.; Vaquette, C.; Fisher, A.G.; Hamlet, S.M.; Xiao, Y.; Hutmacher, D.W.; Ivanovski, S. The influence of cellular source on periodontal regeneration using calcium phosphate coated polycaprolactone scaffold supported cell sheets. *Biomaterials* **2014**, *35*, 113–122. [CrossRef]

35. Vaquette, C.; Fan, W.; Xiao, Y.; Hamlet, S.; Hutmacher, D.W.; Ivanovski, S. A biphasic scaffold design combined with cell sheet technology for simultaneous regeneration of alveolar bone/periodontal ligament complex. *Biomaterials* **2012**, *33*, 5560–5573. [CrossRef] [PubMed]

36. Staples, R.J.; Ivanovski, S.; Vaquette, C. Fibre guiding scaffolds for periodontal tissue engineering. *J. Periodontal Res.* **2020**, *55*, 331–341. [CrossRef] [PubMed]

37. Vaquette, C.; Saifzadeh, S.; Farag, A.; Hutmacher, D.W.; Ivanovski, S. Periodontal tissue engineering with a multiphasic construct and cell sheets. *J. Dental Res.* **2019**, *98*, 673–681. [CrossRef] [PubMed]

38. Vaquette, C.; Mitchell, J.; Fernandez-Medina, T.; Kumar, S.; Ivanovski, S. Resorbable additively manufactured scaffold imparts dimensional stability to extraskeletally regenerated bone. *Biomaterials* **2021**, *269*, 120671. [CrossRef] [PubMed]

39. Han, P.; Ivanovski, S.; Crawford, R.; Xiao, Y. Activation of the canonical Wnt signaling pathway induces cementum regeneration. *J. Bone Miner. Res.* **2015**, *30*, 1160–1174. [CrossRef] [PubMed]

40. Han, P.; Lloyd, T.; Chen, Z.; Xiao, Y. Proinflammatory cytokines regulate cementogenic differentiation of periodontal ligament cells by Wnt/Ca(2+) signaling pathway. *J. Interferon Cytokine Res.* **2016**, *36*, 328–337. [CrossRef]

41. Gholami, L.; Nooshabadi, V.T.; Shahabi, S.; Jazayeri, M.; Tarzemany, R.; Afsartala, Z.; Khorsandi, K. Extracellular vesicles in bone and periodontal regeneration: Current and potential therapeutic applications. *Cell Biosci.* **2021**, *11*, 16. [CrossRef]

42. Gegout, P.Y.; Stutz, C.; Olson, J.; Batool, F.; Petit, C.; Tenenbaum, H.; Benkirane-Jessel, N.; Huck, O. Interests of exosomes in bone and periodontal regeneration: A systematic review. *Adv. Exp. Med. Biol.* **2020**. [CrossRef]

43. Novello, S.; Pellen-Mussi, P.; Jeanne, S. Mesenchymal stem cell-derived small extracellular vesicles as cell-free therapy: Perspectives in periodontal regeneration. *J. Periodontal Res.* **2021**, *56*, 433–442. [CrossRef]

44. Witwer, K.W.; Buzás, E.I.; Bemis, L.T.; Bora, A.; Lässer, C.; Lötvall, J.; Nolte-'t Hoen, E.N.; Piper, M.G.; Sivaraman, S.; Skog, J.; et al. Standardization of sample collection, isolation and analysis methods in extracellular vesicle research. *J. Extracell. Vesicles* **2013**, *2*, 20360. [CrossRef] [PubMed]

45. Greening, D.W.; Xu, R.; Ji, H.; Tauro, B.J.; Simpson, R.J. A protocol for exosome isolation and characterization: Evaluation of ultracentrifugation, density-gradient separation, and immunoaffinity capture methods. In *Proteomic Profiling*; Springer: Cham, Switzerland, 2015; pp. 179–209.

46. Böing, A.N.; Van Der Pol, E.; Grootemaat, A.E.; Coumans, F.A.; Sturk, A.; Nieuwland, R. Single-step isolation of extracellular vesicles by size-exclusion chromatography. *J. Extracell. Vesicles* **2014**, *3*, 23430. [CrossRef] [PubMed]

47. Karttunen, J.; Heiskanen, M.; Navarro-Ferrandis, V.; Das Gupta, S.; Lipponen, A.; Puhakka, N.; Rilla, K.; Koistinen, A.; Pitkänen, A. Precipitation-based extracellular vesicle isolation from rat plasma co-precipitate vesicle-free microRNAs. *J. Extracell. Vesicles* **2019**, *8*, 1555410. [CrossRef] [PubMed]

48. Oliveira-Rodríguez, M.; López-Cobo, S.; Reyburn, H.T.; Costa-García, A.; López-Martín, S.; Yáñez-Mó, M.; Cernuda-Morollón, E.; Paschen, A.; Valés-Gómez, M.; Blanco-López, M.C. Development of a rapid lateral flow immunoassay test for detection of exosomes previously enriched from cell culture medium and body fluids. *J. Extracell. Vesicles* **2016**, *5*, 31803. [CrossRef]

49. Oeyen, E.; Van Mol, K.; Baggerman, G.; Willems, H.; Boonen, K.; Rolfo, C.; Pauwels, P.; Jacobs, A.; Schildermans, K.; Cho, W.C. Ultrafiltration and size exclusion chromatography combined with asymmetrical-flow field-flow fractionation for the isolation and characterisation of extracellular vesicles from urine. *J. Extracell. Vesicles* **2018**, *7*, 1490143. [CrossRef]

50. Konoshenko, M.Y.; Lekchnov, E.A.; Vlassov, A.V.; Laktionov, P.P. Isolation of extracellular vesicles: General methodologies and latest trends. *Biomed. Res. Int.* **2018**, *2018*, 8545347. [CrossRef]

51. Xie, L.; Chen, J.; Ren, X.; Zhang, M.; Thuaksuban, N.; Nuntanaranont, T.; Guan, Z. Alteration of circRNA and lncRNA expression profile in exosomes derived from periodontal ligament stem cells undergoing osteogenic differentiation. *Arch. Oral Biol.* **2021**, *121*, 104984. [CrossRef]

52. Zhang, Z.; Shuai, Y.; Zhou, F.; Yin, J.; Hu, J.; Guo, S.; Wang, Y.; Liu, W. PDLSCs regulate angiogenesis of periodontal ligaments via VEGF transferred by exosomes in periodontitis. *Int. J. Med. Sci.* **2020**, *17*, 558–567. [CrossRef]

53. Chiricosta, L.; Silvestro, S.; Gugliandolo, A.; Marconi, G.D.; Pizzicannella, J.; Bramanti, P.; Trubiani, O.; Mazzon, E. Extracellular vesicles of human periodontal ligament stem cells contain MicroRNAs associated to proto-oncogenes: Implications in cytokinesis. *Front. Genet.* **2020**, *11*, 582. [CrossRef] [PubMed]

54. Zhao, M.; Dai, W.; Wang, H.; Xue, C.; Feng, J.; He, Y.; Wang, P.; Li, S.; Bai, D.; Shu, R. Periodontal ligament fibroblasts regulate osteoblasts by exosome secretion induced by inflammatory stimuli. *Arch. Oral Biol.* **2019**, *105*, 27–34. [CrossRef] [PubMed]

55. Čebatariūnienė, A.; Kriaučiūnaitė, K.; Prunskaitė, J.; Tunaitis, V.; Pivoriūnas, A. Extracellular vesicles suppress basal and Lipopolysaccharide-induced NFκB activity in human periodontal ligament stem cells. *Stem Cells Dev.* **2019**, *28*, 1037–1049. [CrossRef] [PubMed]

56. Pizzicannella, J.; Gugliandolo, A.; Orsini, T.; Fontana, A.; Ventrella, A.; Mazzon, E.; Bramanti, P.; Diomede, F.; Trubiani, O. Engineered extracellular vesicles from human periodontal-ligament stem cells increase VEGF/VEGFR2 expression during bone regeneration. *Front. Physiol.* **2019**, *10*. [CrossRef] [PubMed]

57. Wang, Z.; Maruyama, K.; Sakisaka, Y.; Suzuki, S.; Tada, H.; Suto, M.; Saito, M.; Yamada, S.; Nemoto, E. Cyclic stretch force induces periodontal ligament cells to secrete exosomes that suppress IL-1β production through the inhibition of the NF-κB signaling pathway in macrophages. *Front. Immunol.* **2019**, *10*. [CrossRef]

58. Kang, H.; Lee, M.-J.; Park, S.J.; Lee, M.-S. Lipopolysaccharide-preconditioned periodontal ligament stem cells induce M1 polarization of macrophages through extracellular vesicles. *Int. J. Mol. Sci.* **2018**, *19*, 3843. [CrossRef]

59. Diomede, F.; D'aurora, M.; Gugliandolo, A.; Merciaro, I.; Ettorre, V.; Bramanti, A.; Piattelli, A.; Gatta, V.; Mazzon, E.; Fontana, A. A novel role in skeletal segment regeneration of extracellular vesicles released from periodontal-ligament stem cells. *Int. J. Nanomed.* **2018**, *13*, 3805. [CrossRef]

60. Rajan, T.S.; Giacoppo, S.; Diomede, F.; Ballerini, P.; Paolantonio, M.; Marchisio, M.; Piattelli, A.; Bramanti, P.; Mazzon, E.; Trubiani, O. The secretome of periodontal ligament stem cells from MS patients protects against EAE. *Sci. Rep.* **2016**, *6*, 38743. [CrossRef]

61. Nakao, Y.; Fukuda, T.; Zhang, Q.; Sanui, T.; Shinjo, T.; Kou, X.; Chen, C.; Liu, D.; Watanabe, Y.; Hayashi, C.; et al. Exosomes from TNF-α-treated human gingiva-derived MSCs enhance M2 macrophage polarization and inhibit periodontal bone loss. *Acta Biomater.* **2021**, *122*, 306–324. [CrossRef]

62. Yin, S.; Jia, F.; Ran, L.; Xie, L.; Wu, Z.; Zhan, Y.; Zhang, Y.; Zhang, M. Exosomes derived from idiopathic gingival fibroma fibroblasts regulate gingival fibroblast proliferation and apoptosis. *Oral Dis.* **2020**. [CrossRef]

63. Zhuang, X.-M.; Zhou, B. Exosome secreted by human gingival fibroblasts in radiation therapy inhibits osteogenic differentiation of bone mesenchymal stem cells by transferring miR-23a. *Biomed. Pharmacother.* **2020**, *131*, 110672. [CrossRef] [PubMed]

64. Silvestro, S.; Chiricosta, L.; Gugliandolo, A.; Pizzicannella, J.; Diomede, F.; Bramanti, P.; Trubiani, O.; Mazzon, E. Extracellular vesicles derived from human gingival mesenchymal stem cells: A transcriptomic analysis. *Genes* **2020**, *11*, 118. [CrossRef] [PubMed]

65. Coccè, V.; Franzè, S.; Brini, A.T.; Giannì, A.B.; Pascucci, L.; Ciusani, E.; Alessandri, G.; Farronato, G.; Cavicchini, L.; Sordi, V. in vitro anticancer activity of extracellular vesicles (EVs) secreted by gingival mesenchymal stromal cells primed with paclitaxel. *Pharmaceutics* **2019**, *11*, 61. [CrossRef] [PubMed]

66. Mao, Q.; Nguyen, P.D.; Shanti, R.M.; Shi, S.; Shakoori, P.; Zhang, Q.; Le, A.D. Gingiva-derived mesenchymal stem cell-extracellular vesicles activate schwann cell repair phenotype and promote nerve regeneration. *Tissue Eng. Part A* **2019**, *25*, 887–900. [CrossRef]

67. Diomede, F.; Gugliandolo, A.; Cardelli, P.; Merciaro, I.; Ettorre, V.; Traini, T.; Bedini, R.; Scionti, D.; Bramanti, A.; Nanci, A. Three-dimensional printed PLA scaffold and human gingival stem cell-derived extracellular vesicles: A new tool for bone defect repair. *Stem Cell Res. Ther.* **2018**, *9*, 104. [CrossRef] [PubMed]

68. Faruqu, F.N.; Zhou, S.; Sami, N.; Gheidari, F.; Lu, H.; Al-Jamal, K.T. Three-dimensional culture of dental pulp pluripotent-like stem cells (DPPSCs) enhances Nanog expression and provides a serum-free condition for exosome isolation. *FASEB BioAdv.* **2020**, *2*, 419–433. [CrossRef] [PubMed]

69. Zhou, H.; Li, X.; Yin, Y.; He, X.-T.; An, Y.; Tian, B.-M.; Hong, Y.-L.; Wu, L.-A.; Chen, F.-M. The proangiogenic effects of extracellular vesicles secreted by dental pulp stem cells derived from periodontally compromised teeth. *Stem Cell Res. Ther.* **2020**, *11*. [CrossRef] [PubMed]

70. Ivica, A.; Ghayor, C.; Zehnder, M.; Valdec, S.; Weber, F.E. Pulp-derived exosomes in a fibrin-based regenerative root filling material. *J. Clin. Med.* **2020**, *9*, 491. [CrossRef]

71. Xie, L.; Guan, Z.; Zhang, M.; Lyu, S.; Thuaksuban, N.; Kamolmattayakul, S.; Nuntanaranont, T. Exosomal circLPAR1 promoted osteogenic differentiation of homotypic dental pulp stem cells by competitively binding to hsa-miR-31. *BioMed Res. Int.* **2020**, *2020*, 6319395. [CrossRef]

72. Shen, Z.; Kuang, S.; Zhang, Y.; Yang, M.; Qin, W.; Shi, X.; Lin, Z. Chitosan hydrogel incorporated with dental pulp stem cell-derived exosomes alleviates periodontitis in mice via a macrophage-dependent mechanism. *Bioact. Mater.* **2020**, *5*, 1113–1126. [CrossRef]

73. Li, J.; Ju, Y.; Liu, S.; Fu, Y.; Zhao, S. Exosomes derived from lipopolysaccharide-preconditioned human dental pulp stem cells regulate Schwann cell migration and differentiation. *Connect. Tissue Res.* **2019**, *62*, 277–286. [CrossRef]

74. Ji, L.; Bao, L.; Gu, Z.; Zhou, Q.; Liang, Y.; Zheng, Y.; Xu, Y.; Zhang, X.; Feng, X. Comparison of immunomodulatory properties of exosomes derived from bone marrow mesenchymal stem cells and dental pulp stem cells. *Immunol. Res.* **2019**, *67*, 432–442. [CrossRef]

75. Hu, X.; Zhong, Y.; Kong, Y.; Chen, Y.; Feng, J.; Zheng, J. Lineage-specific exosomes promote the odontogenic differentiation of human dental pulp stem cells (DPSCs) through TGFβ1/smads signaling pathway via transfer of microRNAs. *Stem Cell Res. Ther.* **2019**, *10*, 170. [CrossRef] [PubMed]

76. Huang, C.-C.; Narayanan, R.; Alapati, S.; Ravindran, S. Exosomes as biomimetic tools for stem cell differentiation: Applications in dental pulp tissue regeneration. *Biomaterials* **2016**, *111*, 103–115. [CrossRef] [PubMed]

77. Liu, Y.; Zhuang, X.; Yu, S.; Yang, N.; Zeng, J.; Liu, X.; Chen, X. Exosomes derived from stem cells from apical papilla promote craniofacial soft tissue regeneration by enhancing Cdc42-mediated vascularization. *Stem Cell Res. Ther.* **2021**, *12*, 76. [CrossRef] [PubMed]

78. Zhuang, X.; Ji, L.; Jiang, H.; Liu, Y.; Liu, X.; Bi, J.; Zhao, W.; Ding, Z.; Chen, X. Exosomes derived from stem cells from the apical papilla promote dentine-pulp complex regeneration by inducing specific dentinogenesis. *Stem Cells Int.* **2020**, *2020*, 5816723. [CrossRef]

79. Wang, M.; Li, J.; Ye, Y.; He, S.; Song, J. SHED-derived conditioned exosomes enhance the osteogenic differentiation of PDLSCs via Wnt and BMP signaling in vitro. *Differentiation* **2020**, *111*. [CrossRef]

80. Wei, J.; Song, Y.; Du, Z.; Yu, F.; Zhang, Y.; Jiang, N.; Ge, X. Exosomes derived from human exfoliated deciduous teeth ameliorate adult bone loss in mice through promoting osteogenesis. *J. Mol. Histol.* **2020**, *51*, 455–466. [CrossRef]

81. Luo, P.; Jiang, C.; Ji, P.; Wang, M.; Xu, J. Exosomes of stem cells from human exfoliated deciduous teeth as an anti-inflammatory agent in temporomandibular joint chondrocytes via miR-100-5p/mTOR. *Stem Cell Res. Ther.* **2019**, *10*, 216. [CrossRef]

82. Wu, J.; Chen, L.; Wang, R.; Song, Z.; Shen, Z.; Zhao, Y.; Huang, S.; Lin, Z. Exosomes secreted by stem cells from human exfoliated deciduous teeth promote alveolar bone defect repair through the regulation of angiogenesis and osteogenesis. *ACS Biomater. Sci. Eng.* **2019**, *5*, 3561–3571. [CrossRef] [PubMed]

83. Shi, W.; Guo, S.; Liu, L.; Liu, Q.; Huo, F.; Ding, Y.; Tian, W. Small extracellular vesicles from lipopolysaccharide-preconditioned dental follicle cells promote periodontal regeneration in an inflammatory microenvironment. *ACS Biomater. Sci. Eng.* **2020**, *6*, 5797–5810. [CrossRef] [PubMed]

*nanomaterials*

*Article*

# Influence of Bioinspired Lithium-Doped Titanium Implants on Gingival Fibroblast Bioactivity and Biofilm Adhesion

Aya Q. Alali [1], Abdalla Abdal-hay [1,2], Karan Gulati [1], Sašo Ivanovski [1], Benjamin P. J. Fournier [1,3,*] and Ryan S. B. Lee [1,*]

[1] School of Dentistry, The University of Queensland, Herston, QLD 4006, Australia; a.alali@uqconnect.edu.au (A.Q.A.); abdalla.ali@uq.edu.au (A.A.-h.); k.gulati@uq.edu.au (K.G.); s.ivanovski@uq.edu.au (S.I.)

[2] Department of Engineering Materials and Mechanical Design, Faculty of Engineering, South Valley University, Qena 83523, Egypt

[3] Laboratory of Molecular Oral Physiopathology, INSERM UMRS 1138, Cordeliers Research Center, 75006 Paris, France

* Correspondence: b.fournier@uq.edu.au (B.P.J.F.); r.sblee@uq.edu.au (R.S.B.L.)

**Abstract:** Soft tissue integration (STI) at the transmucosal level around dental implants is crucial for the long-term success of dental implants. Surface modification of titanium dental implants could be an effective way to enhance peri-implant STI. The present study aimed to investigate the effect of bioinspired lithium (Li)-doped Ti surface on the behaviour of human gingival fibroblasts (HGFs) and oral biofilm *in vitro*. HGFs were cultured on various Ti surfaces—Li-doped Ti (Li_Ti), NaOH_Ti and micro-rough Ti (Control_Ti)—and were evaluated for viability, adhesion, extracellular matrix protein expression and cytokine secretion. Furthermore, single species bacteria (*Staphylococcus aureus*) and multi-species oral biofilms from saliva were cultured on each surface and assessed for viability and metabolic activity. The results show that both Li_Ti and NaOH_Ti significantly increased the proliferation of HGFs compared to the control. Fibroblast growth factor-2 (FGF-2) mRNA levels were significantly increased on Li_Ti and NaOH_Ti at day 7. Moreover, Li_Ti upregulated COL-I and fibronectin gene expression compared to the NaOH_Ti. A significant decrease in bacterial metabolic activity was detected for both the Li_Ti and NaOH_Ti surfaces. Together, these results suggest that bioinspired Li-doped Ti promotes HGF bioactivity while suppressing bacterial adhesion and growth. This is of clinical importance regarding STI improvement during the maintenance phase of the dental implant treatment.

**Keywords:** titanium; implants; nanostructure; gingival fibroblasts; biofilm; soft-tissue integration; surface modification

**Citation:** Alali, A.Q.; Abdal-hay, A.; Gulati, K.; Ivanovski, S.; Fournier, B.P.J.; Lee, R.S.B. Influence of Bioinspired Lithium-Doped Titanium Implants on Gingival Fibroblast Bioactivity and Biofilm Adhesion. *Nanomaterials* **2021**, *11*, 2799. https://doi.org/10.3390/nano11112799

Academic Editor: Ion N. Mihailescu

Received: 2 September 2021
Accepted: 14 October 2021
Published: 22 October 2021

**Publisher's Note:** MDPI stays neutral with regard to jurisdictional claims in published maps and institutional affiliations.

## 1. Introduction

Peri-implant diseases of endo-osseous oral implants are mainly initiated by biofilm accumulation and subsequent host immuno-inflammatory responses at the transmucosal (implant abutment-mucosa) interface [1,2]. Peri-implant soft tissues exist as a physical barrier between the oral environment and the implant. However, the peri-implant mucosal seal could be considered fragile in disease, as it lacks the complex supra-crestal connective tissue structures usually found in the natural dentition [3], which may be responsible for the rapid rate of disease progression [4–6]. Hence, various attempts have been made to alter the surface of dental implants and abutments to augment soft tissue integration (STI), as reviewed elsewhere [7].

Surface characteristics of the implant, such as topography and chemistry, play a key role in determining tissue responses [8]. Most of the proposed Ti surface topographical modifications aim to improve osseointegration quantity and quality by altering the surface roughness at microscale levels [9–12], while others focus on creating implant devices with antimicrobial properties by mimicking self-cleansing surfaces found in nature [13–15]. Furthermore, effective bactericidal functions have been achieved by incorporating nanostructures onto implant substrates, such as nanopillars [16,17] or local antibiotic releasing nanotubes fabricated via anodisation [18,19]. With regard to the peri-implant soft tissue's response, previous *in vitro* studies using gingival human fibroblasts demonstrated an increase in the proliferation [20], mechanical stimulation [21], collagen production [22] and attachment [23] to the substrates, indicating the potential for nanostructures to promote connective tissue formation around an implant.

Nature-inspired nano-topographies are commonly reported in the literature, and reproduction of the shape and arrangement of natural nanoscale patterns have been attempted to improve the characteristics of biomaterials, notably in implants research [24–26]. The extracellular matrix (ECM) of biological tissues has been a source of inspiration for several studies [24,26], mainly targeting the arrangement of laminin and collagen nanofibres [27]. Scaffolds with ECM-like features exhibit an increase in the cellular deposition of hydroxyapatite, which aids in promoting the mineralisation required for osseointegration [28]. Moreover, these features influence initial filopodia-surface interactions [29].

Studies have revealed that chemical surface treatments [30], such as hydrothermal alkalinisation [31], electrochemical anodisation [23], and electrochemical oxidization [32], are cost-effective strategies to nano-engineer surfaces on Ti-based implants [31]. Chemically induced nanostructures have been shown to increase gingival fibroblast attachment while inhibiting bacterial adhesion *in vitro* [33]. The incorporation of metal ions, such as zinc [34], magnesium [35] and Li [36], into Ti surfaces enhances various cellular activities in osteoblasts and fibroblasts [37,38]. Our group recently reported the utilisation of hydrothermal transformation to fabricate Li-doped Ti with sustainable Li+ ions release [39]. Lithium belongs to the alkali metal group and is considered a biologically functional ion. It has been shown that Li ions can stimulate bone growth and periodontal ligament cell differentiation through the *Wnt/β*-catenin signalling pathway [40,41]. Abdal-hay et al. [39] investigated the influence of different LiCl concentrations on surface properties of doped Li-Ti. Their results show that a Li-Ti porous layer with nanostructure characteristics was nucleated and formed on the Ti surface. Furthermore, Li-incorporated Ti exhibits improved wettability and mechanical stability compared to untreated Ti surfaces, with an improved effect on osteoblast activity [36,39]. The impacts of Li-incorporated surface modification on gingival fibroblasts, however, have yet to be extensively explored.

An ideal implant surface should modulate cellular responses, leading to the timely establishment and maintenance of osseointegration, soft-tissue integration and prevention of bacterial adhesion. The current study explores the STI and antibacterial functions of ECM-mimicking nanoscale Li-Ti surfaces as the next generation of modified Ti dental implants.

## 2. Materials and Methods

### 2.1. Titanium Surface Modification

A 99.5% Ti flat foil (0.3 mm thickness) was purchased from Nilaco Corporation (Tokyo, Japan). Ti foil was mechanically treated using a gradient of sandpapers to form a micromachining-like surface topography (Control_Ti) [42]. Ti foil was cut into 10 mm × 10 mm squares using diamond EXAKT's saw machine. Next, Ti was etched in an acid mixture (equal volumes of concentrated acids and water $H_2SO_4$: HCl: $H_2O$) at 80 °C for 1 h to remove the natural oxide layer and increase surface roughness, followed by immersion in 200 mL of 5.0 M NaOH aqueous solution at 60 °C for 24 h, and then rinsed with distilled water. To introduce Li ions, the alkali-treated Ti samples were first immersed in lithium chloride (LiCl: 0.025 M), then hydrothermally treated in a Teflon container at 90 °C for 24 h. After Li-containing compound precipitation, the Ti substrates were rinsed in distilled water and dried at 45 °C for 24 h [39]. The substrates were then grouped according to the treatment: (1) lithium-incorporated alkaline-treated Ti (Li_Ti) as a test group, (2) alkaline-treated Ti (NaOH_Ti) as a test group, and (3) mechanically prepared micro-rough Ti (Control_Ti). All surfaces to be tested were sterilised by immersion in 70% ethanol for 8 h followed by air drying for 24 h, and ultraviolet irradiation for 30 min each side. To observe the topography, titanium substrates were mounted on a holder with double-sided conductive tape, coated with 10 nm platinum, and at least 5 substrates of each group were viewed under SEM (SEM, JSM- 7001F, Joel, Tokyo, Japan).

### 2.2. Culture of Human Gingival Fibroblasts

Primary human gingival fibroblast cells cultured at passages 4–6 were used for all experiments. All subjects gave their informed consent for inclusion before they participated in the study. The study was conducted in accordance with the Declaration of Helsinki, and the protocol was approved by the University of Queensland Institutional Human Ethics Research Committee (No. 2019000134). The cells were cultured at 37 °C and 5% $CO_2$ in Dulbecco's modified Eagle's medium (DMEM, Life Technologies, Scoresby, VIC, Australia) supplemented with 10% foetal bovine serum (FBS from Gibco®, Clayton, VIC, Australia) and 1% Penicillin-Streptomycin-Glutamine (Gibco®, Clayton, VIC, Australia). Cells were grown in Corning® T75 Flasks (Thermo Fisher Scientific, Madrid, Spain), and upon 80% confluency, detached using 0.04% trypsin, and then seeded at a density of 5000 cells per Ti substrate in 12-well plates. The LIVE/DEAD assay® (Life Technologies, Scoresby, VIC, Australia) was performed according to the manufacturer's instructions to assess cell viability. At predetermined timepoints, cultured samples were washed twice with phosphate-buffered saline (PBS), then incubated with fluorescein diacetate (FDA/live; 1:200) and propidium iodide (PI/dead) diluted in PBS, for 20 min at 37 °C and 5% $CO_2$. Images of the stained cultures were obtained using confocal microscopy (Nikon Eclipse Ti-E. Nikon Instruments Inc., Melville, NY, USA).

### 2.3. Cell Attachment and Spread Morphology

After 4, 24 h and 7 days incubation, cells on the different Ti substrates were fixed for 20 min with 4% paraformaldehyde in PBS (PFA). After washing twice with PBS, cells were permeabilised with Triton X-100 (0.5%) in PBS for 10 min, followed by incubation in blocking buffer (10% Bovine Serum Albumin, Glycine, tween, and PBS) for 1 h. The primary antibody for collagen I (1:250) was then added for one hour at room temperature. After three PBS washes, secondary antibodies (Goat An-ti-Mouse/Rabbit Alexa fluor 488,568), DAPI Staining Solution (ab228549) (1:500), and Phalloidin-California Red Conjugate (1:1000) were added, and samples were incubated in the dark for 30 min. After a final three PBS washes, images of each sample were obtained using confocal microscopy (Nikon Eclipse Ti-E. Nikon Instruments Inc. USA), and image analysis was performed using Image J (Fiji V1.53 g, National Institutes of Health, Bethesda, MD, USA).

For surface morphology and spreading observation, cultured samples were fixed with 4% PFA for 20 min, washed twice in sodium cacodylate buffer and immersed in glutaraldehyde for 30 min, rinsed twice in sodium cacodylate buffer, dehydrated in multiple concentrations of ethanol (20–100%), then immersed in hexamethyldisilazane (HDMS) for 30 min. Finally, samples were left to fully dry before coating with 10 nm platinum for SEM imaging (JSM-7001F, Jeol, Tokyo, Japan).

### 2.4. Cell Count

Ti substrates were placed in 24 wells containing 350 µL of proteinase K (Invitrogen, Waltham, MA, USA; proteinase K/phosphate buffered EDTA (PBE) 0.5 mg/mL) for DNA content analysis, and were incubated overnight at 56 °C. Following this, 100 µL from each well was aliquoted in triplicate into a black 96-well plate, and 100 µL of the PicoGreen (P11496, Invitrogen, Waltham, MA, USA) working solution was added. Plates were incubated in the dark for 5 min before reading in a fluorescence plate reader (excitation 485 nm, emission 520 nm).

*2.5. Gene Expression by Real-Time Quantitative Polymerase Chain Reaction (RT-qPCR)*

Real-time qPCR was performed to determine changes in expression of selected genes by HGFs on the Ti samples. Briefly, RNA was extracted from HGFs (5 pooled samples, each sample 5000 cell/cm$^2$) using TRIzol following the manufacturer's instructions. Phase separation was performed to generate the aqueous phase, followed by RNA precipitates. cDNA synthesis was completed using a RevertAid First Strand cDNA Synthesis Kit (ThermoFisher Scientific, Scoresby, Australia). mRNA for collagen I, collagen III, CXCL8, IL_1β and FN was measured according to comparative CT values using the StepOnePlusTM Real-Time PCR system (Applied BiosystemsTM, ThermoFisher Scientific, Scoresby, Australia), and normalised against two reference genes, hGAPDH and h18 s. Forward and reverse primer sequences corresponding to each tested gene are listed in Table 1. Fold change analysis was standardised relative to control.

**Table 1.** The experimented genes' symbols and primer sequences in forward 5'-3' and reverse 3'-5', and length in base pair (bp).

| Gene Symbol | Direction | Primer Sequence | Length (bp) |
|---|---|---|---|
| hCOL1A1 | Forward | CCTGCGTGTACCCCACTCA | 115 |
|  | Reverse | ACCAGACATGCCTCTTGTCCTT | 115 |
| hCOL3A1 | Fwd | CCGTTCTCTGCGATGACATAA | 142 |
|  | Rev | CCTTGAGGTCCTTGACCATTAG | 142 |
| hGAPDH | Fwd | TCAGCAATGCATCCTGCAC | 117 |
|  | Rev | TCTGGGTGGCAGTGATGGC | 117 |
| h18S | Fwd | CAGACATTGACCTCACCAAGAG | 99 |
|  | Rev | GAATCTTCTTCAGTCGCTCCAG | 99 |
| hIL_1B | Fwd | GGTGTTCTCCATGTCCTTTGTA | 125 |
|  | Rev | GCTGTAGAGTGGGCTTATCATC | 125 |
| hCXCL8 | Fwd | GAGAGTGATTGAGAGTGGACCAC | 112 |
|  | Rev | CACAACCCTCTGCACCCAGTTT | 112 |
| hFN1 | Fwd | CACAGTCAGTGTGGTTGCCT | 68 |
|  | Rev | CTGTGGACTGGGTTCCAATCA | 68 |

*2.6. Extracellular Matrix Expression by Luminex*

A Magnetic Luminex Screening Assay with a Human Premixed Multi-Analyte Kit (LXSAHM, R&D Systems Luminex®, Minneapolis, MN, USA) was utilised to assay conditioned media from human gingival fibroblast cultures for proteins of interest. The customised 5-plex panel included primary growth and inflammatory analytes: fibroblast growth factor (FGF)-2, matrix metalloproteinase (MMP)1, MMP8, vascular endothelial growth factor (VEGF)-A, and platelet-derived growth factor (PDGF)-BB, and Multiplex-ELISA was performed according to the manufacturer's instructions. Triplicate supernatant samples were assayed in duplicate. Culture media were used as a negative control for all the samples.

*2.7. Ethics Approval and Saliva Collection*

Methicillin-sensitive *Staphylococcus aureus* (MSSA) was obtained from the American Type Culture Collection (ATCC; 25923, Manassas, VA, USA) for growing mono-species biofilms. Saliva from healthy volunteers was used for growing polymicrobial salivary biofilms. It was conducted in accordance with the Declaration of Helsinki, and the protocol was approved by the University of Queensland Institutional Human Ethics Research Committee (No. 2019001113). Informed consent was obtained from all subjects before sample collection. Unstimulated saliva from six healthy individuals was collected using a protocol previously reported [43]. In brief, volunteers were requested to provide approximately 2.0 mL of unstimulated saliva by spitting it into a 50 mL centrifuge tube. The volunteers had good gingival health, as evidenced by oral examination, and had not consumed antimicrobials and were not regularly using antimicrobial mouth rinses. The collected saliva was pooled, mixed with equal amounts of 70% glycerol stock solution and vortexed. The resultant mix was aliquoted into 1.5 mL Eppendorf tubes and stored at $-80\,^\circ$C until further processing.

### 2.8. Biofilm Development

#### 2.8.1. Single Species Biofilms

*S. aureus* was inoculated into 10 mL of brain heart infusion broth (BHI) in a centrifuge tube using Culti-Loops™, cultured overnight, then the tubes were centrifuged, and the supernatant was discarded. The sedimented bacteria were resuspended in sterile phosphate-buffered saline. The turbidity of the suspension was measured spectrophotometrically (Thermo Scientific™ GENESYS 10S UV-Vis spectrophotometer). The turbidity of the suspension was adjusted to approximately $1 \times 10^7$ CFU/mL of *S. aureus*, which was subsequently used for culturing purposes.

#### 2.8.2. Multispecies Biofilms

Similarly, 1.0 mL of unstimulated saliva was mixed with 9 mL of BHI broth for overnight culturing. The inoculum was adjusted to $1 \times 10^7$ CFU/mL as above.

#### 2.8.3. Biofilm Culture and Development

One millilitre of the bacteria was mixed with 8.0 mL BHI and 1.0 mL defibrinated sheep's blood, and kept in an anaerobic gas box inside a shaker (80 rpm) at 37 Celsius overnight to allow bacterial growth. The following day, concentrations of bacteria were determined spectrophotometrically. Approximately $1 \times 10^7$ CFU/mL of *S. aureus* or salivary biofilm were cultured separately over sterile Li_Ti, NaOH_Ti, and control substrates ($n = 3$) placed in a sterile 24-well tissue culture (Corning CLS3524, Thermo Fisher Scientific, Scoresby, Australia). At the predetermined time points, samples were washed twice with phosphate-buffered saline (pH 7.2) prior to further experiments.

### 2.9. Bacterial Metabolic Activity

An XTT (2,3-Bis-(2-methoxy-4-nitro-5-sulfophenyl)-2H-tetrazolium-5-carboxanilide) kit (Sigma-Aldrich, Castle-Hill, NSW, Australia) was used to test bacterial metabolic activity. In this process, 200 µg/mL of XTT was mixed with 25 µM of menadione. Ti substrates ($n = 4$) were washed with PBS, immersed in 300 µL of the working solution, and then incubated at 37 $^\circ$C for 4 h. Three technical replicates of 100 µL were transferred to a 6-well plate and read at 492 nm absorbance using a Tecan infinite 200 pro spectrophotometer described previously [44,45].

### 2.10. Biofilm Viability Staining

Triplicate Ti samples were washed twice with phosphate-buffered saline (pH 7.2) at 24 and 72 h of culture before assessment of biofilm viability using a Filmtracer™ LIVE/DEAD™ Biofilm Viability Kit (Invitrogen, ThermoFisher Scientific, Scoresby, VIC, Australia) as previously described [46]. Following a 20 min incubation at room temperature (25 $^\circ$C), the biofilms were washed once for removal of unbound stain and two-dimensional images of the biofilms captured using the confocal microscopy. Subsequently, 3D images were reconstructed with a step size of 2.0 µm.

### 2.11. Statistical Analysis

GraphPad Prism version 9.0.0 (Windows, GraphPad Software, San Diego, CA, USA) was used for all data analysis. All data are presented as mean and SD. The difference between the control, NaOH_Ti, and Li_Ti groups was analysed using two-way ANOVA with Tukey's multiple comparisons test. The fold change for qPCR values was analysed using the $2^{-\Delta\Delta Ct}$ method. A $p$-value of <0.05 was considered statistically significant.

## 3. Results

### 3.1. Surface Characterisation of Ti Substrates

The surface topography of Ti substrates was characterised using SEM and the images are presented in Figure 1.

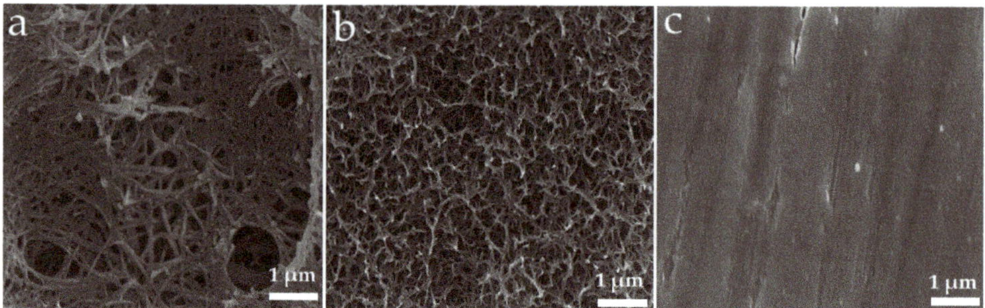

**Figure 1.** Top view SEM images showing the surface of Ti substrates. (**a**) Lithium-incorporated Ti (Li_Ti), (**b**) alkaline-treated Ti (without Li) (NaOH_Ti) and (**c**) mechanically micro-machined Ti (Control_Ti). All scale bars represent 1 um.

### 3.2. HGF Viability and Early Proliferation

Live/dead staining of HGFs over the sample groups showed no cytotoxicity signs after 1 and until 5 days of culture. DNA content was quantified after 4 and 24 h of culture to assess HGFs proliferation. More cells were present for both the Li_Ti and NaOH_Ti surfaces at 4 h compared to the control (untreated Ti) (Figure 2).

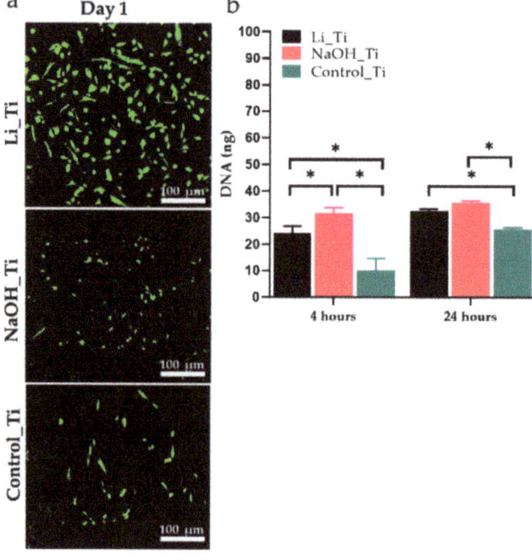

**Figure 2.** Viability and proliferation of human gingival fibroblasts. (**a**) Confocal microscopy images of Live-Dead staining over Li_Ti, NaOH_Ti and Control_Ti substrates at day 1, (**b**) analysis of PicoGreen assay for DNA content. * $p < 0.05$.

### 3.3. HGF Attachment and Morphology

Three-dimensional confocal microscopy images were used to view and analyse the HGF nuclei and actin filaments (Figure 3a–f) at 4 and 24 h post-seeding. Most cells were attached at 24 h in all groups, with no significant differences in nuclei count (Figure 4a). The measurements for the length and the aspect ratio (major axis of a cell/minor axis) were performed using ImageJ software (1.53f51, Wayne Rasband, Bethesda, MD, USA)

(Figure 4b,c). Scanning electron microscopy images (Figure 3g–i) taken at 24 h confirmed an elongated and narrow cellular arrangement in the control group (spindle shape), compared to a wider, more branched appearance (stellate cells) of the HGFs in the treated Ti groups: Li_Ti and NaOH_Ti.

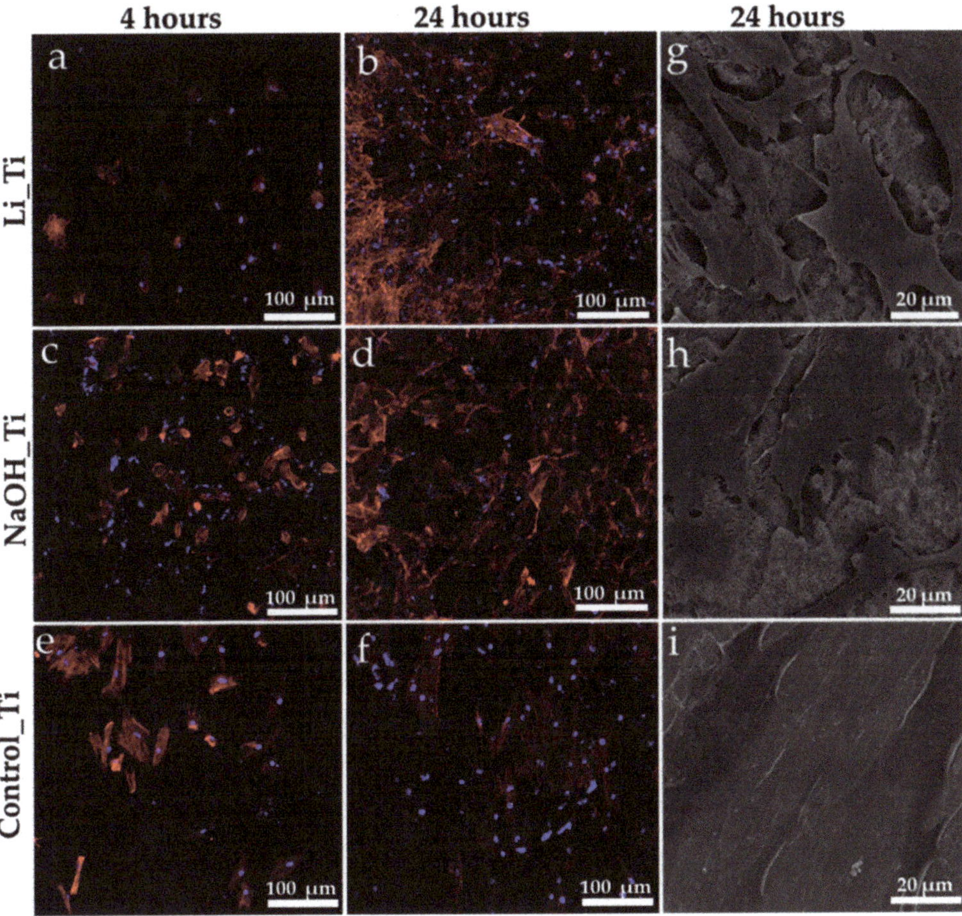

**Figure 3.** Human gingival fibroblasts' attachment. (**a–f**) Nuclei (blue) and actin F (red) staining using confocal microscopy images of HGFs at 4 and 24 h, (**g–i**) SEM images of the HGF over Ti groups, ×1000 magnification.

### 3.4. HGF Proliferation and Gene Expression of after 7 Days of Culture

After a longer incubation period (7 days), cells produced a denser and more irregular filament network on Li_Ti samples than other groups (Figure 5a). Both Li_Ti and NaOH_Ti surfaces induced significantly higher HGF proliferation than the control group (Figure 5b). Real-time PCR analysis (Figure 5c) demonstrated significantly increased expression of collagen I in both treated Ti groups compared to the control, and approximately an 8-fold increase in collagen I expression by HGFs on the Li_Ti surface compared to NaOH_Ti. Similarly, fibronectin was significantly increased in both treated Ti groups, with a six-fold increase in the Li_Ti compared to the NaOH_Ti group. The expression of collagen III, CXCL8 (interleukin 8) and IL1β (interleukin-1-beta) was higher in the Li_Ti and NaOH_Ti than in control, although not reaching statistical significance.

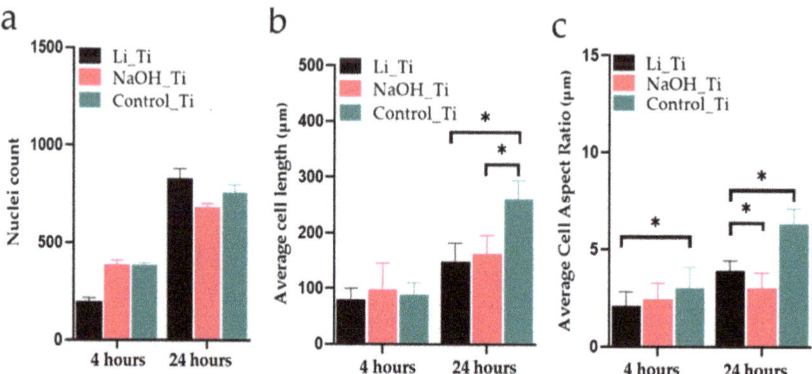

**Figure 4.** Cell morphology analysis at 4 and 24 h culture. (**a**) Nuclei counts, (**b**) cell length, and (**c**) length-to-width ratio (Aspect Ratio). Three-dimensional confocal microscopy images in Figure 3 were used for the analysis, * $p < 0.05$.

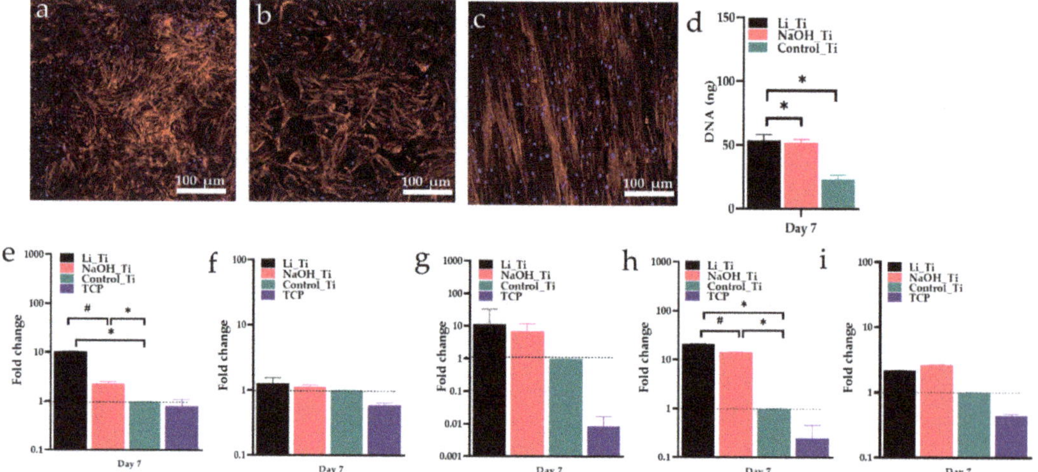

**Figure 5.** Human gingival fibroblasts characterisation 7 days after culture. (**a–c**) Confocal microscopy images of HGF at day 7 showing nuclei (blue) and Actin F (red) staining, (**d**) PicoGreen assay for DNA content, (**e–i**) HGF gene expression showing the fold change of COL-I, COL-III, CXCL8, fibronectin and IL1β. Dotted lines refers to the reference value * $p < 0.05$; # $p < 0.01$.

### 3.5. Analysis of Selected HGF-Secreted Proteins

Multiplex ELISA of conditioned culture media at 7 days (Figure 6) showed a significant increase in the concentration of FGF-2 in HGF cultures with Li_Ti and NaOH_Ti substrates, compared to the untreated control Ti (Figure 6a). Moreover, a significant decrease in MMP8 (Figure 6b) and VEGF (Figure 6e) was shown on days 3 and 7 compared to the control. No significant change was detected in MMP1 (Figure 6c) or PDGF-BB (Figure 6d). A summary of the *in vitro* assessments on HGFs bioactivity is illustrated in Table 2.

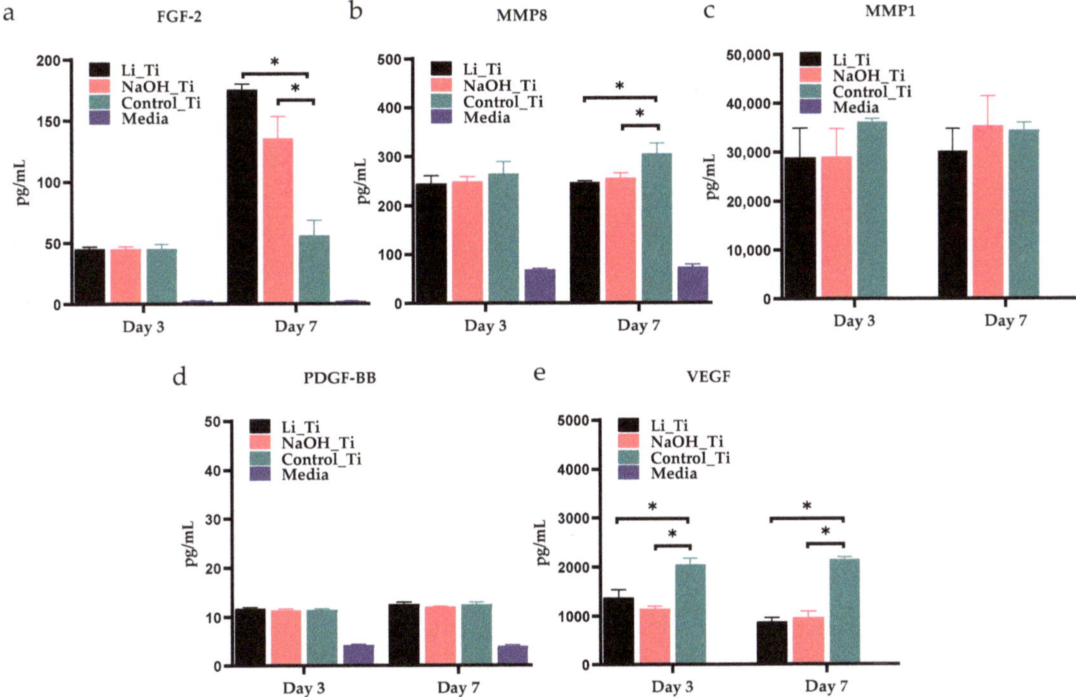

**Figure 6.** Multiplex ELISA quantification of conditioned media concentrations of FGF-2 (**a**), MMP-8 (**b**), MMP-1 (**c**), PDGF-BB (**d**) and VEGF (**e**) from HGF cultures with Ti substrates. * $p < 0.05$.

**Table 2.** Summary of the bioactivity assessments of varied Ti implants.

| Figure | Test/Assay | Time Points | Description | Inference |
|---|---|---|---|---|
| Figure 2a | Livedead staining | D1 | Viability of cells | Live cells were observed in all groups ( no signs of cytotoxicity) |
| Figure 2b | Picogreen | D1 | Early cell proliferation measured by DNA content | Some significance in DNA content was observed |
| Figure 3a–f | Immunofluorescence staining (DAPI, phalloidin) | 4 h, D1 | Early visualization of nuclei and actin filaments | Generated images (at least 3 samples per group) were used for the analysis showed in Figure 4 |
| Figure 3g–i | Scanning electron microscopy | D1 | Detailed information of the surface and attached cells | Closer visualization of cellular morphology |
| Figure 5a–c | Immunofluorescence staining (DAPI, Phalloidin) | D7 | 1 week old visualization of nuclei and actin filaments | Some difference of the filaments density was observed |
| Figure 5d | Picogreen | D7 | 1 week old cell proliferation measured by DNA content | Significance of DNA content between groups |
| Figure 5e–i | Real time PCR | D7 | Quantification of mRNA levels of selected primers | Significant increase in the expression of COL 1 and Fibronectin between Li_Ti and NaOH_Ti |
| Figure 6 | Multiplex-ELISA | D7 | Quantification of protein concentrations in the culture media | Significant difference in protein concentration between treated titanium groups vs. control titanium |

### 3.6. Analysis of Bacterial Metabolic Activity

For single-species biofilms, the metabolic activity of *S. aureus* in the Li_Ti group was the lowest after 1 and 3 days of culture and exhibited a significant difference to the alkaline group on the first-day post-culture. NaOH_Ti demonstrated slightly more bacterial activity than the control group on the first day. (Figure 7a). Metabolic activity of the salivary biofilms was significantly lower in Li_Ti than the control on days 1 and 3, and also showed remarkably fewer active bacteria than NaOH_Ti on day 3. (Figure 7c).

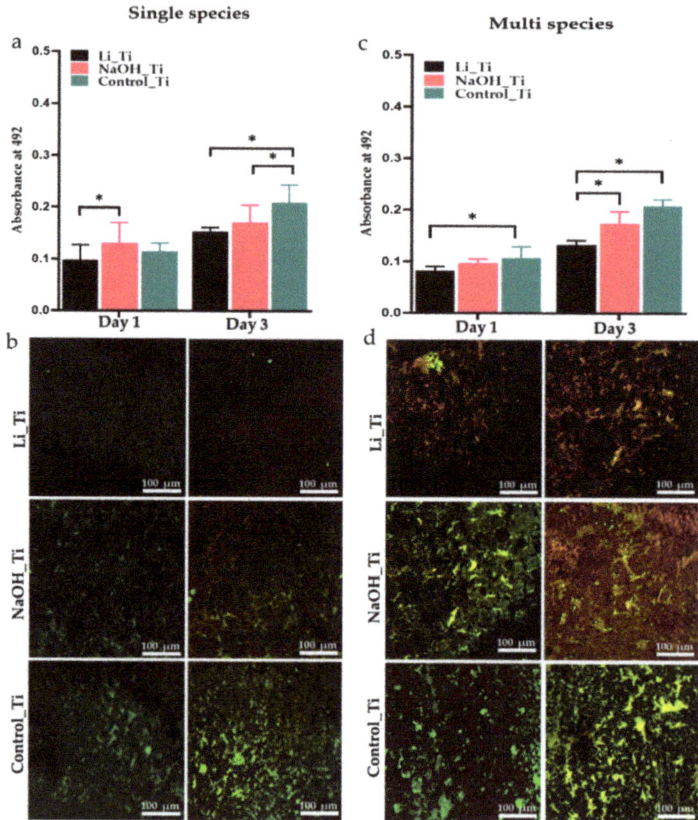


**Figure 7.** Metabolic activity (**a**,**c**), and live (green)/dead (red) staining (**b**,**d**) of single species (*S. aureus*) and multispecies (saliva) biofilm on three Ti substrates (Li_Ti, NaOH_Ti and Control_Ti) after 1 and 3 days of culture. * $p < 0.05$ by two-way ANOVA with Tukey's multiple comparison test.

### 3.7. Biofilm Viability

Three-dimensional sections from stained samples were imaged under confocal microscopy to view live and dead single species and salivary biofilms over 1- and 3-days post-culture. Live/dead staining of single species (*S. aureus*) biofilms showed very few living bacterial cells in Li_Ti compared to the other groups on day 3 (Figure 7b). Similarly, salivary biofilms exhibited fewer viable bacteria for both Li_Ti and NaOH_Ti than the control surface, with the Li_Ti surface being the least favourable for viable bacteria, as shown by the red-stained areas 3 days post-culture compared to the control surface (Figure 7d).

## 4. Discussion

This study aimed to explore the effects of ECM-mimicking lithium-doped Ti nanostructure [39], on human gingival fibroblasts and oral biofilms. Previous studies have focused on the interaction between nanostructures of Ti surface and osteogenic cells in the context of osseous healing and osseointegration [9–12]. The current study focused on biocompatibility and anti-microbial properties of the Li-doped Ti surface from the peri-implant STI perspective.

### 4.1. Gingival Fibroblasts Response to the Surface

We hypothesized that the ECM-mimicking surface, doped with lithium ions [39], could positively influence the interaction of oral soft connective tissue cells. HGFs were chosen for the study, as they are the primary constituent cells in peri-implant connective tissue, responsible for forming the soft tissue seal against the oral environment [6]. HGFs produce adhesion proteins and ECM molecules essential in the soft tissue healing process, tissue attachment and formation at the transmucosal level [47].

Previous studies have shown that HGFs display enhanced proliferation on nanomodi-fied substrates compared to micro-textured or smooth Ti groups [12,20,23]. The current results of increased viability actively formed an actin cytoskeleton, and the DNA produc-tion at a higher rate in HGFs cultured with nano-textured Ti substrates (NaOH_Ti and Li_Ti) (Figures 2 and 3) corroborate these studies. The biocompatibility of the Li_Ti surface has been demonstrated previously by it promoting adhesion and growth of other cell types such as osteoblasts [39]. The nanowire-like mesh on the Ti surface with high resemblance to the collagen fibril arrangement in the ECM of native bone tissues considerably increased osteoblast viability, metabolism, adhesion, and proliferation. Isoshima et al. [48] used Li ions to create positive charges on the Ti surface for increased hydrophilicity, resulting in increased osteoblast attachment to the lithium charged surface, further supporting the potential benefit of these approaches for future clinical applications.

ECM formation is an important biological event for cellular attachment during the early phase of healing. After adhesion to the ECM surface, fibroblasts produce adhesion proteins such as collagen and fibronectin to ensure its structural support [49–51]. Collagen I is the main collagen type constituting the periodontal and peri-implant connective tissue structure [52]. In previous studies, Ti surfaces tuned with nanopores influenced the gene expression of collagen I [20,53]. Here, COL-I gene expression was shown to be significantly upregulated on the Li_Ti surface compared to both the NaOH_Ti and control-Ti surfaces (Figure 5c). A significant increase in the expression of fibronectin for the Li_Ti surface was also observed. Elevated fibronectin levels as an indicator of effective adhesion are well-established in the literature [54–57]. Indeed, it is the one glycoprotein produced by fibroblasts that regulates the adhesion process [58], acting as a "glue" for cell attachment. Together, these findings strongly support promotion of fibroblast metabolic activity by Li-induced surface nano-topography.

In our secretome analysis, the levels of VEGF and MMP-8 produced by HGFs cultured with both Li_Ti and NaOH_Ti substrates were significantly reduced compared to control (Figure 6). In previous tissue degradation models, VEGF inhibition was related to the reduction in collagen degradation [59], suggesting that the modified surfaces in the current study could potentially reduce collagenase activities [60]. Moreover, the level of FGF-2 secretion was significantly increased in the Li_Ti and NaOH_Ti cultures. FGF-2 is well known for its function in soft tissue healing and regeneration [61–63]. It is thus plausible that both the nano-scale topographies used in the present study could positively influence collagen production while reducing the expression of metalloproteinases.

### 4.2. Bacterial Activity over the Surface

The biofilm is considered the primary aetiological factor in the development and progression of peri-implant disease [4]. Complete eradication of pathogenic microbes in the oral environment is neither feasible nor realistic; however, considerable effort has been placed into developing surfaces that can restrain bacterial adhesion and growth, hence disturbing biofilm formation [64,65]. This is of clinical importance as the implant-transmucosal interface is where biofilm initially forms.

Our bacterial study was conducted using a single strain of bacteria (*S. aureus*) and multi-species bacteria collected from saliva. *S. aureus* is a commonly found bacterium on the skin and plays an essential role in the causation of medical device/implant-related biofilm infections [66,67]. The behaviour of mono-species biofilms in the lab is more predictable and controlled compared to polymicrobial biofilms. Hence, we initially chose to grow mono-species biofilms. However, most biofilm infections are polymicrobial, especially oral infections. So, the antimicrobial properties of the surfaces were evaluated against polymicrobial biofilms by using pooled saliva [68,69]. Quantitative data from our XTT experiment indicated that the bioinspired Li_Ti surface significantly reduced the early bacterial activity of both single and multi-species biofilm. Similarly, viability data for both microbial environments were consistent, showing a significant reduction in the bacterial growth in the Li_Ti group. Our results are in line with previous studies [17,18,33], in which nanoscale modifications on Ti substrates showed either bacteriostatic or bactericidal ability. Moghanian [70] reported that increased Li concentration in bioactive glass led to a prominent decrease in *Staphylococcus aureus* activity. This was compatible with our bacterial activity study, where the Li-containing substrate exhibited increased suppression of *S. aureus* activity compared to the alkaline group (Figure 7a). Moreover, the metabolic activity of the multi-species salivary biofilm was significantly reduced on the Li_Ti surface compared to NaOH_Ti (Figure 7c). Importantly, the present study is the first to investigate the antibacterial effect of Li on a multi-species biofilm model.

In addition to HGFs, peri-implant soft tissue is composed of other cell types including epithelial cells and innate immune cells, and hence the current work's sole focus on HGF response to the modified Ti substrates may be considered a study limitation. It would be of importance to investigate all cell responses to the Li_Ti modified surface from a clinical perspective. Our bacterial culture study being conducted under a static condition is a second limitation. The flow of saliva in the oral cavity, as simulated in a dynamic model, may influence bacterial activity and survival on Ti surfaces not accounted for in our current static model. Nevertheless, the current work is the first to demonstrate the antibacterial characteristics of the nano-modified Ti surface by using multi-species biofilm, rather than single species bacteria alone.

Biocompatibility and antibacterial effects of the nano-modified Li_Ti surface should be further investigated in an in vivo environment, preferably in an oral environment, to provide a better understanding of the biological and microbiological mechanisms of the surface, therefore allowing the exploitation of its potential for clinical application.

## 5. Conclusions

A dental implant surface capable of augmenting the function of gingival fibroblasts and reducing the adhesion of bacteria may enable the early establishment of STI and increase long-term survival. Remarkably, the Li-doped Ti (Li_Ti) surface resulted in up-regulated expression of COL-I and fibronectin compared to the Ti nanostructure without lithium (NaOH_Ti). In addition, the Li_Ti surface promoted an increase in the concentration of growth factors (FGF2), while significantly reducing collagenase (MMP8) and VEGF secretion compared to the control Ti surface. Concerning its effects on bioactivity, the bioinspired Li_Ti surface can augment HGF cellular attachment, proliferation, collagen formation, and extracellular matrix deposition. As for antibacterial activity, both treated Ti (nanoscale modified topographies) surfaces significantly reduced bacterial adhesion and growth compared to the untreated smooth machine polished (control) Ti surface. These antibacterial effects were more evident at day 3 for the Li_Ti surface compared to the control group. As such, it may be concluded that the bioinspired Li-doped Ti surface can promote HGF bioactivity while suppressing bacterial adhesion and growth. This is of clinical importance in terms of improved STI during the maintenance phase of implant treatment. Further *in vivo* studies are warranted to investigate Li-doped surfaces' effects on the host immune responses and tissue formation quality.

**Author Contributions:** A.Q.A.: methodology, analysis, writing—original draft preparation, A.A.-h.: Methodology, review and editing, K.G.: Methodology, review and editing, S.I. Project funding, review and editing. B.P.J.F.: Conceptualisation, validation, review and editing, R.S.B.L.: Conceptualisation, validation, review and editing. All authors have read and agreed to the published version of the manuscript.

**Funding:** This research was funded by the Australian Dental Research Foundation ADRF (2804-2020) 2020–2021, UQ Colgate-Palmolive Student Research Grants, 2020 and UQ Postgraduate Research Support Funding, 2019 and 2020. K.G. is supported by the National Health and Medical Research Council (NHMRC) Early Career Fellowship (APP1140699).

**Institutional Review Board Statement:** The study was conducted according to the guidelines of the Declaration of Helsinki, and approved by the University of Queensland Institutional Human Ethics Research Committee (No. 2019000134 and No. 2019001113).

**Informed Consent Statement:** Informed consent was obtained from all subjects involved in the study.

**Data Availability Statement:** The data are not publicly available.

**Acknowledgments:** The authors would like to thank Pingping Han and Kexin Jiao (School of Dentistry, University of Queensland) for their help in performing RT-PCR. The authors would also like to thank Stephen Hamlet (School of Dentistry and Oral Health, Griffith University) for his guidance during Multiplex ELISA. The authors would also like to thank Srinivas Ramachandra for his help with the bacterial testing.

**Conflicts of Interest:** The authors declare no conflict of interest.

## References


1. Roos, J.; Sennerby, L.; Lekholm, U.L.F.; Jemt, T.; Gröndahl, K.; Albrektsson, T. A qualitative and quantitative method for evaluating implant success: A 5-year retrospective analysis of the Branemark implant. *Int. J. Oral Maxillofac. Implants* **1997**, *12*, 1–20.
2. Becker, W.; Becker, B.E.; Newman, M.G.; Nyman, S. Clinical and microbiologic findings that may contribute to dental implant failure. *Int. J. Oral Maxillofac. Implant.* **1990**, *5*, 1–17.
3. Ivanovski, S.; Lee, R. Comparison of peri-implant and periodontal marginal soft tissues in health and disease. *Periodontology 2000* **2018**, *76*, 116–130. [CrossRef] [PubMed]
4. Berglundh, T.; Armitage, G.; Araujo, M.G.; Avila-Ortiz, G.; Blanco, J.; Camargo, P.M.; Chen, S.; Cochran, D.; Derks, J.; Figuero, E.; et al. Peri-implant diseases and conditions: Consensus report of workgroup 4 of the 2017 world workshop on the classification of periodontal and peri-implant diseases and conditions. *J. Periodontol.* **2018**, *89*, S313–S318. [CrossRef] [PubMed]
5. Maksoud, M.A. Manipulation of the peri-implant tissue for better maintenance: A periodontal perspective. *J. Oral Implant.* **2003**, *29*, 120–123. [CrossRef]
6. Guo, T.; Gulati, K.; Arora, H.; Han, P.; Fournier, B.; Ivanovski, S. Race to invade: Understanding soft tissue integration at the transmucosal region of titanium dental implants. *Dent. Mater.* **2021**, *37*, 816–831. [CrossRef]
7. Guo, T.; Gulati, K.; Arora, H.; Han, P.; Fournier, B.; Ivanovski, S. Orchestrating soft tissue integration at the transmucosal region of titanium implants. *Acta Biomater.* **2021**, *124*, 33–49. [CrossRef]
8. Kim, H.; Murakami, H.; Chehroudi, B.; Textor, M.; Brunette, D.M. Effects of surface topography on the connective tissue attachment to subcutaneous implants. *Int. J. Oral Maxillofac. Implant.* **2006**, *21*, 354–365.
9. Hao, J.; Li, Y.; Li, B.; Wang, X.; Li, H.; Liu, S.; Liang, C.; Wang, H. Biological and mechanical effects of micro-nanostructured titanium surface on an osteoblastic cell line *In Vitro* and osteointegration In Vivo. *Appl. Biochem. Biotechnol.* **2017**, *183*, 280–292. [CrossRef]
10. Gui, N.; Xu, W.; Myers, D.; Shukla, R.; Tang, H.; Qian, M. The effect of ordered and partially ordered surface topography on bone cell responses: A review. *Biomater. Sci.* **2017**, *6*, 250–264. [CrossRef]
11. Smeets, R.; Stadlinger, B.; Schwarz, F.; Beck-Broichsitter, B.; Jung, O.; Precht, C.; Kloss, F.; Gröbe, A.; Heiland, M.; Ebker, T. Impact of dental implant surface modifications on osseointegration. *Biol. Med. Res. Int.* **2016**, *2016*, 1–16. [CrossRef] [PubMed]
12. Souza, J.C.; Sordi, M.B.; Kanazawa, M.; Ravindran, S.; Henriques, B.; Silva, F.; Aparicio, C.; Cooper, L.F. Nano-scale modification of titanium implant surfaces to enhance osseointegration. *Acta Biomater.* **2019**, *94*, 112–131. [CrossRef] [PubMed]
13. Hasan, J.; Jain, S.; Chatterjee, K. Nanoscale topography on black titanium imparts multi-biofunctional properties for orthopedic applications. *Sci. Rep.* **2017**, *7*, 41118. [CrossRef] [PubMed]
14. Bhadra, C.M.; Truong, V.K.; Pham, V.T.H.; Al Kobaisi, M.; Seniutinas, G.; Wang, J.; Juodkazis, S.; Crawford, R.; Ivanova, E.P. Antibacterial titanium nano-patterned arrays inspired by dragonfly wings. *Sci. Rep.* **2015**, *5*, 16817. [CrossRef] [PubMed]
15. Modaresifar, K.; Azizian, S.; Ganjian, M.; Fratila-Apachitei, L.E.; Zadpoor, A.A. Bactericidal effects of nanopatterns: A systematic review. *Acta Biomater.* **2019**, *83*, 29–36. [CrossRef] [PubMed]
16. Bandara, C.D.; Singh, S.; Afara, I.O.; Wolff, A.; Tesfamichael, T.; Ostrikov, K.; Oloyede, A. Bactericidal effects of natural nanotopography of dragonfly wing on *Escherichia coli*. *ACS Appl. Mater. Interfaces* **2017**, *9*, 6746–6760. [CrossRef] [PubMed]

17. Jaggessar, A.; Mathew, A.; Wang, H.; Tesfamichael, T.; Yan, C.; Yarlagadda, P.K. Mechanical, bactericidal and osteogenic behaviours of hydrothermally synthesised $TiO_2$ nanowire arrays. *J. Mech. Behav. Biomed. Mater.* **2018**, *80*, 311–319. [CrossRef] [PubMed]

18. Ivanova, E.P.; Hasan, J.; Webb, H.; Truong, V.K.; Watson, G.; Watson, J.; Baulin, V.; Pogodin, S.; Wang, J.; Tobin, M.; et al. Natural bactericidal surfaces: Mechanical rupture of pseudomonas aeruginosa cells by cicada wings. *Small* **2012**, *8*, 2489–2494. [CrossRef] [PubMed]

19. Chopra, D.; Gulati, K.; Ivanovski, S. Understanding and optimising the antibacterial functions of anodized nano-engineered titanium implants. *Acta Biomater.* **2021**, *127*, 80–101. [CrossRef] [PubMed]

20. Ferrà-Cañellas, M.D.M.; Llopis-Grimalt, M.A.; Monjo, M.; Ramis, J.M. Tuning nanopore diameter of titanium surfaces to improve human gingival fibroblast response. *Int. J. Mol. Sci.* **2018**, *19*, 2881. [CrossRef] [PubMed]

21. Gulati, K.; Moon, H.J.; Kumar, P.S.; Han, P.; Ivanovski, S. Anodized anisotropic titanium surfaces for enhanced guidance of gingival fibroblasts. *Mater. Sci. Eng. C* **2020**, *112*, 110860. [CrossRef]

22. Guida, L.; Oliva, A.; Basile, M.A.; Giordano, M.; Nastri, L.; Annunziata, M. Human gingival fibroblast functions are stimulated by oxidized nano-structured titanium surfaces. *J. Dent.* **2013**, *41*, 900–907. [CrossRef]

23. Gulati, K.; Moon, H.-J.; Li, T.; Kumar, P.S.; Ivanovski, S. Titania nanopores with dual micro-/nano-topography for selective cellular bioactivity. *Mater. Sci. Eng. C* **2018**, *91*, 624–630. [CrossRef]

24. Ziegler, N.; Sengstock, C.; Mai, V.; Schildhauer, T.A.; Köller, M.; Ludwig, A. Glancing-angle deposition of nanostructures on an implant material surface. *Nanomaterials* **2019**, *9*, 60. [CrossRef] [PubMed]

25. Liu, K.; Jiang, L. Bio-inspired design of multiscale structures for function integration. *Nano Today* **2011**, *6*, 155–175. [CrossRef]

26. Izquierdo-Barba, I.; García-Martín, J.M.; Alvarez, R.; Palmero, A.; Esteban, J.; Pérez-Jorge, C.; Arcos, D.; Vallet-Regí, M. Nanocolumnar coatings with selective behavior towards osteoblast and *Staphylococcus aureus* proliferation. *Acta Biomater.* **2015**, *15*, 20–28. [CrossRef] [PubMed]

27. Ng, R.; Zang, R.; Yang, K.K.; Liu, N.; Yang, S.-T. Three-dimensional fibrous scaffolds with microstructures and nanotextures for tissue engineering. *RSC Adv.* **2012**, *2*, 10110–10124. [CrossRef]

28. Wu, S.; Liu, X.; Yeung, K.W.Y.; Yang, X. Biomimetic porous scaffolds for bone tissue engineering. *Mater. Sci. Eng. R Rep.* **2014**, *80*, 1–36. [CrossRef]

29. Dalby, M.J.; Riehle, M.O.; Johnstone, H.; Affrossman, S.; Curtis, A.S.G. Investigating the limits of filopodial sensing: A brief report using SEM to image the interaction between 10 nm high nano-topography and fibroblast filopodia. *Cell Biol. Int.* **2004**, *28*, 229–236. [CrossRef] [PubMed]

30. Pachauri, P.; Bathala, L.R.; Sangur, R. Techniques for dental implant nanosurface modifications. *J. Adv. Prosthodont.* **2014**, *6*, 498–504. [CrossRef] [PubMed]

31. Morgan, D.L. Alkaline Hydrothermal Treatment of Titanate Nanostructures. Ph.D. ThesIs, Queensland University of Technology, Brisbane City, QLD, Australia, 2010.

32. Dorkhan, M.; Yucel-Lindberg, T.; Hall, J.; Svensäter, G.; Davies, J.R. Adherence of human oral keratinocytes and gingival fibroblasts to nano-structured titanium surfaces. *BMC Oral Health* **2014**, *14*, 75. [CrossRef]

33. Miao, X.; Wang, D.; Xu, L.; Wang, J.; Zeng, D.; Lin, S.; Huang, C.; Liu, X.; Jiang, X. The response of human osteoblasts, epithelial cells, fibroblasts, macrophages and oral bacteria to nanostructured titanium surfaces: A systematic study. *Int. J. Nanomed.* **2017**, *12*, 1415–1430. [CrossRef] [PubMed]

34. Zhang, L.; Guo, J.; Yan, T.; Han, Y. Fibroblast responses and antibacterial activity of Cu and Zn co-doped $TiO_2$ for percutaneous implants. *Appl. Surf. Sci.* **2018**, *434*, 633–642. [CrossRef]

35. Okawachi, H.; Ayukawa, Y.; Atsuta, I.; Furuhashi, A.; Sakaguchi, M.; Yamane, K.; Koyano, K. Effect of titanium surface calcium and magnesium on adhesive activity of epithelial-like cells and fibroblasts. *Biointerphases* **2012**, *7*, 27. [CrossRef] [PubMed]

36. Liu, W.; Chen, D.; Jiang, G.; Li, Q.; Wang, Q.; Cheng, M.; He, G.; Zhang, X. A lithium-containing nanoporous coating on entangled titanium scaffold can enhance osseointegration through Wnt/β-catenin pathway. *Nanomed. Nanotechnol. Biol. Med.* **2018**, *14*, 153–164. [CrossRef]

37. Wang, L.; Luo, Q.; Zhang, X.; Qiu, J.; Qian, S.; Liu, X. Co-implantation of magnesium and zinc ions into titanium regulates the behaviors of human gingival fibroblasts. *Bioact. Mater.* **2021**, *6*, 64–74. [CrossRef]

38. Galli, C.; Piemontese, M.; Lumetti, S.; Manfredi, E.; Macaluso, G.M.; Passeri, G. GSK3b-inhibitor lithium chloride enhances activation of Wnt canonical signaling and osteoblast differentiation on hydrophilic titanium surfaces. *Clin. Oral Implant. Res.* **2013**, *24*, 921–927. [CrossRef] [PubMed]

39. Abdal-Hay, A.; Gulati, K.; Fernandez-Medina, T.; Qian, M.; Ivanovski, S. In Situ hydrothermal transformation of titanium surface into lithium-doped continuous nanowire network towards augmented bioactivity. *Appl. Surf. Sci.* **2020**, *505*, 144604. [CrossRef]

40. Tang, G.H.; Xu, J.; Chen, R.J.; Qian, Y.F.; Shen, G. Lithium delivery enhances bone growth during midpalatal expansion. *J. Dent. Res.* **2011**, *90*, 336–340. [CrossRef]

41. Han, P.; Wu, C.; Chang, J.; Xiao, Y. The cementogenic differentiation of periodontal ligament cells via the activation of Wnt/β-catenin signalling pathway by $Li^+$ ions released from bioactive scaffolds. *Biomaterials* **2012**, *33*, 6370–6379. [CrossRef]

42. Li, T.; Gulati, K.; Wang, N.; Zhang, Z.; Ivanovski, S. Bridging the gap: Optimized fabrication of robust titania nanostructures on complex implant geometries towards clinical translation. *J. Colloid Interface Sci.* **2018**, *529*, 452–463. [CrossRef]

43. Han, P.; Ivanovski, S. Effect of saliva collection methods on the detection of periodontium-related genetic and epigenetic biomarkers-a pilot study. *Int. J. Mol. Sci.* **2019**, *20*, 4729. [CrossRef] [PubMed]

44. Philip, N.; Bandara, H.; Leishman, S.J.; Walsh, L.J. Inhibitory effects of fruit berry extracts on *Streptococcus mutans* biofilms. *Eur. J. Oral Sci.* **2019**, *127*, 122–129. [CrossRef]

45. Koban, I.; Matthes, R.; Hübner, N.O.; Welk, A.; Sietmann, R.; Lademann, J.; Kramer, A.; Kocher, T. XTT assay of Ex Vivo saliva biofilms to test antimicrobial influences. *GMS Krankenhaushygiene Interdisziplinär* **2012**, *7*, Doc06.

46. Kim, H.-J.; Cho, M.-Y.; Lee, E.-S.; Jung, H.I.; Kim, B.-I. Effects of short-time exposure of surface pre-reacted glass-ionomer eluate on dental microcosm biofilm. *Sci. Rep.* **2020**, *10*, 1–8. [CrossRef] [PubMed]

47. Dean, J.W.; Blankenship, J.A. Migration of gingival fibroblasts on fibronectin and laminin. *J. Periodontol.* **1997**, *68*, 750–757. [CrossRef] [PubMed]

48. Isoshima, K.; Ueno, T.; Arai, Y.; Saito, H.; Chen, P.; Tsutsumi, Y.; Hanawa, T.; Wakabayashi, N. The change of surface charge by lithium ion coating enhances protein adsorption on titanium. *J. Mech. Behav. Biomed. Mater.* **2019**, *100*, 103393. [CrossRef] [PubMed]

49. Newman, M.G.; Takei, H.H.; Carranza, F.A. *Carranza's Clinical Periodontology*; Elsevier: Amsterdam, The Netherlands, 2002.

50. Roman-Malo, L.; Bullon, B.; De Miguel, M.; Bullon, P. Fibroblasts collagen production and histological alterations in hereditary gingival fibromatosis. *Diseases* **2019**, *7*, 39. [CrossRef]

51. Meyle, J. Cell adhesion and spreading on different implant surfaces. In Proceedings of the 3rd European Workshop on Periodontology, Waldenburg, Switzerland, 1999.

52. Knowles, G.C.; McKeown, M.; Sodek, J.; McCulloch, C.A. Mechanism of collagen phagocytosis by human gingival fibroblasts: Importance of collagen structure in cell recognition and internalization. *J. Cell Sci.* **1991**, *98*, 551–558. [CrossRef]

53. Wang, X.; Lu, T.; Wen, J.; Xu, L.; Zeng, D.; Wu, Q.; Cao, L.; Lin, S.; Liu, X.; Jiang, X. Selective responses of human gingival fibroblasts and bacteria on carbon fiber reinforced polyetheretherketone with multilevel nanostructured $TiO_2$. *Biomaterials* **2016**, *83*, 207–218. [CrossRef]

54. Zhang, P.; Katz, J.; Michalek, S.M. Glycogen synthase kinase-3beta (GSK3beta) inhibition suppresses the inflammatory response to Francisella infection and protects against tularemia in mice. *Mol. Immunol.* **2009**, *46*, 677–687. [CrossRef]

55. Hormia, M.; Könönen, M. Immunolocalization of fibronectin and vitronectin receptors in human gingival fibroblasts spreading on titanium surfaces. *J. Periodontal Res.* **1994**, *29*, 146–152. [CrossRef]

56. Zhou, Z.; Dai, Y.; Liu, B.-B.; Xia, L.-L.; Liu, H.-B.; Vadgama, P.; Liu, H.-R. Surface modification of titanium plate enhanced fibronectin-mediated adhesion and proliferation of MG-63 cells. *Trans. Nonferrous Met. Soc. China* **2014**, *24*, 1065–1071. [CrossRef]

57. Qi, H.; Shi, M.; Ni, Y.; Mo, W.; Zhang, P.; Jiang, S.; Zhang, Y.; Deng, X. Size-confined effects of nanostructures on fibronectin-induced macrophage inflammation on titanium implants. *Adv. Health Mater.* **2021**, e2100994. [CrossRef] [PubMed]

58. Bauer, J.S.; Schreiner, C.L.; Giancotti, F.G.; Ruoslahti, E.; Juliano, R.L. Motility of fibronectin receptor-deficient cells on fibronectin and vitronectin: Collaborative interactions among integrins. *J. Cell Biol.* **1992**, *118*, 217. [CrossRef]

59. Ohshima, M.; Yamaguchi, Y.; Ambe, K.; Horie, M.; Saito, A.; Nagase, T.; Nakashima, K.; Ohki, H.; Kawai, T.; Abiko, Y.; et al. Fibroblast VEGF-receptor 1 expression as molecular target in periodontitis. *J. Clin. Periodontol.* **2015**, *43*, 128–137. [CrossRef]

60. Checchi, V.; Maravic, T.; Bellini, P.; Generali, L.; Consolo, U.; Breschi, L.; Mazzoni, A. The role of matrix metalloproteinases in periodontal disease. *Int. J. Environ. Res. Public Health* **2020**, *17*, 4923. [CrossRef] [PubMed]

61. Maddaluno, L.; Urwyler, C.; Werner, S. Fibroblast growth factors: Key players in regeneration and tissue repair. *Development* **2017**, *144*, 4047–4060. [CrossRef] [PubMed]

62. An, S.; Huang, X.; Gao, Y.; Ling, J.; Huang, Y.; Xiao, Y. FGF-2 induces the proliferation of human periodontal ligament cells and modulates their osteoblastic phenotype by affecting Runx2 expression in the presence and absence of osteogenic inducers. *Int. J. Mol. Med.* **2015**, *36*, 705–711. [CrossRef]

63. Takayama, S.I.; Yoshida, J.; Hirano, H.; Okada, H.; Murakami, S. Effects of basic fibroblast growth factor on human gingival epithelial cells. *J. Periodontol.* **2002**, *73*, 1467–1473. [CrossRef]

64. Wang, L.; Xie, X.; Li, C.; Liu, H.; Zhang, K.; Zhou, Y.; Chang, X.; Xu, H.H. Novel bioactive root canal sealer to inhibit endodontic multispecies biofilms with remineralising calcium phosphate ions. *J. Dent.* **2017**, *60*, 25–35. [CrossRef] [PubMed]

65. Afkhami, F.; Pourhashemi, S.J.; Sadegh, M.; Salehi, Y.; Fard, M.J.K. Antibiofilm efficacy of silver nanoparticles as a vehicle for calcium hydroxide medicament against Enterococcus faecalis. *J. Dent.* **2015**, *43*, 1573–1579. [CrossRef] [PubMed]

66. Ribeiro, M.; Monteiro, F.J.; Ferraz, M.P. Infection of orthopedic implants with emphasis on bacterial adhesion process and techniques used in studying bacterial-material interactions. *Biomatter* **2012**, *2*, 176–194. [CrossRef]

67. Garbacz, K.; Jarzembowski, T.; Kwapisz, E.; Daca, A.; Witkowski, J. Do the oral *Staphylococcus aureus* strains from denture wearers have a greater pathogenicity potential? *J. Oral. Microbiol.* **2018**, *11*, 1536193. [CrossRef]

68. Willems, H.M.; Xu, Z.; Peters, B.M. Polymicrobial biofilm studies: From basic science to biofilm control. *Curr. Oral Health Rep.* **2016**, *3*, 36–44. [CrossRef] [PubMed]

69. Bowen, W.H.; Burne, R.A.; Wu, H.; Koo, H. Oral biofilms: Pathogens, matrix, and polymicrobial interactions in microenvironments. *Trends Microbiol.* **2018**, *26*, 229–242. [CrossRef]

70. Moghanian, A.; Firoozi, S.; Tahriri, M.; Sedghi, A. A comparative study on the *In Vitro* formation of hydroxyapatite, cytotoxicity and antibacterial activity of 58S bioactive glass substituted by Li and Sr. *Mater. Sci. Eng. C* **2018**, *91*, 349–360. [CrossRef]

*nanomaterials*

*Article*

# Correlation between LncRNA Profiles in the Blood Clot Formed on Nano-Scaled Implant Surfaces and Osseointegration

Long Bai [1,2,3,4,†], Peiru Chen [5,†], Bin Tang [2], Ruiqiang Hang [2,*] and Yin Xiao [3,4,*]

[1] Key Laboratory for Ultrafine Materials of Ministry of Education, The State Key Laboratory of Bioreactor Engineering, East China University of Science and Technology, Shanghai 200237, China; bailong@ecust.edu.cn
[2] Laboratory of Biomaterial Surfaces & Interfaces, Institute of New Carbon Materials, Taiyuan University of Technology, Taiyuan 030000, China; tangbin@tyut.edu.cn
[3] Institute of Health and Biomedical Innovation, Queensland University of Technology, Brisbane 4059, Australia
[4] Australia-China Centre for Tissue Engineering and Regenerative Medicine, Queensland University of Technology, Brisbane 4059, Australia
[5] Beijing Proteome Research Center, State Key Laboratory of Proteomics, National Center for Protein Sciences (Beijing), Institute of Lifeomics, Beijing 102206, China; chenpeiru12@126.com
* Correspondence: hangruiqiang@tyut.edu.cn (R.H.); yin.xiao@qut.edu.au (Y.X.)
† The authors contribute equally to this work.

**Citation:** Bai, L.; Chen, P.; Tang, B.; Hang, R.; Xiao, Y. Correlation between LncRNA Profiles in the Blood Clot Formed on Nano-Scaled Implant Surfaces and Osseointegration. *Nanomaterials* **2021**, *11*, 674. https://doi.org/10.3390/nano11030674

Academic Editor: Karan Gulati

Received: 1 February 2021
Accepted: 6 March 2021
Published: 9 March 2021

**Publisher's Note:** MDPI stays neutral with regard to jurisdictional claims in published maps and institutional affiliations.

**Abstract:** Implant surfaces with a nanoscaled pattern can dominate the blood coagulation process resulting in a defined clot structure and its degradation behavior, which in turn influence cellular response and the early phase of osseointegration. Long non-coding (Lnc) RNAs are known to regulate many biological processes in the skeletal system; however, the link between the LncRNA derived from the cells within the clot and osseointegration has not been investigated to date. Hence, the sequence analysis of LncRNAs expressed within the clot formed on titania nanotube arrays (TNAs) with distinct nano-scaled diameters (TNA 15 of 15 nm, TNA 60 of 60 nm, TNA 120 of 120 nm) on titanium surfaces was profiled for the first time. LncRNA LOC103346307, LOC103352121, LOC108175175, LOC103348180, LOC108176660, and LOC108176465 were identified as the pivotal players in the early formed clot on the nano-scaled surfaces. Further bioinformatic prediction results were used to generate co-expression networks of LncRNAs and mRNAs. Gene Ontology and Kyoto Encyclopedia of Genes and Genomes pathway analyses revealed that distinct nano-scaled surfaces could regulate the biological functions of target mRNAs in the clot. LOC103346307, LOC108175175, and LOC108176660 upregulated mRNAs related to cell metabolism and Wnt, TGF-beta, and VEGF signaling pathways in TNA 15 compared with P-Ti, TNA 60, and TNA 120, respectively, whereas LOC103352121, LOC103348180, and LOC108176465 downregulated mRNAs related to bone resorption and inflammation through negatively regulating osteoclast differentiation, TNF, and NF-kappa signaling pathways. The results indicated that surface nano-scaled characteristics can significantly influence the clot-derived LncRNAs expression profile, which affects osseointegration through multiple signaling pathways of the targeted mRNAs, thus paving a way for better interpreting the link between the properties of a blood clot formed on the nano-surface and de novo bone formation.

**Keywords:** implant; nano-scaled surface; blood clot; LncRNA; osseointegration; bone regeneration

## 1. Introduction

Osseointegration indicates a direct anchorage of a biomedical metal implant onto the host bone, allowing the newly formed bone to be attached directly to the surface of the implant [1]. Osseointegrated implants show promise to replace damaged joint tissues, alleviate pain, and restore bone function. However, implant loosening contributes to more than half of replacement failures due to the poor osseointegration between host bone and implant [2]. Fulfilled osseointegration requires rapid bone formation with a qualified volume, ensuring the incorporation and longevity of the implant [3]. The commercial pure titanium-based implant failed to satisfy this requirement due to its bioinert lacking bioactivity [4]. Surface modification in nano-scale on titanium-based implant has been verified to promote osseointegration [5,6]. However, the underlying mechanism pertaining to the nano-scaled surface-mediated osseointegration is not well-understood. Osseointegration is a sophisticated process that initiates immediately with clot formation on the implant surface. The blood clot has been recognized as a natural healing scaffold consisting of fibrin fiber structure and myriad immune cells including T cells, B cells, macrophages, and neutrophils [7,8]. The clot exerts a pivotal role in the manipulation of osseointegration, as it can modulate the early immune response and osteogenesis/angiogenesis through osteoimmunomodulation [8]. Efforts have been conducted to demonstrate the role of mRNAs within the clot on osseointegration in our recent study [8]. However, more is to be explored based on the recent advances in the potential role of the Long non-coding RNAs (LncRNAs) on cell differentiation, especially these within the clot on the osseointegration.

LncRNAs, whose length of the transcripts ranges from 200 nt to 100 kb, are emerging pivotal factors in the regulation of gene transcription and thus affect various aspects of cellular homeostasis, including proliferation, survival, migration, and genomic stability [9]. Numerous studies have demonstrated that the expression profile of LncRNAs is related to tissue regeneration and disease development, thus the LncRNAs profiling study will help to unravel their underlying functions [10]. However, the role of LncRNAs in the skeletal system and the regulation of osseointegration remain largely unclear to date. Specifically, no report was made on the expression profiles of LncRNAs in the early bone healing clot. LncRNA is multifaceted and varies differently from its locations, binding sites, and acting modes when exerting its biological function [11]. The regulating role of LncRNAs is not solitary but instead occurs through a large complex network that involves mRNAs, miRNAs, and proteins [12]. Based on our previous study [8], we demonstrated the nano-scaled surfaces could significantly influence the osteoimmunological reaction, angiogenesis, and osseointegration. In this study, we further investigated the expression of LncRNAs in blood clot formed on different nano-scaled surfaces and their potential impact on the osseointegration.

The nanostructured titania nanotube array (TNA) was chosen as the nano-scaled surfaces on the titanium implant due to the advantageous features of proven biocompatibility, thermal stability, and corrosion resistance [13]. Previous studies have demonstrated that surface modification of TNAs conferred the pristine titanium implant with enhanced osteogenesis, angiogenesis, and potentially induced favorable osseointegration [13,14]. However, osseointegration is a sophisticated process that involved multiple cells as aforementioned and the inconsistency of relevant in vitro/vivo studies requires a deeper investigation and clarification. Herein, TNAs with three distinct diameters (15, 60, and 120 nm) were fabricated aiming to unravel whether the distinct nano-surfaces can influence the LncRNAs profiles within the clot. Moreover, we specifically focused on the potential effect of the LncRNAs expression on the early phases of osseointegration, that is, whether the different LncRNAs profiles can a distinct osteoimmunomodulation effect on confer the nano-surfaces.

## 2. Materials and Methods

### 2.1. Surface Modification and Characterization of the Ti Implant

Ti implants shaped in rod (99.6% purity, the length of 5.0 mm, the diameter of 3.0 mm) were introduced for the in vivo study. The implants were ultrasonically cleaned in acetone, ethanol, and ultrapure water sequentially before the anodization. The titania nanotube arrays with different diameters on the surfaces were fabricated via an electrochemical cell (IT6120, ITECH, Shanghai, China) with a two-electrode configuration. The implants were used as the anode electrode while the platinum foil was set as the counter electrode. Electrochemical treatments were carried out in ethylene glycol solution containing 0.5 wt% ammonium fluoride ($NH_4F$), 5 vol% methanol, and 5 vol% distilled water. The applied potentials were 5 V for 2 h, 30 V for 1 h, and 60 V for 10 min separately at room temperature. Afterwards, the as-prepared implants with distinct nano-scaled surfaces were ultrasonic-cleaned in ethanol for 10 min and air-dried. The surface morphology of the implants was characterized by a field-emission scanning electron microscope (SEM, JSM-7001F, Tokyo, Japan).

### 2.2. In Vitro Observation of the Platelet Activation on the Nano-Scaled Surfaces

Blood was collected in 3.8% sodium citrate (9:1, *v/v*) from healthy aspirin-free donors following informed consent at the Australian Red Cross Blood Bank. All procedures were carried out under approval by the University Human Research Ethics Committee at the Queensland University of Technology (1500000918). Citrated blood was centrifuged at 1200 rpm for 10 min and the platelet-rich plasma (PRP) was collected using the 1.5 mL EP tubes. To investigate the platelet activation, the implants were placed in a 96-well plate (Corning, St. Louis, MO, USA) and 50 µL PRP was dropped on each specimen. The plate was incubated at 37 °C for 30 min. Then the activation of the platelets on the implants was observed by SEM after gradient elution via ethanol.

### 2.3. In Vivo Clot Observation on the Nano-Scaled Surfaces

The animal surgery was done consistent with protocols approved by the Animal Research Committee of Taiyuan University of Technology (TYUT202001003). A total of 8 New Zealand rabbits (male, 9 to 10 months) were used. The animals were sedated with 2 mg/kg intramuscular midazolam, and general anesthesia was conducted with an intramuscular injection of 50 mg/kg ketamine and 15 mg/kg xylazine. With animals under local anesthesia with oxybuprocaine 0.4%, four cylindrical titanium implants covered with nanotubes were placed on each distal surface of the bilateral femoral condyles of an animal. After 24 h of implantation, the animals were euthanized, and the implants were harvested immediately and then fixed in 4% PFA. The clot morphology was observed by SEM after the process with the standard procedures.

### 2.4. Histological Analysis

After 8 weeks of implantation, the implants were dehydrated and embedded in the resin. EXAKT saw was introduced to process the specimens into several sections with 200 µm in thickness. Afterwards, 50 µm sections were grinded then polished through the EXAKT grinder (EXAKT 400 CS, Norderstedt, Germany). The sections were then stained with Toluidine blue. Three histological sections were evaluated via the Leica microscope and quantified by Image J.

### 2.5. LncRNAs Profile of the Clot on the Nano-Scaled Surfaces

Three days after implantation, the other 4 animals were euthanized, and the clot was collected from the surface of the specimens for total RNA extraction using Trizol (Invitrogen). Subsequently, total RNA was qualified and quantified using a NanoDrop and Agilent 2100 bioanalyzer (Thermo Fisher Scientific, MA, USA). For the Lnc-RNA profile detection, ribosomal RNA (rRNA) was removed using target-specific oligos and RNase H reagents were used to deplete both cytoplasmic (5S rRNA, 5.8S rRNA, 18S rRNA, and 28S rRNA) and mitochondrial ribosomal RNA (12S rRNA and 16S rRNA) from total RNA preparations. Following SPRI beads purification, the RNA was fragmented into small pieces using divalent cations under elevated temperature. The cleaved RNA fragments were copied into the first-strand cDNA using reverse transcriptase and random primers, followed by second-strand cDNA synthesis using DNA Polymerase I and RNase H. This process would remove the RNA template and synthesizes a replacement strand, incorporating dUTP in place of dTTP to generate ds cDNA. These cDNA fragments then had the addition of a single 'A' base and subsequent ligation of the adapter. After UDG treatment, the incorporation of dUTP quenched the second strand during amplification. The products were enriched with PCR to create the final cDNA library. The libraries were assessed quality and quantity using two methods: check the distribution of the size of the fragments using the Agilent 2100 bioanalyzer, and quantify the library using real-time quantitative PCR (QPCR) (TaqMan Probe, St. Louis, MO, USA). The qualified libraries were sequenced pair end on the BGISEQ-500 System. The sequencing platform of BGI-500 (BGI, Shenzhen, China) was used to obtain the LncRNA gene expression profiles. Quality control checks were performed to confirm sequencing saturation and gene mapping distribution. Fragments per Kilobase of transcript per Million mapped reads (FPKM) value were used to express relative gene abundance. Different LncRNAs in comparison with the nanotubes with a diameter of 15 nm were analyzed. RNAplex was used to reveal the potential targeted mRNAs of the LncRNAs. The target mRNAs were then subjected to enrichment analysis of GO functions and KEGG pathways.

### 2.6. Statistical Analysis

The quantitative data were displayed as means $\pm$ standard deviation (SD). The data were statistically analyzed using the software SPSS. Statistical analysis was performed using one-way ANOVA methodology. Significance and high significance were indicated at $p$ values < 0.05 and 0.01, respectively.

## 3. Results and Discussion

The surface morphology of P-Ti and TNAs is shown in Figure 1a. Highly ordered TNAs with distinct diameters are obtained after one-step anodization. The average diameter of TNAs is 15, 60, and 120 nm, thus are denoted as TNA 15, TNA 60, and TNA 120 respectively.

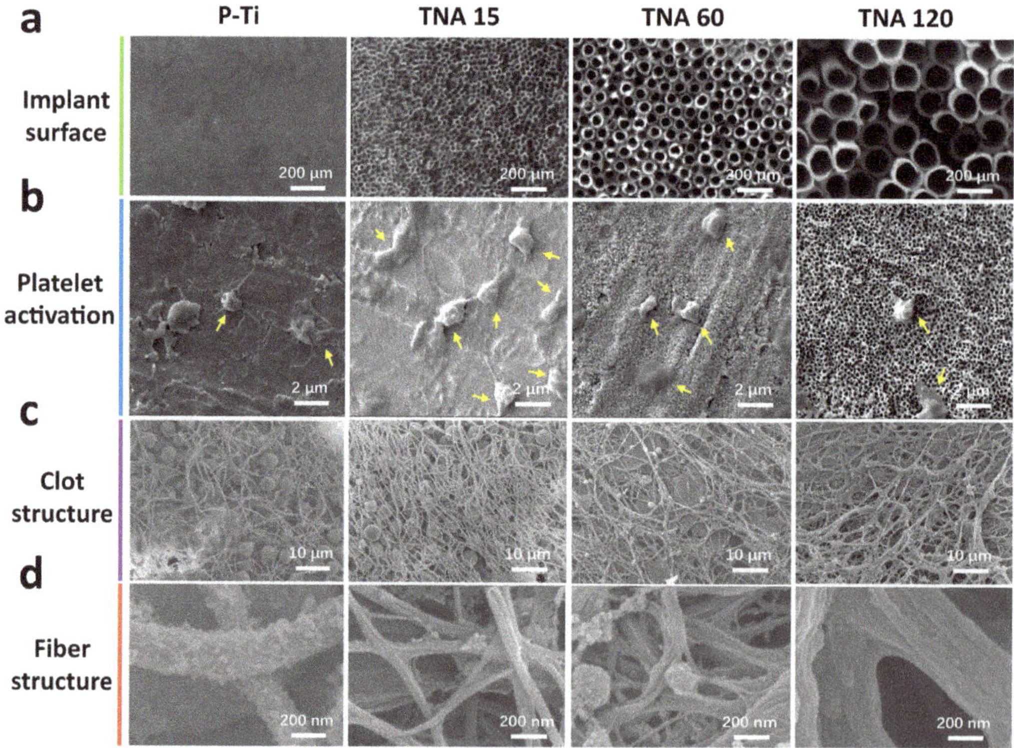


**Figure 1.** (**a**) Scanning electron microscope (SEM) images of the surface of P-Ti and TNAs. (**b**) Platelet activation (yellow arrows) on the surfaces. (**c,d**) Clot and fiber structures on the surfaces.

Figure 1b displays the platelet morphology adhered to the surfaces of the implants after 30 min of incubation. TNA 15 attracts more platelet adhesion and enables a significant activation morphology manifested by a huge extension area with abundant lamellipodia and filopodia in comparison with that on other groups.

After implanted for 3 days, the clot morphology on the surfaces of the implants is shown in Figure 1c and detailed information is displayed in Figure 1d. Clots with a much more compact and thinner fiber network formed on TNA 15, while a structure with larger pores with thicker fibers was observed on other groups, especially on TNA 120.

Figure 2 discloses the detailed information of the in vivo clot features on TNA 15 after 24 h of implantation. Abundant activated immune cells can be observed within the clot (Figure 2a); meanwhile, numerous platelets with a huge degree of activation are also seen near the immune cells (Figure 2b). These results are consistent with the aforementioned in vitro outcome that TNA 15 enables a significant activation of immune cells and platelet activity [8].

## Immune cells in clot of TNA 15

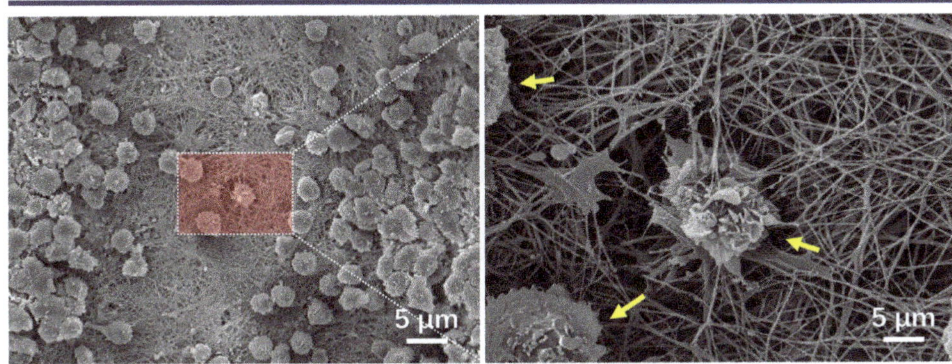


## Platelets in clot of TNA 15

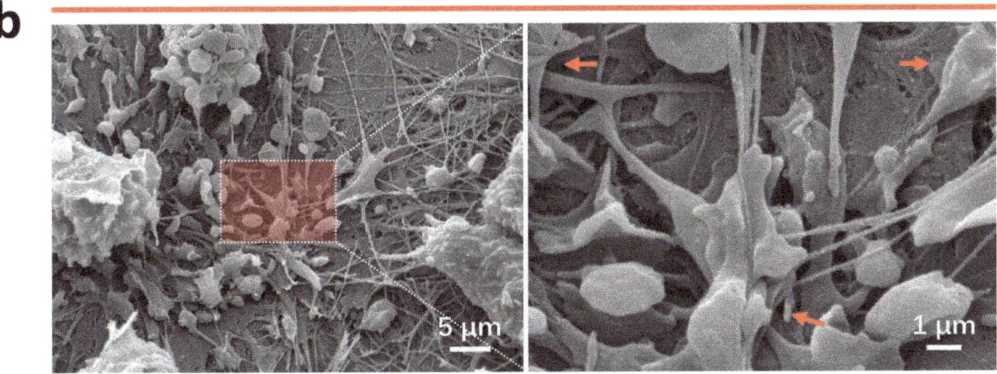


**Figure 2.** (**a**) SEM images of the immune cell (yellow arrows) morphology within the clot on the specimens. (**b**) SEM images of the platelet activations (red arrows) within the clot on the specimens. The right image is magnified from the left one.

In vivo osseointegration results of the modified implant are shown in Figure 3. Figure 3a displays the representative images of the bone-to-implant interface and peri-implant bone tissue stained with the toluidine blue. Similarly, significantly elevated BIC can be observed near the surface of TNA 15 from the images and further verified via the quantitative results in Figure 3b while compared with that of other groups [8]. Accordingly, TNA 15 was chosen as a reference group in the following study to investigate the regulation effect of the LncRNA profile within the clot.

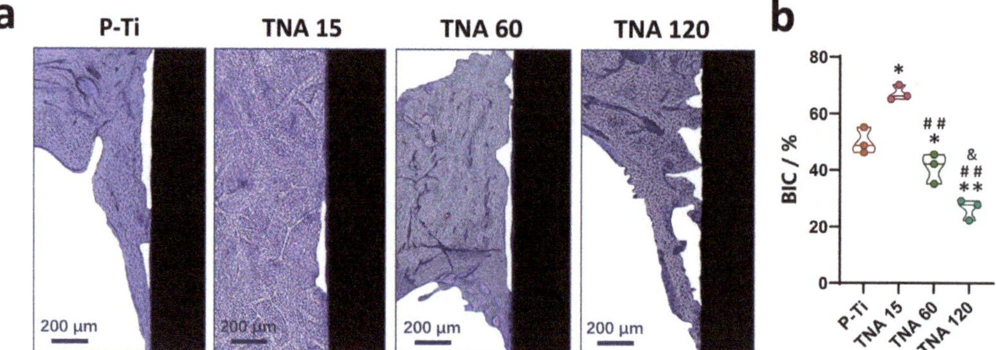

**Figure 3.** (**a**) Representative histological images of peri-implant bone tissue. (**b**) Quantitative results of BIC. * $p < 0.05$ compared to P-Ti, & $p < 0.05$ compared to TNA 60, ** $p < 0.01$ compared to P-Ti, ## $p < 0.01$ compared to TNA 15.

Figure 4 depicts the different LncRNAs expression profiles among groups. A total of 508 LncRNAs are significantly upregulated and 61 LncRNAs are downregulated of TNA 15 when compared with P-Ti (Figure 4a,b). 250 LncRNAs are significantly upregulated and 28 LncRNAs are downregulated of TNA 60 when compared with TNA 15 (Figure 4c,d). Similarly, 92 LncRNAs are significantly upregulated and 205 LncRNAs are downregulated of TNA 120 when compared with TNA 15 (Figure 4e,f).

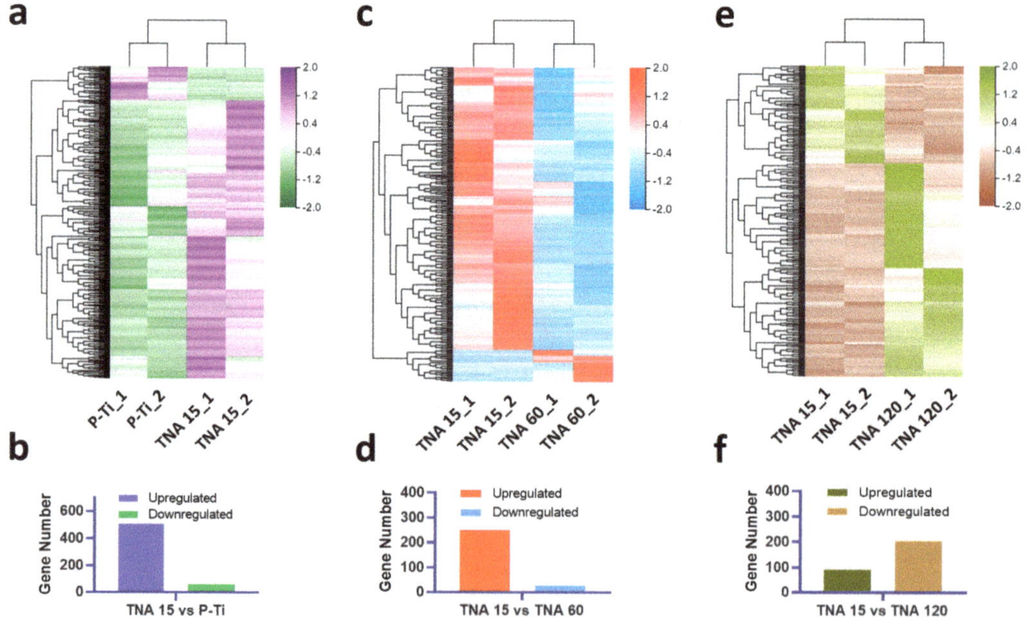

**Figure 4.** Visualization of different LncRNA expression profiles with heatmap. (**a**) P-Ti vs. TNA 15. (**b**) TNA 60 vs. TNA 15. (**c**) TNA 120 vs. TNA 15. Quantitative results of the LncRNA expression profiles. (**d**) P-Ti vs. TNA 15. (**e**) TNA 60 vs. TNA 15. (**f**) TNA 120 vs. TNA 15.

Figure 5 indicates key LncRNAs involved in the regulation of LncRNA-mRNA among groups. LOC103346307, LOC108175175, and LOC108176660 targeted most of the upregulated mRNAs in TNA 15 compared with P-Ti, TNA 60, and TNA 120, respectively, whereas LOC103352121, LOC103348180, and LOC108176465 targeted most of the downregulated mRNAs.

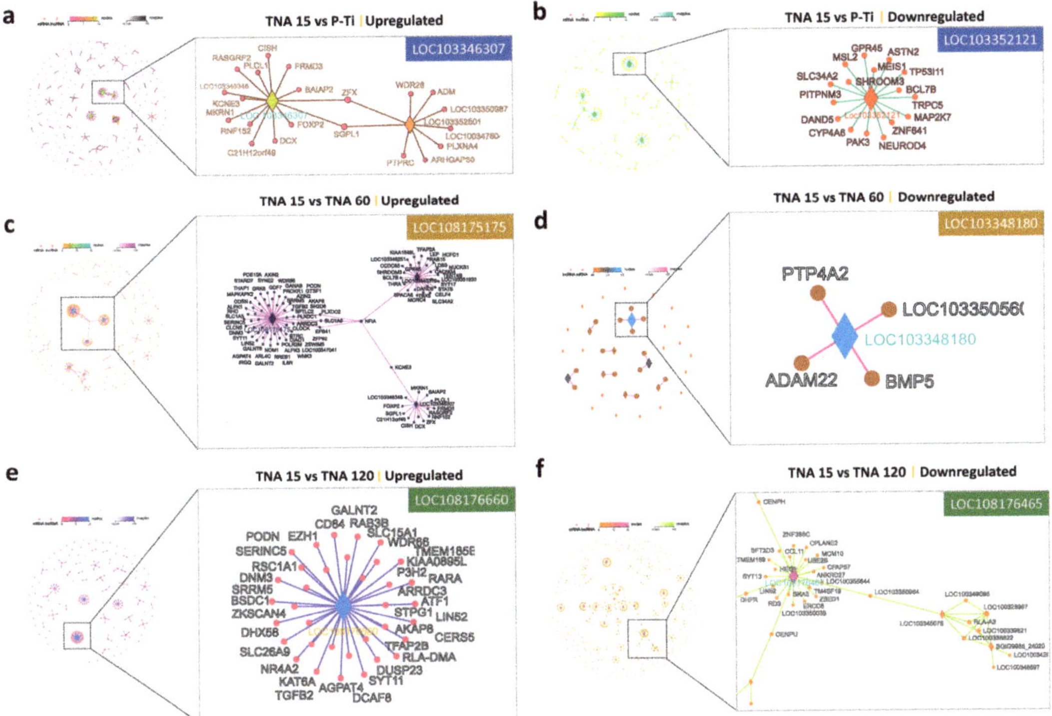

**Figure 5.** (**a**) Key LncRNA identified within the upregulated profile in TNA 15 vs. P-Ti. (**b**) Key LncRNA identified within the downregulated profile in TNA 15 vs. P-Ti. (**c**) Key LncRNA identified within the upregulated profile in TNA 15 vs. TNA 60. (**d**) Key LncRNA identified within the downregulated profile in TNA 15 vs. TNA 60. (**e**) Key LncRNA identified within the upregulated profile in TNA 15 vs. TNA 120. (**f**) Key LncRNA identified within the downregulated profile in TNA 15 vs. TNA 120.

LncRNAs targeted mRNAs were introduced to KEGG pathway enrichment analysis (Figure 6) and the detailed gene expression in the pathways is shown in Figure 7. Figure 6a,b indicate that upregulated mRNAs in group TNA 15 vs. P-Ti were significantly enriched in growth metabolism-related signaling pathways such as the Focal adhesion, PI3K-Akt signaling pathway, Regulation of actin cytoskeleton, cAMP signaling pathway, Jak-STAT signaling pathway, and platelet activation, while the downregulated mRNAs are significantly enriched in inflammation-related signaling pathways such as the Chemokine signaling pathway, TNF signaling pathway, Toll-like receptor signaling pathway, C-type lectin receptor signaling pathway, T cell receptor signaling pathway, and IL-17 signaling pathway. Figure 6c,d indicate that upregulated mRNAs in group TNA 15 vs. TNA 60 were significantly enriched in growth metabolism-related signaling pathways such as the Wnt signaling pathway, Hippo signaling pathway, PI3K-Akt signaling pathway, Signaling pathways regulating pluripotency of stem cells, cAMP signaling pathway, and Jak-STAT signaling pathway, while the downregulated mRNAs were significantly enriched in inflammation-related signaling pathways such as the mTOR signaling pathway, p53 signaling pathway, Apoptosis, Fc gamma R-mediated phagocytosis, Toll-like receptor signaling pathway, and B cell receptor signaling pathway. Figure 6e,f indicate that upregulated mRNAs in group TNA 15 vs. TNA 60 were significantly enriched in growth metabolism-related signaling pathways such as the Wnt signaling pathway, Adherens junction, TGF-beta signaling pathway, Signaling pathways regulating pluripotency of stem cells, Hippo signaling pathway, and MAPK signaling pathway, while the downregulated mRNAs were significantly enriched in inflammation-related signaling pathways such as the Chemokine signaling pathway, Fc gamma R-mediated phagocytosis, T cell receptor signaling pathway, Cellular senescence, and Inflammatory mediator regulation of TRP.

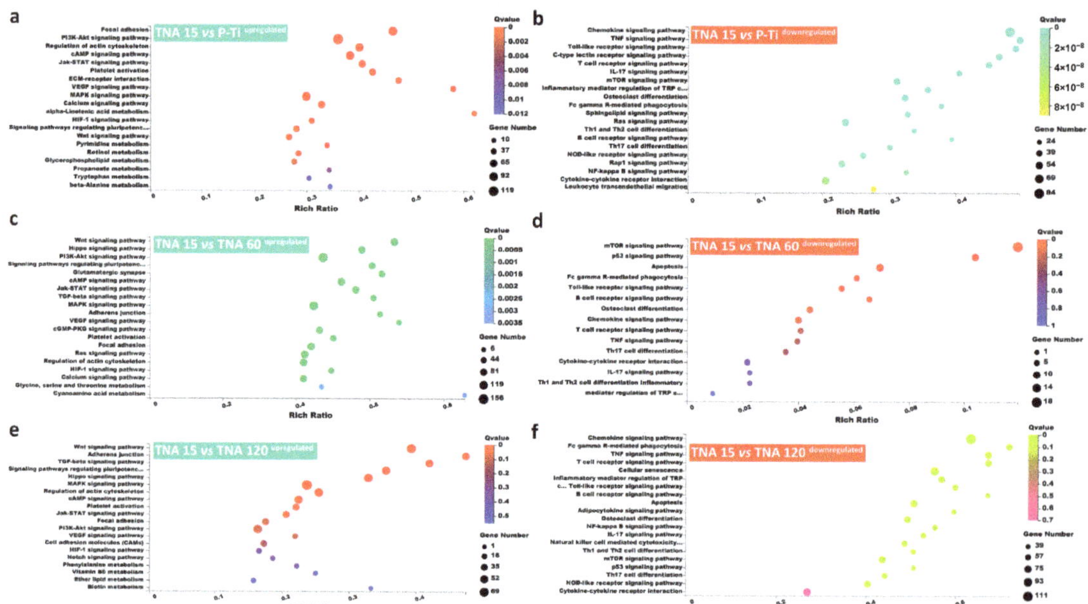

**Figure 6.** (**a**) KEGG pathway enrichment analysis of the upregulated LncRNAs targeted mRNAs within the comparison TNA 15 vs. P-Ti. (**b**) KEGG pathway enrichment analysis of the downregulated LncRNAs targeted mRNAs within the comparison TNA 15 vs. P-Ti. (**c**) KEGG pathway enrichment analysis of the upregulated LncRNAs targeted mRNAs within the comparison TNA 60 vs. P-Ti. (**d**) KEGG pathway enrichment analysis of the downregulated LncRNAs targeted mRNAs within the comparison TNA 60 vs. P-Ti. (**e**) KEGG pathway enrichment analysis of the upregulated LncRNAs targeted mRNAs within the comparison TNA 120 vs. P-Ti. (**f**) KEGG pathway enrichment analysis of the downregulated LncRNAs targeted mRNAs within the comparison TNA 120 vs. P-Ti.

Notably, WNT, TGF-beta, and VEGF signaling pathways had more synchronized appearances in TNA groups when compared with others. During osseointegration, Wnt signaling, which plays a pivotal role in MSC lineage commitment and progression, was implicated in proximal-distal outgrowth and dorsoventral limb patterning and, later, in osteogenesis mostly through the canonical pathway but also involving noncanonical elements [15]. Inhibition of WNT signaling resulted in low bone mass in osteoporosis-pseudoglioma syndrome [15]. Additionally, a total of 23 genes were significantly enhanced when compared with P-Ti. TGF-beta signaling is another crucial pathway evidenced to regulate bone mass and quality and loss of TGF-beta signaling also reduce bone matrix mineralization [16]. Besides the enhancement of de novo bone formation, angiogenesis at the bone-implant interface is recognized as a prerequisite for a fulfilled osseointegration [17]. VEGF signaling is capable of activating eNOS that is responsible for the production of vascular nitric oxide (NO), which consequently contributes to neovascularization [18].

Similarly, downregulation of osteoclast differentiation, TNF, and NF-kappa signaling pathways was identified in TNA groups when compared with others.

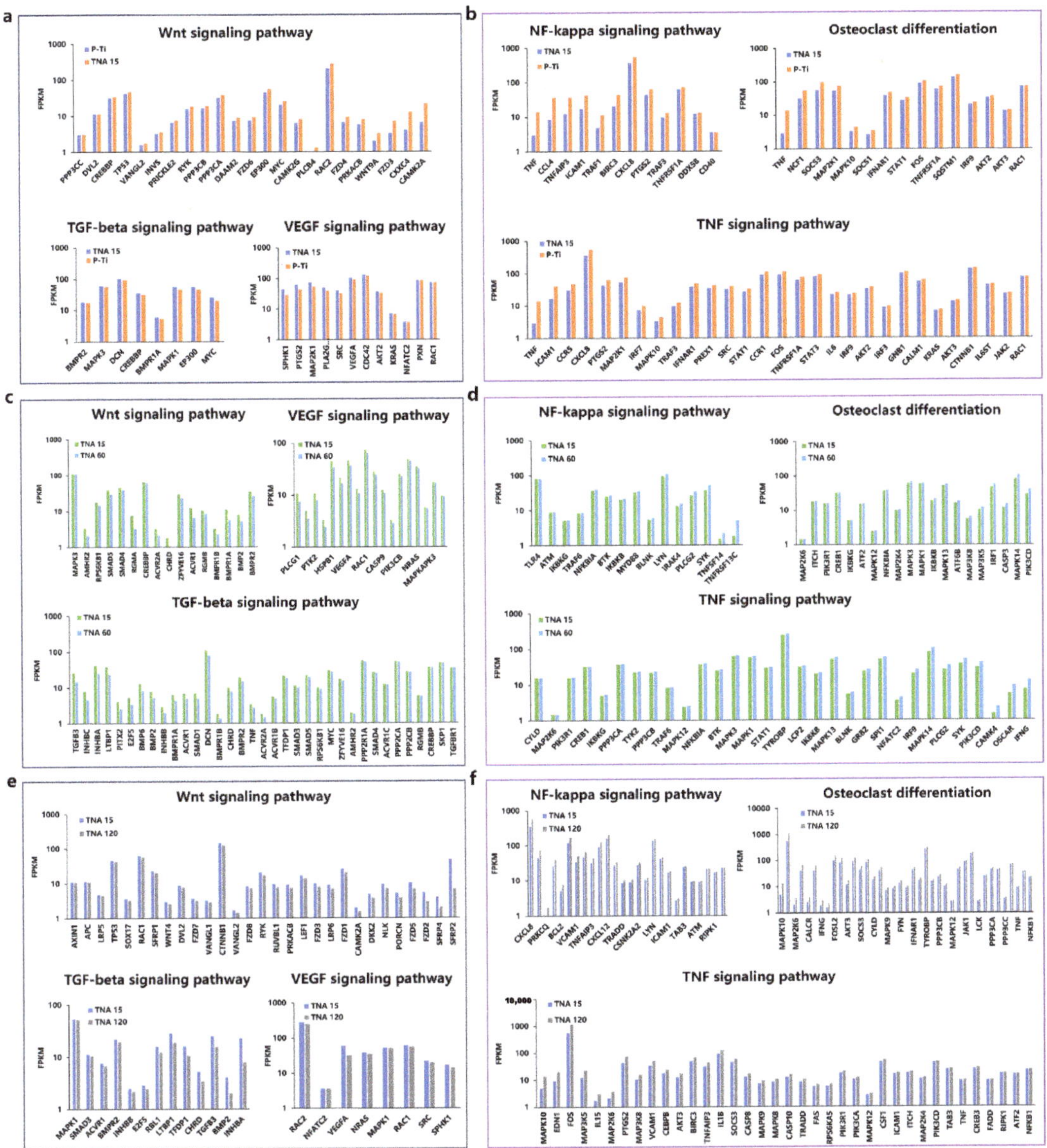

**Figure 7.** (**a**) Upregulated genes in the Wnt, TGF-beta, and VEGF signaling pathways within the comparison TNA 15 vs. P-Ti. (**b**) Downregulated genes in the osteoclast differentiation, TNF, NF-kappa signaling pathways within the comparison TNA 15 vs. P-Ti. (**c**) Upregulated genes in the WNT, TGF-beta, and VEGF signaling pathways within the comparison TNA 15 vs. TNA 60. (**d**) Downregulated genes in the osteoclast differentiation, TNF, NF-kappa signaling pathways within the comparison TNA 15 vs. TNA 60. (**e**) Upregulated genes in the Wnt, TGF-beta, and VEGF signaling pathways within the comparison TNA 15 vs. TNA 120. (**f**) Downregulated genes in the osteoclast differentiation, TNF, and NF-kappa signaling pathways within the comparison TNA 15 vs. TNA 120.

Osteoclasts are bone-resorbing multinucleated cells derived from hematopoietic precursors that are formed in the bone marrow through osteoclast differentiation of macrophages [19]. Osteoclasts break down de novo bone formation by secreting proteases, a process is known as bone resorption via degrading type I collagen and promote osteoclastogenesis [20]. The TNF signaling pathway functions in vivo to increase osteoclast precursors, as well as indirectly increasing osteoclastogenesis through augmentation of RANK expression on osteoclast precursors, plays a major role in promoting osteoclastogenesis and bone resorption [21]. The NF kappa signaling pathway is a family of transcription factors that have been comprehensively studied in the promotion of osteoclasts function [22]. Increasing evidence suggested that the pathway activation decreases osteogenic differentiation and suppresses de novo bone formation [23]. Additionally, the NF kappa signaling pathway plays an important regulatory role in the developmental process of inflammation in combination with its downstream inflammatory cytokines interleukin-6 (IL-6), interleukin-1β (IL-1β), and TNF-α, while suppression of inflammation is beneficial to osseointegration [24]. Accordingly, the LncRNAs identified herein contribute to the fulfilled osseointegration of TNA 15 through activation of Wnt, TGF-beta, and VEGF signaling pathways and the suppression of osteoclast differentiation, TNF, and NF-kappa signaling pathways.

In our previous study, TNA 15 was shown to be highly promising for enhancing osseointegration through manipulating a fulfilled osteoimmune microenvironment by using the specialized thinner, porous blood clot fibrous network and the releasing of growth factors (PDGF-AB, TGF-beta) [8]. Herein, we depicted the results from a different viewpoint and deepened the current understanding of the osseointegration (Figure 8). Firstly, the study unprecedently indicated that TNA 15 can manipulate the expressions of LncRNAs within the clot, and thus highlighted that single nano-surfaces can significantly regulate LncRNAs profiles. Then, it was found that among hundreds of identified LncRNAs in each group, there are always several key LncRNAs that play a pivotal role in osteoimmunomodulation through regulation of multiple signaling pathways with their targeting mRNAs. Moreover, the study advanced the current comprehension of LncRNAs which have a powerful function in regulating de novo bone formation, and thus, provided a new paradigm for surface modification of the implantable materials to enhance osseointegration. The further study shall pay attention to deepen the understanding of the LncRNA modulated osteoimmune response during bone formation. Additionally, an in vitro study using larger size specimens is of interest to verify the in vivo effect of the blood clot-derived LncRNAs.

**Figure 8.** Illustration of the correlation between LncRNAs in the Blood Clot formed on Nano-Scaled Implant Surfaces and Osseointegration.

## 4. Conclusions

We demonstrated that distinct nano-surfaces are capable of regulating LncRNAs expression within the clot, and 6 key LncRNAs were identified in each comparison exerting a pivotal role in manipulating the expression of the targeted mRNA. In the TNA15 group, the LncRNAs targeted mRNAs subsequently manipulate a favorable osteogenesis microenvironment through upregulation of Wnt, TGF-beta, and VEGF signaling pathways and suppression of osteoclast differentiation, TNF, and NF-kappa signaling pathways, which resulted in promoted osseointegration. The findings firstly indicated the link of nano-surfaces on the LncRNAs expression in the blood clot and demonstrated that LncRNAs may strongly impact the osseointegration through the targeted mRNAs.

**Author Contributions:** Conceptualization, L.B. and P.C.; methodology, L.B. and P.C.; software, L.B. and P.C.; validation, L.B.; formal analysis, L.B. and P.C.; investigation, L.B.; writing—original draft preparation, L.B. and P.C.; writing—review and editing, B.T., R.H. and Y.X.; visualization, L.B.; supervision, R.H. and Y.X.; project administration, R.H. and Y.X.; funding acquisition, L.B., R.H. and Y.X. All authors have read and agreed to the published version of the manuscript.

**Funding:** This work is supported by the Fund for Shanxi "1331 Project" Key Innovative Research Team (PY201809), the National Natural Science Foundation of China (81901898), the Prince Charles Hospital Research Foundation, ITI Research Grant 1260, and Australia-China Centre for Tissue Engineering and Regenerative Medicine at QUT. The first author appreciates the financial support from the China Scholarship Council (CSC, 201606930017).

**Institutional Review Board Statement:** The study was approved by the Animal Research Committee of Taiyuan University of Technology (TYUT202001003) and the University Human Research Ethics Committee at the Queensland University of Technology (1500000918).

**Informed Consent Statement:** Not applicable.

**Data Availability Statement:** The data presented in this study are available on request from the corresponding author.

**Conflicts of Interest:** The authors declare no conflict of interest.

## References

1.  Guglielmotti, M.B.; Olmedo, D.G.; Cabrini, R.L. Research on implants and osseointegration. *Periodontology 2000* **2019**, *79*, 178–189. [CrossRef]
2.  Apostu, D.; Lucaciu, O.; Berce, C.; Lucaciu, D.; Cosma, D. Current methods of preventing aseptic loosening and improving osseointegration of titanium implants in cementless total hip arthroplasty: A review. *J. Int. Med. Res.* **2018**, *46*, 2104–2119. [CrossRef]
3.  Agarwal, R.; García, A.J. Biomaterial strategies for engineering implants for enhanced osseointegration and bone repair. *Adv. Drug Delivery Rev.* **2015**, *94*, 53–62. [CrossRef]
4.  Bosshardt, D.D.; Chappuis, V.; Buser, D. Osseointegration of titanium, titanium alloy and zirconia dental implants: Current knowledge and open questions. *Periodontology 2000* **2017**, *73*, 22–40. [CrossRef]
5.  Souza, J.C.M.; Sordi, M.B.; Kanazawa, M.; Ravindran, S.; Henriques, B.; Silva, F.S.; Aparicio, C.; Cooper, L.F. Nano-scale modification of titanium implant surfaces to enhance osseointegration. *Acta Biomater.* **2019**, *94*, 112–131. [CrossRef] [PubMed]
6.  Bai, L.; Hang, R.; Gao, A.; Zhang, X.; Huang, X.; Wang, Y.; Tang, B.; Zhao, L.; Chu, P.K. Nanostructured titanium–silver coatings with good antibacterial activity and cytocompatibility fabricated by one-step magnetron sputtering. *Appl. Surf. Sci.* **2015**, *355*, 32–44. [CrossRef]
7.  Shahneh, F.; Alexandra, G.; Klein, M.; Frauhammer, F.; Bopp, T.; Schäfer, K.; Raker, V.; Becker, C. Specialized regulatory T cells control venous blood clot resolution through SPARC. *Blood* **2020**. [CrossRef] [PubMed]
8.  Bai, L.; Zhao, Y.; Chen, P.; Zhang, X.; Huang, X.; Du, Z.; Crawford, R.; Yao, X.; Tang, B.; Hang, R.; et al. Targeting Early Healing Phase with Titania Nanotube Arrays on Tunable Diameters to Accelerate Bone Regeneration and Osseointegration. *Small* **2020**. [CrossRef]
9.  Wu, Z.; Liu, X.; Liu, L.; Deng, H.; Zhang, J.; Xu, Q.; Cen, B.; Ji, A. Regulation of lncRNA expression. *Cell. Mol. Biol. Lett.* **2014**, *19*, 561–575. [CrossRef]
10. Wapinski, O.; Chang, H.Y. Long noncoding RNAs and human disease. *Trends Cell Biol.* **2011**, *21*, 354–361. [CrossRef]
11. Fatica, A.; Bozzoni, I. Long non-coding RNAs: New players in cell differentiation and development. *Nat. Rev. Genet.* **2014**, *15*, 7–21. [CrossRef]
12. Young, R.S.; Ponting, C.P. Identification and function of long non-coding RNAs. *Essays Biochem.* **2013**, *54*, 113–126. [CrossRef]

13. Bai, L.; Wu, R.; Wang, Y.; Wang, X.; Zhang, X.; Huang, X.; Qin, L.; Hang, R.; Zhao, L.; Tang, B. Osteogenic and angiogenic activities of silicon-incorporated $TiO_2$ nanotube arrays. *J. Mater. Chem. B* **2016**, *4*, 5548–5559. [CrossRef]

14. Bai, L.; Yang, Y.; Mendhi, J.; Du, Z.; Hao, R.; Hang, R.; Yao, X.; Huang, N.; Tang, B.; Xiao, Y. The effects of TiO2 nanotube arrays with different diameters on macrophage/endothelial cell response and ex vivo hemocompatibility. *J. Mater. Chem. B* **2018**, *6*, 6322–6333. [CrossRef] [PubMed]

15. Baron, R.; Kneissel, M. WNT signaling in bone homeostasis and disease: From human mutations to treatments. *Nat. Med.* **2013**, *19*, 179–192. [CrossRef] [PubMed]

16. Dole, N.S.; Mazur, C.M.; Acevedo, C.; Lopez, J.P.; Monteiro, D.A.; Fowler, T.W.; Gludovatz, B.; Walsh, F.; Regan, J.N.; Messina, S.; et al. Osteocyte-Intrinsic TGF-β Signaling Regulates Bone Quality through Perilacunar/Canalicular Remodeling. *Cell Rep.* **2017**, *21*, 2585–2596. [CrossRef]

17. Langen, U.H.; Pitulescu, M.E.; Kim, J.M.; Enriquez-Gasca, R.; Sivaraj, K.K.; Kusumbe, A.P.; Singh, A.; Di Russo, J.; Bixel, M.G.; Zhou, B.; et al. Cell-matrix signals specify bone endothelial cells during developmental osteogenesis. *Nat. Cell Biol.* **2017**, *19*, 189–201. [CrossRef] [PubMed]

18. Ahmad, S.; Hewett, P.W.; Wang, P.; Al-Ani, B.; Cudmore, M.; Fujisawa, T.; Haigh, J.J.; le Noble, F.; Wang, L.; Mukhopadhyay, D.; et al. Direct evidence for endothelial vascular endothelial growth factor receptor-1 function in nitric oxide–mediated angiogenesis. *Circ. Res.* **2006**, *99*, 715–722. [CrossRef] [PubMed]

19. Park, J.H.; Lee, N.K.; Lee, S.Y. Current Understanding of RANK Signaling in Osteoclast Differentiation and Maturation. *Mol. Cells* **2017**, *40*, 706–713. [CrossRef]

20. Boyle, W.J.; Simonet, W.S.; Lacey, D.L. Osteoclast differentiation and activation. *Nature* **2003**, *423*, 337–342. [CrossRef]

21. Zhao, B. TNF and Bone Remodeling. *Curr. Osteoporos. Rep.* **2017**, *15*, 126–134. [CrossRef]

22. Chang, J.; Wang, Z.; Tang, E.; Fan, Z.; McCauley, L.; Franceschi, R.; Guan, K.; Krebsbach, P.H.; Wang, C.-Y. Inhibition of osteoblastic bone formation by nuclear factor-κB. *Nat. Med.* **2009**, *15*, 682–689. [CrossRef] [PubMed]

23. Krum, S.A.; Chang, J.; Miranda-Carboni, G.; Wang, C.-Y. Novel functions for NFκB: Inhibition of bone formation. *Nat. Rev. Rheumatol.* **2010**, *6*, 607–611. [CrossRef] [PubMed]

24. Bai, L.; Du, Z.; Du, J.; Yao, W.; Zhang, J.; Weng, Z.; Liu, S.; Zhao, Y.; Liu, Y.; Zhang, X.; et al. A multifaceted coating on titanium dictates osteoimmunomodulation and osteo/angio-genesis towards ameliorative osseointegration. *Biomaterials* **2018**, *162*, 154–169. [CrossRef] [PubMed]

 *nanomaterials*

*Communication*

# *Towards Clinical Translation*: Optimized Fabrication of Controlled Nanostructures on Implant-Relevant Curved Zirconium Surfaces

**Divya Chopra, Karan Gulati * and Sašo Ivanovski ***

The University of Queensland, School of Dentistry, Herston, QLD 4006, Australia; d.chopra@uq.net.au
* Correspondence: k.gulati@uq.edu.au (K.G.); s.ivanovski@uq.edu.au (S.I.)

**Abstract:** Anodization enables fabrication of controlled nanotopographies on Ti implants to offer tailorable bioactivity and local therapy. However, anodization of Zr implants to fabricate $ZrO_2$ nanostructures remains underexplored and are limited to the modification of easy-to-manage flat Zr foils, which do not represent the shape of clinically used implants. In this pioneering study, we report extensive optimization of various nanostructures on implant-relevant micro-rough Zr curved surfaces, bringing this technology closer to clinical translation. Further, we explore the use of sonication to remove the top nanoporous layer to reveal the underlying nanotubes. Nano-engineered Zr surfaces can be applied towards enhancing the bioactivity and therapeutic potential of conventional Zr-based implants.

**Keywords:** zirconium; zirconia; dental implants; nanopores; electrochemical anodization

**Citation:** Chopra, D.; Gulati, K.; Ivanovski, S. *Towards Clinical Translation*: Optimized Fabrication of Controlled Nanostructures on Implant-Relevant Curved Zirconium Surfaces. *Nanomaterials* **2021**, *11*, 868. https://doi.org/10.3390/nano11040868

Academic Editor: May Lei Mei

Received: 25 February 2021
Accepted: 25 March 2021
Published: 29 March 2021

**Publisher's Note:** MDPI stays neutral with regard to jurisdictional claims in published maps and institutional affiliations.



## 1. Introduction

Zirconium (Zr) is a valve metal that is very stable with a high dielectric constant, and hence it is a suitable material choice for the nuclear and microelectronic industries [1]. Further, Zr and its alloys are extensively used in the field of optics, magnetics, chemical sensors, and biomedical implants [2]. Due to their favourable characteristics (physical, chemical, and biological), Zr-based implants are gaining popularity in the dental and orthopaedic markets [3,4]. For this application, the favourable biocompatibility of Zr is mainly attributed to its surface oxide film ($ZrO_2$). Further, Zr (metal, grey colour) with a $ZrO_2$ (ceramic, white colour) surface has a toughness comparable to metals, and hence is suitable for a variety of biomedical applications [5]. Clinically, ceramic structures have shown a higher risk of fracture due to the nature of the material. However, oxidized Zr surfaces offer the potential to decrease wear and tear as the bulk of the material is metal, and not a monolithic ceramic [6]. It is noteworthy that oxidized Zr is not a ceramic but the transition of metal to ceramic. Studies have established that $ZrO_2$ not only promotes osseointegration but also demonstrates reduced cytotoxicity as compared to Ti-based implants [2]. Moreover, $ZrO_2/Zr$ presents greater mechanical strength and low ion release when compared to Ti [7]. Overall, as compared to Ti, $ZrO_2$ based implants offers many advantages including superior aesthetics, with favourable biological, mechanical, and optical properties [8].

In the last few decades, the potential of Zr and its alloys in the field of dental implants has gained increasing attention [9–11]. Although there are some in vivo studies that demonstrate the biocompatibility of Zr, the surface modification and related bioactivity assessment of Zr-based implants needs in-depth investigation [2]. It is noteworthy that in compromised patient conditions (e.g., diabetic and osteoporotic), 'normal' bioactivity may not be sufficient to encourage bone-implant integration, and hence enhanced bioactivity is needed. In that light, surface modifications of Zr-based implants to form an oxide layer have been performed via various physical, chemical, and electrochemical means [12–14].

Further, electrochemical anodization (EA) has been regarded as an effective strategy to fabricate $ZrO_2$ with nanoscale surface roughness and the ability to incorporate bioactive ions (Ca or P) [15].

It is well established that the bioactivity of modified implant surfaces follows the trend nano- > micro- > macro-scale [16]. As a result, research has shifted towards the fabrication of controlled nanotopographies on Zr-based implants (including Zr, Ti–Zr alloys etc.). Various strategies have been employed for nano-engineering Zr implants, such as electrochemical anodization (EA) [17], plasma treatment [18], micro-arc oxidation [19], hydrothermal treatment [20], chemical co-precipitation, and sol–gel method [21]. Among these, EA stands out due to its cost-effectiveness, scalability, and control over the characteristics of the fabricated nanostructures [22]. Briefly, EA involves immersion of a target substrate (Zr) as an anode and a counter electrode (cathode) in a suitable electrolyte (containing water and fluoride ions), and a supply of constant voltage/current. Upon attainment of optimized conditions, self-ordering of $ZrO_2$ nanotubes occurs on the surface of the anode. Relevant to biomedical applications, EA to fabricate self-ordered $ZrO_2$ nanotubes has gained attention, with various attempts made to optimize the EA fabrication [23–25]. Further, the augmented bioactivity and osteogenic ability of $ZrO_2$ nanotubes has also been demonstrated [26–28].

With respect to anodized nano-engineered zirconium implants, key fabrication challenges remain unaddressed:

1. Fabrication optimization has only been restricted to planar Zr flat foil that is easy to manage. However, clinically used orthopaedic and dental implants are based on curved surfaces and edges, thereby limiting the clinical translation of conventional anodized Zr flat foil.
2. Dental implants generally use microscale roughness which, to date, is regarded as a 'gold standard' for ensuring osseointegration. Thus, preserving rather than removal of this micro-roughness (which is routinely performed to fabricate nanotubes) is needed along with superimposition of nanostructures (dual micro–nano).

To further optimize the fabrication of anodic nanostructures on Zr-based implants, in this study, we explore EA optimization of Zr wires as models for curved clinically relevant implant architectures. Briefly, EA parameters, including voltage and time, were varied to fabricate oxide nanocrystals, nanopores, and nanotubes on the Zr wires (Figure 1). This study bridges the gap between the fabrication of controlled nanostructures on clinically relevant Zr surfaces, with the objective of facilitating future clinical translation. We also report on the use of sonication to reveal the underlying nanostructures by removing the superficial nanoporous oxide film. Optimized fabrication of controlled nanotopographies on implant substrates that preserves the underlying micro-roughness can be paradigm shifting in the domain of Zr-based biomedical applications.

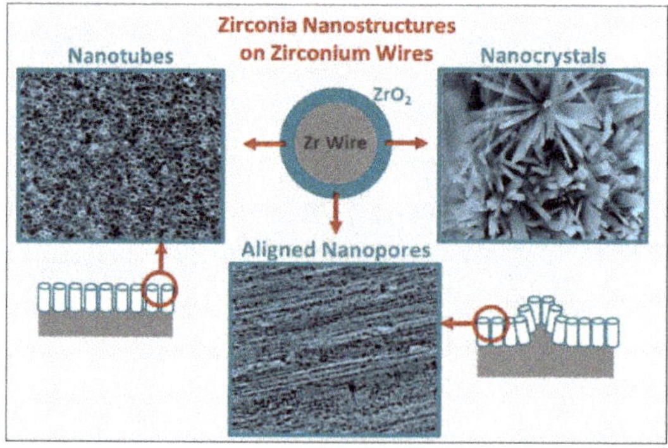


**Figure 1.** Schematic representation of various nanostructures fabricated on clinical implant relevant zirconium wire.

## 2. Experimental Section

### 2.1. Materials and Chemicals

Zirconium wire with 0.5 mm diameter [annealed, 99.2% purity (metal basis excluding Hf), 4.5% Hf max] was obtained from Alfa Aesar (Lancashire, UK) and used as received. High-purity $(NH_4)_2SO_4$, $NH_4F$, and methanol were purchased from Sigma Aldrich (North Ryde, Australia).

### 2.2. Electrochemical Anodization (EA)

Prior to EA, as-received Zr wires were cut into 10 cm lengths and sonicated in ethanol to remove any surface contaminants. EA was carried out in a custom-designed two-electrode electrochemical cell at room temperature using a DC power source (Keithley, Cleveland, OH, USA) with the current precisely monitored [29,30]. EA was performed using as-received Zr wire as the anode (5 mm exposed in the electrolyte) and non-targeted Zr wire as a cathode in an electrolyte with 1 M $(NH_4)_2SO_4$ + 0.5 wt% $NH_4F$. Anodization was performed at 20–100 V for 10–120 min, with current vs. time precisely recorded (Power Supply App, Keithley KickStart Software, Solon, OH, USA). Anodization voltage and time was decided based on current literature and prior optimizations studies using Ti wires [29]. Briefly, current density was calculated (current/area of anode) and plotted against time to visualize key features identifying anodization [29]. To remove the anodic oxide layer, anodized samples were sonicated in methanol for various time intervals to reveal the underlying features.

### 2.3. Surface Characterization

Surface topography characterization of the nanostructures was performed using scanning electron microscopy (JSM 7001F, JEOL, Tokyo, Japan). Before imaging, samples were mounted on an SEM holder using double-sided conductive tape and coated with a 5 nm thick layer of platinum. Images with a range of scan sizes at normal incidence and a 30° angle were acquired from the top surfaces.

## 3. Results and Discussion

Figure S1 (Supplementary Information) shows the SEM image of as-received Zr wire with clearly visible micro-machined features (micro-rough). There is an obvious resemblance to conventional dental implants/abutments with respect to the micro-scale features, which for dental implants, ensures osseointegration. This micro-roughness is regarded as the 'gold-standard' in dentistry and, hence, its removal to fabricate nanostructures could prove detrimental [31]. We have previously demonstrated that dual micro–nano features with nanopores superimposed on micro-machined Ti can be fabricated using an optimized EA procedure [32]. Fabrication of controlled nanostructures with preserved underlying micro-features on Zr implants can result in a paradigm shift in achieving enhanced bioactivity from nano-engineering, without compromising the benefits obtained from micro-roughness. In that light, we optimized the anodization of Zr implants using Zr wire as a model for Zr dental/orthopaedic implants with curved surfaces and micro-machined lines.

Figure 2 shows low-magnification SEM images of the anodized wire, demonstrating an even coverage of the anodic $ZrO_2$ film, with clearly visible cracks. We have previously reported similar cracks on $TiO_2$ films formed on anodized Ti wire [32]. Briefly, these instabilities of the anodic layer could be attributed to the electric field concentrations at the topographical peaks of the substrate—which, in this case, is an irregular micro-rough curved surface [29,30]. The surface heterogeneity (micro-roughness) upon EA can also result in thicker oxide at the convex part and thinner oxide at the concave part [33]. It is noteworthy that these surface inconsistencies do not compromise the mechanical stability of the nano-engineered surface and can be used to accommodate drugs or enhance cellular adhesion [34]. Further, we have also explored strategies, including electrolyte ageing and surface polishing, to reduce anodic layer cracks [30].

In summary, these cracks or pits are unavoidable on the anodized curved substrates and are attributed to (Figure 3):

(1)  Curved substrate: radial/perpendicular growth of nanotubes outwards [29].
(2)  Internal stresses: due to uneven electric field distribution.
(3)  Mechanical stress: due to volume expansion and limited space for growth.
(4)  Weak spots: electrolyte penetration resulting in unstable/fragile anodic layers [35].
(5)  Substrate: micro-roughness further exacerbates the stresses/weak spots [36].
(6)  Nanotube collapse (or bundling): especially for longer tubes.

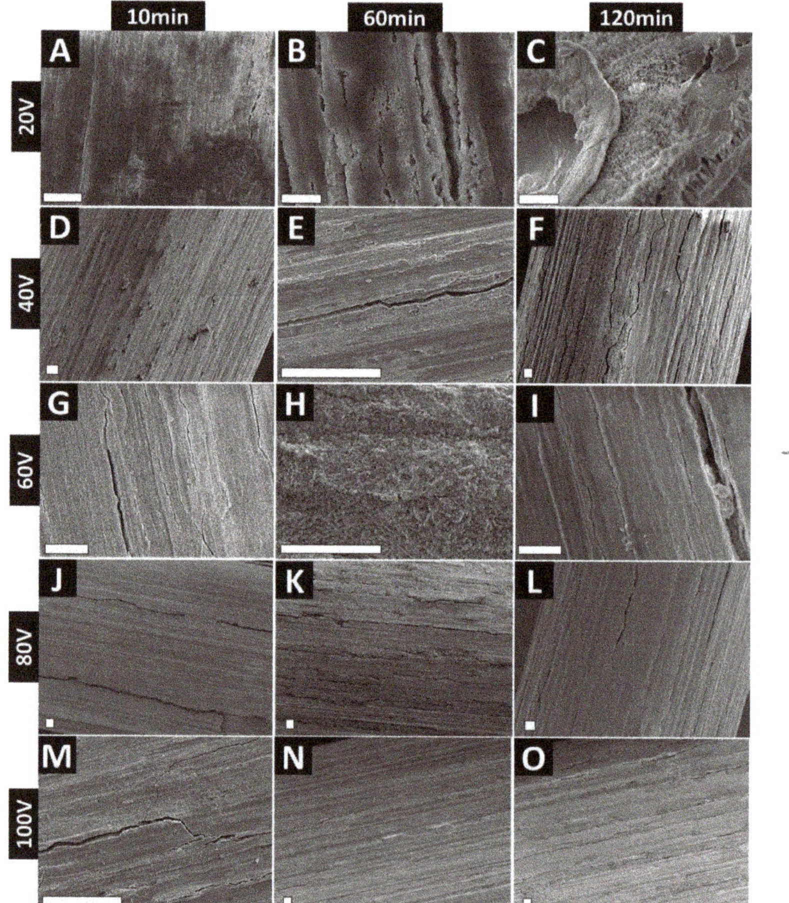


**Figure 2.** Top-view SEM images of anodized Zr wires at various voltages and times. (**A–C**) 20 V; (**D–F**) 40 V; (**G–I**) 60 V; (**J–L**) 80 V and (**M–O**) 100 V. Scale bars represent 20 μm.

Various strategies can be employed to reduce such cracks or instabilities on anodized metal surfaces (however, these remain poorly explored for Zr anodization) [37]:

(1)  Use of appropriately aged electrolyte—mostly applicable to anodization with organic electrolytes (like ethylene glycol) [38,39].
(2)  Polishing the substrate prior to anodization using mechanical, chemical or electropolishing treatments (will reduce/remove micro-roughness) [37].
(3)  Reducing water content, voltage/current, or anodization time (may reduce diameter/length of anodized nanostructures due to reduced growth rates).

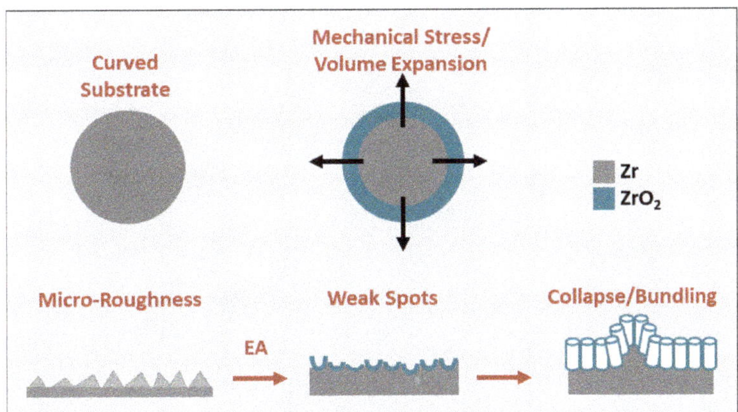



**Figure 3.** Representation of the formation of cracks or pits on the surface of anodized curved surfaces.

There have been attempts at exploring electrolyte ageing for Zr EA, with confirmation of the transition of nanoporous to nanotubular topography for EA performed in glycerol-based electrolyte [38,40]. It is noteworthy that such surface defects can also be reduced or minimized by electropolishing of the substrate, as shown elsewhere [29,41]. However, any polishing will 'consume' the underlying micro-roughness, removing this desirable feature of the implant and potentially compromising the positive osseointegrating property of the micro-roughness. Further, the cracks on the anodic film have been shown to survive drug loading and release in in vitro, ex vivo, and in vivo settings [34,42,43]. Indeed, cracks allow for higher drug loading amounts and enhance the overall surface roughness (at the microscale), allowing for higher cellular adhesion and anchoring points. Figure 2 also shows clear evidence of the anodic film with preserved micro-machined lines, with the anodic film aligned parallel to the lines on the underlying substrate. Cracks corresponding to voltage and time are also evident from Figure 2. Similar to Ti wire EA, cracks and instabilities increase with voltage and time of EA.

High-magnification images of the anodized Zr wires are presented in Figure 4. At 20 V, a bare oxide layer with no distinguishable features is visible for 10–60 min of EA. For 120 min EA at 20 V, delamination of the oxide film reveals the presence of underlying nanocrystal-like features (Figure 4C). Using 40 V 60 min yielded alignment of the nanoporous layer onto the underlying micro-roughness (Figure 4E). However, for 40 V at 120 m, some evidence of the underlying nanotubular structures is visible, covered by the oxide film (Figure 4F). Further, clear evidence of nanopore formation is visible for 60 V at 10 m (diameter ~46 nm) and 60 m (diameter ~52 nm) (Figure 4G–I). In summary, for all of the 60 V anodized samples, we observed nanopore formation throughout the surface of the wire, with the irregular sponge-like patches of the ZrO$_2$ layer (which was prominent for 60 V 10 m samples). It is worth noting that the nanopores on the Zr wire are aligned in the direction of the underlying microfeatures of the substrate. Our group has shown that aligned TiO$_2$ nanopores on Ti can be used to mechanically stimulate cells [44,45]. Briefly, the activity of primary gingival fibroblasts and osteoblasts on aligned TiO$_2$ nanopores was enhanced and the cells aligned parallel to the nanopores, indicating a strong mechanotransduction effect [45]. Additionally, as clear nanopores are visible, loading and release of various therapies may be enabled, which has never been demonstrated for ZrO$_2$ nanopores and, hence, warrants further investigation. We have previously shown that that TiO$_2$ nanopores are mechanically superior to conventional as well as mechanically enhanced (via various physical/chemical techniques) nanotubes (shown for TiO$_2$ nanotubes) [37]. Additionally, for EA at 80–100 V for 10–60 m (Figure 4J–O), nanopore-like surface features were observed, which were aligned in the direction of the underlying substrate micro-roughness.

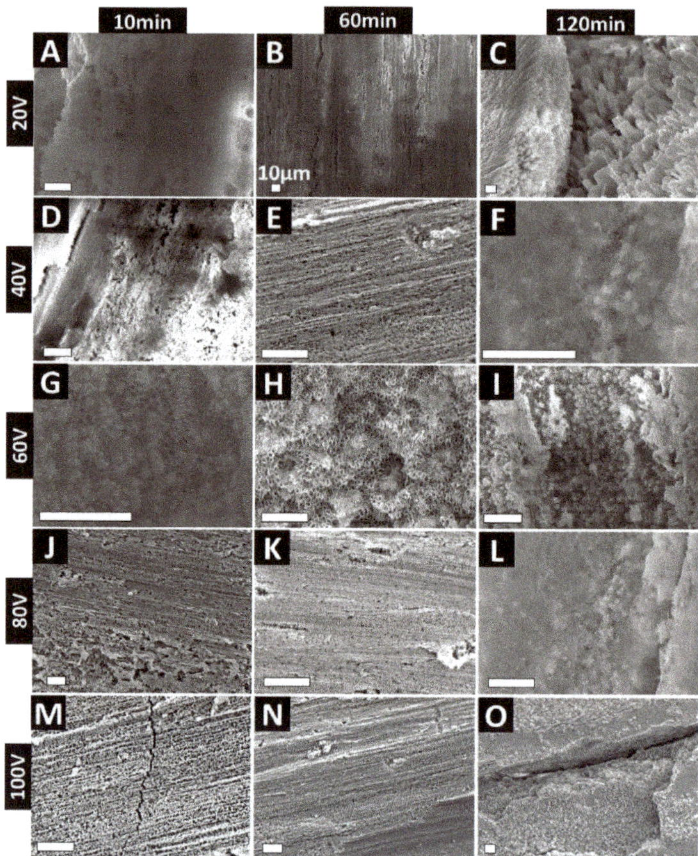

**Figure 4.** High-magnification SEM images showing various ZrO$_2$ nanostructures formed on Zr wires at different voltage and times. (**A–C**) 20 V; (**D–F**) 40 V; (**G–I**) 60 V; (**J–L**) 80 V and (**M–O**) 100 V. Unmarked scale bars represent 1 µm.

To elucidate the mechanism of formation of the various ZrO$_2$ nanostructures, we undertook a detailed analysis of the current density (J) vs. time (t) plots, as presented in Figure S2 (Supplementary Information). The first 15 s of J vs. t plots provide information with respect to the first two phases of Zr EA: (1) formation of compact barrier layer (BL) and (2) pit formation [46]. There are significant differences between the J values at different voltages, and the presented data provide information about the time to reach equilibrium (teq) and barrier oxide layer (BL) thickness. The delay in reaching equilibrium equates to a thicker BL, strong adherence to the underlying substrate, and a stable anodic film, attributed to reduced compressive stress at the ZrO$_2$–Zr interface [47]. teq and J are highest for the 60 V EA, which corresponds to a previous study showing that improved ordering is obtained for higher growth rates (or higher J values) [48]. This explains the findings from Figure 4G–I, which shows that the most nanoporous structures were obtained for 60 V EA. Based on J values corresponding to 60 V, an increased 'outward expansion pressure' for fast growth also explains the abovementioned. For EA performed at higher voltages (80 and 100 V), it can be assumed that the BL will be severely etched (higher field results in increased inward O$^{2-}$ migration) and the electric field polarises the Zr–O bond and damages the tubular structures [46]. As previously reported, besides the internal growth-induced stresses, electric field-induced stresses can also result in compromised stability of the anodized ZrO$_2$ film [29,32].

Next, in order to expose the nanostructures covered by the $ZrO_2$ film or nanopores, we sonicated the anodized wires at various times from 5–60 min. The resultant nanostructures are presented in Figure 5. It was found that dependent on the overall anodized film stability, higher sonication times disrupted the nanostructures. Five-minute sonication for the 20 V 120 min samples exposed the underlying nanocrystal-like topography, which was found to cover the underlying substrate (Figure 5B,C). For 60 V 10 m, 15 min sonication partially removed the nanoporous layer, while 30 m completely removed the nanopores, revealing the $ZrO_2$ nanotubes (Figure 5E,F). The survival of the nanotubes even at 30 m sonication confirms the mechanical stability and robustness of the dual micro–nanostructures onto the underlying wire substrate. This correlates with previous studies whereby the microfeatures of the underlying substrates allowed for increased interfacial contact area between the anodic film and the substrate [35,40–42]. This increased area reduces the mechanical stress and volume expansion during anodic film growth and hence improves overall mechanical stability. Next, 10 min sonication of the aligned nanopores on 100 V 10 min wire revealed the $ZrO_2$ nanotubes (Figure 5H,I) underneath. A similar effect was also observed for 80 V 10 m anodized wires, as shown in Figure S3 (Supplementary Information). It is noteworthy that for 60 V 10 m, the anodic structures survived the extended sonication time (15–30 m, Figure 5F), though for higher voltages (80 and 100 V, Figure 5I and Figure S3C), a small duration (5–10 min) exposed the underlying structures. We have previously shown that increased EA voltage is associated with higher growth rates on curved substrates, and hence reduced structural integrity of the anodized nanostructures (as compared to low-voltage-anodized structures) [29].

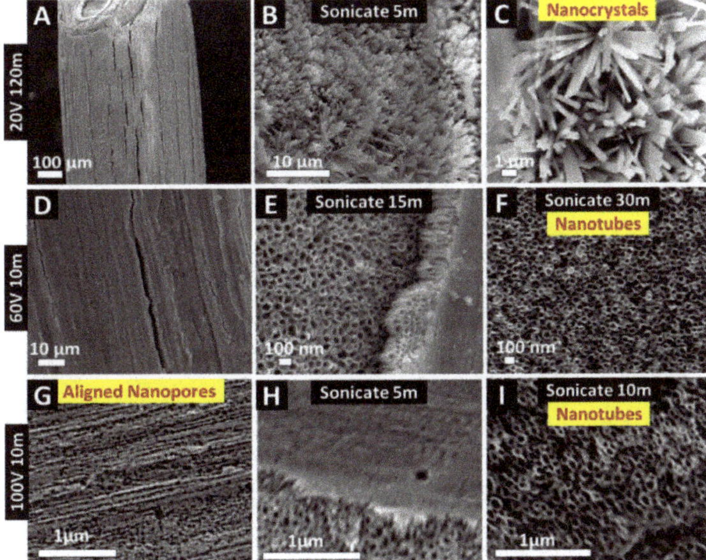


**Figure 5.** Top-view SEM images showing the influence of sonication of anodized Zr wire for various durations to remove superficial nanoporous oxide layer and expose underlying nanostructures. (**A–C**) 20 V 120 min anodized wire for 5 min sonication reveals nanocrystal-like features; (**D–F**) 15–30 min sonication of 60 V 10 min Zr wire reveals nanotubes; and (**G–I**) 5–10 min sonication removes oxide film and exposes underlying nanotubes on 100 V 10 min anodized wire. Survival of nanotubes on Zr wire post-sonication confirms mechanical stability and strong adherence to the underlying substrate.

In summary, this study highlights the fabrication of stable nanotopographies on clinically relevant Zr surfaces—ensuring clinical translatability of electrochemically anodized Zr implants. The innovation of the study is the fact that it is a pioneering attempt at the fabrication of complex $ZrO_2$ nanostructures on Zr curved surfaces via EA, while preserving the 'gold standard' micro-roughness to fabricate dual micro–nanostructures. Further, such controlled dual micro–nanostructures on Zr implants have the potential to augment cell activity and local therapy. Previous studies suggest that such aligned dual micro–nanostructures can mechanically stimulate cells [44]. Therefore, future studies will focus on the evaluation of soft- and hard-tissue integration on the surface of nano-engineered Zr implants, with the current study providing important data that bridges the gap to clinical translation by evaluating clinically relevant implant surfaces. It is noteworthy that bioactivity and therapeutic evaluations of such curved 3D implant substrates (Zr wires) is difficult to achieve in conventional 2D cell culture in vitro, which is more suitable for flat/planar substrates. We have previously undertaken extensive bioactivity evaluations of nano-engineered Ti wires in a 3D cell culture system in vitro [42], animal tissues ex vivo [43], and animal implantation in vivo [34]. However, inclusion of such detailed assessments is outside the scope of the current paper that is focussed on fabrication optimization.

## 4. Conclusions

With the objective of bridging the gap between nano-engineered zirconia and the dental implant industry, this study showcases the fabrication of various controlled nanotopographies on Zr wire substrates (as a model for dental implants) via electrochemical anodization (EA). In a pioneering approach, by tuning EA voltage and time, EA of micromachined Zr wire enabled the fabrication of aligned nanopores, nanotubes, and nanocrystals. We also showed the impact of removing the top layer of oxide/nanopores to reveal the underlying nanotubes. Preserving the underlying micro-roughness and superimposition of controlled $ZrO_2$ nanostructures holds great promise towards improving the bioactivity and therapeutic potential of conventional Zr-based dental and orthopaedic implants.

**Supplementary Materials:** The following are available online at https://www.mdpi.com/article/10.3390/nano11040868/s1, Figure S1: SEM images of as-received Zr wires, Figure S2: current density vs. time plots for anodization of Zr wire, Figure S3: SEM images confirming the influence of sonication on 80 V 10 min anodized Zr wire.

**Author Contributions:** Conceptualization, D.C., K.G. and S.I.; methodology, D.C.; software, D.C.; investigation, D.C.; resources, K.G. and S.I.; writing—original draft preparation, D.C.; writing—review and editing, D.C., K.G. and S.I.; supervision, S.I.; project administration, S.I.; funding acquisition, K.G. and S.I. All authors have read and agreed to the published version of the manuscript.

**Funding:** Divya Chopra is supported by the UQ Graduate School Scholarships (UQGSS) funded by the University of Queensland. Karan Gulati is supported by the National Health and Medical Research Council (NHMRC) Early Career Fellowship (APP1140699).

**Acknowledgments:** The authors acknowledge the facilities, and the scientific and technical assistance, of the Australian Microscopy & Microanalysis Research Facility at the Centre for Microscopy and Microanalysis, The University of Queensland.

**Conflicts of Interest:** The authors declare no conflict of interest.

## References

1. Zhang, Y.; Chen, H.-X.; Duan, L.; Fan, J.-B. The Electronic Structures, Elastic Constants, Dielectric Permittivity, Phonon Spectra, Thermal Properties and Optical Response of Monolayer Zirconium Dioxide: A First-Principles Study. *Thin Solid Film.* **2021**, *721*, 138549. [CrossRef]
2. Schünemann, F.H.; Galárraga-Vinueza, M.E.; Magini, R.; Fredel, M.; Silva, F.; Souza, J.C.M.; Zhang, Y.; Henriques, B. Zirconia Surface Modifications for Implant Dentistry. *Mater. Sci. Eng. C* **2019**, *98*, 1294–1305. [CrossRef]
3. Rupp, F.; Liang, L.; Geis-Gerstorfer, J.; Scheideler, L.; Hüttig, F. Surface Characteristics of Dental Implants: A Review. *Dent. Mater.* **2018**, *34*, 40–57. [CrossRef] [PubMed]
4. Guo, T.; Gulati, K.; Arora, H.; Han, P.; Fournier, B.; Ivanovski, S. Race to Invade: Understanding Soft Tissue Integration at the Transmucosal Region of Titanium Dental Implants. *Dent. Mater.* **2021**, in press. [CrossRef] [PubMed]
5. Tuna, T.; Wein, M.; Swain, M.; Fischer, J.; Att, W. Influence of Ultraviolet Photofunctionalization on the Surface Characteristics of Zirconia-Based Dental Implant Materials. *Dent. Mater.* **2015**, *31*, e14–e24. [CrossRef] [PubMed]

6.  Honda, J.; Komine, F.; Kusaba, K.; Kitani, J.; Matsushima, K.; Matsumura, H. Fracture Loads of Screw-Retained Implant-Supported Zirconia Prostheses after Thermal and Mechanical Stress. *J. Prosthodont. Res.* **2020**, *64*, 313–318. [CrossRef] [PubMed]

7.  AlFarraj, A.A.; Aldosari, A.; Sukumaran, A.; Al Amri, M.D.; van Oirschot, A.J.A.B.; Jansen, J.A. A Comparative Study of the Bone Contact to Zirconium and Titanium Implants after 8 Weeks of Implantation in Rabbit Femoral Condyles. *Odontology* **2018**, *106*, 37–44. [CrossRef] [PubMed]

8.  Sivaraman, K.; Chopra, A.; Narayan, A.I.; Balakrishnan, D. Is Zirconia a Viable Alternative to Titanium for Oral Implant? A Critical Review. *J. Prosthodont. Res.* **2018**, *62*, 121–133. [CrossRef] [PubMed]

9.  Patil, N.A.; Kandasubramanian, B. Biological and mechanical enhancement of zirconium dioxide for medical applications. *Ceram. Int.* **2020**, *46*, 4041–4057. [CrossRef]

10.  Gomez, S.A.; Schreiner, W.; Duffó, G.; Ceré, A.S. Surface Characterization of Anodized Zirconium for Biomedical Applications. *Appl. Surf. Sci.* **2011**, *257*, 6397–6405.

11.  Wang, Y.B.; Zheng, Y.F.; Wei, S.C.; Li, M. In Vitro Study on Zr-Based Bulk Metallic Glasses as Potential Biomaterials. *J. Biomed. Mater. Res. Part B Appl. Biomater.* **2011**, *96*, 34–46. [CrossRef]

12.  Hobbs, L.W.; Rosen, V.B.; Mangin, S.P.; Treska, M.; Hunter, G. Oxidation Microstructures and Interfaces in the Oxidized Zirconium Knee. *Int. J. Appl. Ceram. Technol.* **2005**, *2*, 221–246. [CrossRef]

13.  Uchida, M.; Kim, H.M.; Miyaji, F.; Kokubo, T.; Nakamura, T. Apatite Formation on Zirconium Metal Treated with Aqueous Naoh. *Biomaterials* **2002**, *23*, 313–317. [CrossRef]

14.  Gomez, S.A.; Ballarre, J.; Orellano, J.C.; Duffó, G.; Cere, S. Surface Modification of Zirconium by Anodisation as Material for Permanent Implants: In Vitro and in Vivo Study. *J. Mater. Sci. Mater. Med.* **2013**, *24*, 161–169. [CrossRef]

15.  Jović, V.D.; Jović, B.M. The Influence of the Conditions of the $ZrO_2$ Passive Film Formation on Its Properties in 1 M NaOH. *Corros. Sci.* **2008**, *50*, 3063–3069. [CrossRef]

16.  Gulati, K.; Hamlet, S.M.; Ivanovski, S. Tailoring the Immuno-Responsiveness of Anodized Nano-Engineered Titanium Implants. *J. Mater. Chem. B* **2018**, *6*, 2677–2689. [CrossRef]

17.  de la Hoz, M.F.T.; Katunar, M.R.; González, A.; Sanchez, A.G.; Díaz, A.O.; Ceré, S. Effect of Anodized Zirconium Implants on Early Osseointegration Process in Adult Rats: A Histological and Histomorphometric Study. *Prog. Biomater.* **2019**, *8*, 249–260. [CrossRef] [PubMed]

18.  Bacchelli, B.; Giavaresi, G.; Franchi, M.; Martini, D.; de Pasquale, V.; Trirè, A.; Fini, M.; Giardino, R.; Ruggeri, A. Influence of a Zirconia Sandblasting Treated Surface on Peri-Implant Bone Healing: An Experimental Study in Sheep. *Acta Biomater.* **2009**, *5*, 2246–2257. [CrossRef] [PubMed]

19.  Zhang, L.; Zhu, S.; Han, Y.; Xiao, C.; Tang, W. Formation and Bioactivity of Ha Nanorods on Micro-Arc Oxidized Zirconium. *Mater. Sci. Eng. C* **2014**, *43*, 86–91. [CrossRef]

20.  Quan, R.; Yang, D.; Yan, J.; Li, W.; Wu, X.; Wang, H. Preparation of Graded Zirconia–Cap Composite and Studies of Its Effects on Rat Osteoblast Cells in Vitro. *Mater. Sci. Eng. C* **2009**, *29*, 253–260. [CrossRef]

21.  Li, X.; Deng, J.; Lu, Y.; Zhang, L.; Sun, J.; Wu, F. Tribological Behavior of $ZrO_2/Ws_2$ Coating Surfaces with Biomimetic Shark-Skin Structure. *Ceram. Int.* **2019**, *45*, 21759–21767. [CrossRef]

22.  Gulati, K.; Kogawa, M.; Maher, S.; Atkins, G.; Findlay, D.; Losic, D. Titania Nanotubes for Local Drug Delivery from Implant Surfaces. In *Electrochemically Engineered Nanoporous Materials*; Springer: Berlin/Heidelberg, Germany, 2015; pp. 307–355.

23.  Tsuchiya, H.; Macak, j.; Taveira, L.; Schmuki, P. Fabrication and Characterization of Smooth High Aspect Ratio Zirconia Nanotubes. *Chem. Phys. Lett.* **2005**, *410*, 188–191. [CrossRef]

24.  Katunar, M.R.; Sanchez, A.G.; Coquillat, A.S.; Civantos, A.; Campos, E.M.; Ballarre, J.; Vico, T.; Baca, M.; Ramos, V.; Cere, S. In vitro and in vivo characterization of anodised zirconium as a potential material for biomedical applications. *Mater. Sci. Eng. C* **2017**, *75*, 957–968. [CrossRef] [PubMed]

25.  Zhao, J.; Xu, R.; Wang, X.; Li, Y. In Situ Synthesis of Zirconia Nanotube Crystallines by Direct Anodization. *Corros. Sci.* **2008**, *50*, 1593–1597. [CrossRef]

26.  Guo, L.; Zhao, J.; Wang, X.; Xu, R.; Lu, Z.; Li, Y. Bioactivity of Zirconia Nanotube Arrays Fabricated by Electrochemical Anodization. *Mater. Sci. Eng. C* **2009**, *29*, 1174–1177. [CrossRef]

27.  Frandsen, C.J.; Brammer, K.S.; Noh, K.; Connelly, L.S.; Oh, S.; Chen, L.H.; Jin, S. Zirconium Oxide Nanotube Surface Prompts Increased Osteoblast Functionality and Mineralization. *Mater. Sci. Eng. C* **2011**, *31*, 1716–1722. [CrossRef]

28.  Zhang, L.; Han, Y. Enhanced Bioactivity of Self-Organized $ZrO_2$ Nanotube Layer by Annealing and Uv Irradiation. *Mater. Sci. Eng. C* **2011**, *31*, 1104–1110. [CrossRef]

29.  Gulati, K.; Santos, A.; Findlay, D.; Losic, D. Optimizing Anodization Conditions for the Growth of Titania Nanotubes on Curved Surfaces. *J. Phys. Chem. C* **2015**, *119*, 16033–16045. [CrossRef]

30.  Gulati, K.; Maher, S.; Chandrasekaran, S.; Findlay, D.M.; Losic, D. Conversion of Titania ($TiO_2$) into Conductive Titanium (Ti) Nanotube Arrays for Combined Drug-Delivery and Electrical Stimulation Therapy. *J. Mater. Chem. B* **2016**, *4*, 371–375. [CrossRef]

31.  Gulati, K.; Ivanovski, S. Dental Implants Modified with Drug Releasing Titania Nanotubes: Therapeutic Potential and Developmental Challenges. *Expert Opin. Drug Deliv.* **2017**, *14*, 1009–1024. [CrossRef]

32.  Gulati, K.; Li, T.; Ivanovski, S. Consume or Conserve: Microroughness of Titanium Implants toward Fabrication of Dual Micro–Nanotopography. *ACS Biomater. Sci. Eng.* **2018**, *4*, 3125–3131. [CrossRef] [PubMed]

33. Zhao, J.; Wang, X.; Xu, R.; Meng, F.; Guo, L.; Li, Y. Fabrication of High Aspect Ratio Zirconia Nanotube Arrays by Anodization of Zirconium Foils. *Mater. Lett.* **2008**, *62*, 4428–4430. [CrossRef]
34. Kaur, G.; Willsmore, T.; Gulati, K.; Zinonos, I.; Wang, Y.; Kurian, M.; Hay, S.; Losic, D.; Evdokiou, A. Titanium Wire Implants with Nanotube Arrays: A Study Model for Localized Cancer Treatment. *Biomaterials* **2016**, *101*, 176–188. [CrossRef] [PubMed]
35. Proost, J.; Vanhumbeeck, J.; van Overmeere, Q. Instability of Anodically Formed $TiO_2$ Layers (Revisited). *Electrochim. Acta* **2009**, *55*, 350–357. [CrossRef]
36. Fan, M.; la Mantia, F. Effect of Surface Topography on the Anodization of Titanium. *Electrochem. Commun.* **2013**, *37*, 91–95. [CrossRef]
37. Li, T.; Gulati, K.; Wang, N.; Zhang, Z.; Ivanovski, S. Understanding and Augmenting the Stability of Therapeutic Nanotubes on Anodized Titanium Implants. *Mater. Sci. Eng. C* **2018**, *88*, 182–195. [CrossRef] [PubMed]
38. Muratore, F.; Hashimoto, T.; Skeldon, P.; Thompson, G.E. Effect of Ageing in the Electrolyte and Water on Porous Anodic Films on Zirconium. *Corros. Sci.* **2011**, *53*, 2299–2305. [CrossRef]
39. Guo, T.; Oztug, N.A.K.; Han, P.; Ivanovski, S.; Gulati, K. Old Is Gold: Electrolyte Aging Influences the Topography, Chemistry, and Bioactivity of Anodized $TiO_2$ Nanopores. *ACS Appl. Mater. Interfaces* **2021**, *13*, 7897–7912. [CrossRef] [PubMed]
40. Li, T.; Gulati, K.; Wang, N.; Zhang, Z.; Ivanovski, S. Bridging the Gap: Optimized Fabrication of Robust Titania Nanostructures on Complex Implant Geometries Towards Clinical Translation. *J. Colloid Interface Sci.* **2018**, *529*, 452–463. [CrossRef]
41. Pilling, N.B. The Oxidation of Metals at High Temperature. *J. Inst. Met.* **1923**, *29*, 529–582.
42. Gulati, K.; Kogawa, M.; Prideaux, M.; Findlay, D.M.; Atkins, G.J.; Losic, D. Drug-Releasing Nano-Engineered Titanium Implants: Therapeutic Efficacy in 3D Cell Culture Model, Controlled Release and Stability. *Mater. Sci. Eng. C* **2016**, *69*, 831–840. [CrossRef]
43. Rahman, S.; Gulati, K.; Kogawa, M.; Atkins, G.J.; Pivonka, P.; Findlay, D.M.; Losic, D. Drug Diffusion, Integration, and Stability of Nanoengineered Drug-Releasing Implants in Bone Ex-Vivo. *J. Biomed. Mater. Res. Part A* **2016**, *104*, 714–725. [CrossRef]
44. Gulati, K.; Moon, H.G.; Kumar, P.T.S.; Han, P.; Ivanovski, S. Anodized Anisotropic Titanium Surfaces for Enhanced Guidance of Gingival Fibroblasts. *Mater. Sci. Eng. C* **2020**, *112*, 110860. [CrossRef]
45. Gulati, K.; Moon, H.G.; Li, T.; Kumar, P.T.S.; Ivanovski, S. Titania Nanopores with Dual Micro-/Nano-Topography for Selective Cellular Bioactivity. *Mater. Sci. Eng. C* **2018**, *91*, 624–630. [CrossRef]
46. Ismail, S.; Ahmad, Z.A.; Berenov, A.; Lockman, Z. Effect of Applied Voltage and Fluoride Ion Content on the Formation of Zirconia Nanotube Arrays by Anodic Oxidation of Zirconium. *Corros. Sci.* **2011**, *53*, 1156–1164. [CrossRef]
47. Zhou, X.; Nguyen, N.T.; Özkan, S.; Schmuki, P. Anodic $TiO_2$ Nanotube Layers: Why Does Self-Organized Growth Occur—A Mini Review. *Electrochem. Commun.* **2014**, *46*, 157–162. [CrossRef]
48. Jessensky, O.; Müller, F.; Gösele, U. Self-Organized Formation of Hexagonal Pore Arrays in Anodic Alumina. *Appl. Phys. Lett.* **1998**, *72*, 1173–1175. [CrossRef]

 *nanomaterials*

*Article*

# Evaluation of Surface Change and Roughness in Implants Lost Due to Peri-Implantitis Using Erbium Laser and Various Methods: An In Vitro Study

Aslihan Secgin-Atar [1], Gokce Aykol-Sahin [2], Necla Asli Kocak-Oztug [1], Funda Yalcin [1], Aslan Gokbuget [3] and Ulku Baser [1,*]

[1] Periodontology Department, Faculty of Dentistry, Istanbul University, 34452 Istanbul, Turkey; aslihansecginatar@gmail.com (A.S.-A.); asli.kocak@istanbul.edu.tr (N.A.K.-O.); fyalcin@istanbul.edu.tr (F.Y.)

[2] Department of Periodontology, Faculty of Dentistry, Istanbul Okan University, 34959 Istanbul, Turkey; gokceaykol@gmail.com

[3] PGG Private Practice, 34365 Istanbul, Turkey; aslangokbuget@gmail.com

* Correspondence: baserulk@istanbul.edu.tr; Tel.: +90-5053568644

**Abstract:** The aim of our study was to obtain similar surface properties and elemental composition to virgin implants after debridement of contaminated titanium implant surfaces covered with debris. Erbium-doped:yttrium, aluminum, and garnet (Er:YAG) laser, erbium, chromium-doped:yttrium, scandium, gallium, and garnet (Er,Cr:YSGG) laser, curette, and ultrasonic device were applied to contaminated implant surfaces. Scanning electron microscopy (SEM) images were taken, the elemental profile of the surfaces was evaluated with energy dispersive X-ray spectroscopy (EDX), and the surface roughness was analyzed with profilometry. Twenty-eight failed implants and two virgin implants as control were included in the study. The groups were designed accordingly; titanium curette group, ultrasonic scaler with polyetheretherketone (PEEK) tip, Er: YAG very short pulse laser group (100 μs, 120 mJ/pulse 10 Hz), Er: YAG short-pulse laser group (300 μs, 120 mJ/pulse, 10 Hz), Er: YAG long-pulse laser group (600 μs, 120 mJ/pulse, 10 Hz), Er, Cr: YSGG1 laser group (1 W 10 Hz), Er, Cr: YSGG2 laser group (1.5 W, 30 Hz). In each group, four failed implants were debrided for 120 s. When SEM images and EDX findings and profilometry results were evaluated together, Er: YAG long pulse and ultrasonic groups were found to be the most effective for debridement. Furthermore, the two interventions have shown the closest topography of the sandblasted, large grit, acid-etched implant surface (SLA) as seen on virgin implants.

**Keywords:** dental implants; titanium; laser therapy; peri-implantitis; debridement

**Citation:** Secgin-Atar, A.; Aykol-Sahin, G.; Kocak-Oztug, N.A.; Yalcin, F.; Gokbuget, A.; Baser, U. Evaluation of Surface Change and Roughness in Implants Lost Due to Peri-Implantitis Using Erbium Laser and Various Methods: An In Vitro Study. *Nanomaterials* **2021**, *11*, 2602. https://doi.org/10.3390/nano11102602

Academic Editor: Ion N. Mihailescu

Received: 2 September 2021
Accepted: 28 September 2021
Published: 2 October 2021

**Publisher's Note:** MDPI stays neutral with regard to jurisdictional claims in published maps and institutional affiliations.



## 1. Introduction

Peri-implantitis is a pathological condition characterized by inflammation of the peri-implant mucosa and progressive destruction of the supporting bone as a result of biofilm formation on the implant surface [1–3]. Peri-implantitis is diagnosed by increased peri-implant probing depth, bleeding on probing in the peri-implant pocket, and detecting bone loss around implant with radiography [4]. Etiology of peri-implant infection comprises various factors such as the macro design of the implant, the level of surface roughness, and the condition of the hard/ soft tissue surrounding the implant [5]. However, bacterial plaque formation on the implant surface is the most important factor in the etiology of peri-implantitis [6].

There are multiple risk factors affecting the formation of peri-implanter disease, such as a history of periodontitis, poor oral hygiene habits, and smoking [7]. One or more of these risk factors may cause the degradation of the biocompatibility between the host and the implant surface [8]. The development of an adherent layer of plaque (biofilm) on the implant can result in the formation of calcified deposits similar to calculus and appears to be critical for the development of peri-implant diseases [9–11]. Therefore, the primary goal of treatment is to eliminate all microbial and calcified deposits from the surface of the implant and to supply appropriate conditions for re-osseointegration [8,10].

Despite the fact that periodontitis and peri-implantitis have similar features, they differ in some directions. These variations cause the treatment of the two diseases to own different ideas from every other [12,13]. Treatment methods such as tooth and root surface cleaning, which are effectively used in the treatment of teeth with periodontitis, cannot be used in the same way on rough and grooved implant surfaces [14]. In periodontal inflammation, it is possible to treat the infected cementum surface by cleaning it with various mechanical or chemical methods, whereas, in peri-implantitis treatment, successful surface debridement may not be fully achieved due to the complex implant surface topography. This creates an extraordinary shelter for bacterial attachment and colonization.

Increasing the amount of bone-implant contact and re-osseointegration of implants are vital for successful peri-implantitis treatment [15]. Numerous in vitro studies have analyzed the effects of debridement instruments used in peri-implantitis treatment which showed serious impairment on the titanium topography. Firstly, the traces that may occur on the implant surface impair cell adhesion to the titanium surface and damage proper wound healing. Furthermore, surface defects lead to changes in the implant surface characteristics, which may alter the bone integration of the implant [16–18].

Many in vitro studies evaluated the effects of decontamination and debridement instruments on implants by electron spectroscopy. According to the results of these studies, the most prominent pollutant was carbon. Carbon indicates the presence of any non-biocompatible substance (calcified and organic) that could change the re-osseointegration of implants. In particular, studies on failed implant surfaces have shown varying degrees of carbon [19,20]. Titanium-bone connection is altered by the deposition of organic molecules on implant surfaces, and it was concluded that the removal of hydrocarbon gradient is a crucial step in achieving the bioactivation and osseointegration of titanium [21,22]. It was also suggested that the implant surface must be treated to intensify the surface wettability and energy [23].

The high prevalence of peri-implantitis, ranging from 16 to 47.1%, has also prompted scientists to explore a range of therapeutic applications for implant surface decontamination [24,25]. These include mechanical debridement methods such as curettes, rubber cups, ultrasonic devices and powder spray systems, chemical disinfection methods such as chlorhexidine, tetracycline, metronidazole, citric acid application, and various surgical and non-surgical treatment methods including antibiotics, photodynamic therapy, lasers, and combinations of all these treatments [26,27]. However, the optimal procedure or procedures are still not fully specified [28,29].

Different laser systems have been tested for potential use in implant surfaces including erbium-doped:yttrium, aluminum, and garnet (Er:YAG), and erbium, chromium-doped:yttrium, scandium, gallium, and garnet (Er,Cr:YSGG) lasers [30]. Due to the high absorption by water, the 2.940 nm wavelength of Er:YAG lasers seemed to be capable of effectively removing bacterial deposits from either smooth or rough titanium implants without damaging their surfaces [31,32]. Although both Erbium lasers have high absorption by water, the Er,Cr:YSGG laser operating at a wavelength of 2.780 nm has a lower absorption coefficient in water than the Er:YAG laser [33].

Although the use of lasers is currently on the agenda, there is no standard recommendation regarding laser type, irradiation, or settings protocol for the treatment of peri-implantitis [34]. Before implementing a patient-level irradiation protocol, in vitro studies are required to establish ideal settings.

Our aim is to obtain similar surface properties and elemental composition to virgin implants using erbium lasers (Er:YAG and Er,Cr:YSGG) and mechanical methods (curette, ultrasonic device) in the debridement of implants lost due to peri-implantitis by evaluating the surface change of the implants treated with these methods by scanning electron microscopy (SEM), energy dispersive X-ray spectroscopy (EDX), and profilometry.

## 2. Materials and Methods

### 2.1. Sample Collection

Twenty-eight implants (DE2™, SLA G4) explanted because of advanced peri-implantitis, with debris on their micro-structured surface, and two virgin titanium dental implants (DE2™, SLA G4) representing the positive control group were included in this present study. Implants were autoclaved and immersed in acrylic blocs from a 3 mm apical part to facilitate the precise application of the instruments and standardization. SEM images, EDX, and profilometry measurements were done before and after debridement. Twenty-eight failed implants were allocated into seven test groups, and two virgin implants represented the control group. Test groups were divided based on treatment methods of implant surfaces and each test group had 4 failed implants. All assessments were done in two-time points, before and after the debridement. All implant surface debridement procedures were performed by the same investigator. This study was approved by the Ethics Commission of Istanbul University Faculty of Dentistry (Approved number 2020/22).

### 2.2. Interventions

Treatment procedures of contaminated titanium implant surfaces for each group are shown in Table 1.

**Table 1.** Treatment procedures.

| Groups | Instruments | Parameters |
|---|---|---|
| Ti-Cur ($n$ = 4) | Titanium curette [1] | 120 s |
| US-PEEK ($n$ = 4) | Ultrasonic scaler [2] with PEEK [3] tip | 120 s |
| ErL-VSP ($n$ = 4) | ER:YAG laser [4] with R02-C | VSP (100 μs); 120 mJ/pulse; 10 Hz; Air 6 Water 4; 120 s; 19.04 J/cm$^2$ |
| ErL-SP ($n$ = 4) | ER:YAG laser [4] with R02-C | SP (300 μs); 120 mJ/pulse; 10 Hz; Air 6 Water 4; 120 s; 19.04 J/cm$^2$ |
| ErL-LP ($n$ = 4) | ER:YAG laser [4] with R02-C | LP (600 μs); 120 mJ/pulse; 10 Hz; Air 6 Water 4; 120 s; 19.04 J/cm$^2$ |
| ErCrL-1 ($n$ = 4) | Er,Cr:YSGG laser [5] with RFPT5 14 mm fiber tip | 1 W, 10 Hz (100 mJ/pulse); Air 40 Water 50; 120 s; 38.46 J/cm$^2$ |
| ErCrL-2 ($n$ = 4) | Er,Cr:YSGG laser [5] with RFPT5 14 mm fiber tip | 1.5 W, 30 Hz (50 mJ/pulse); Air 40 Water 50; 120 s; 19.23 J/cm$^2$ |
| Control ($n$ = 2) | No intervention was made on the implants | |

[1] LM ErgoMix™, Pargas, Finland. [2] Woodpecker®, Guilin, China. [3] Scorpion™ İnsert CLiP Fine Ultrasonic Implant Scaler Kit™, Romagnat, France. [4] Fotona® Fidelis Plus II, Ljubljana, Slovenia. [5] WaterLase iPlus®, Foothill Ranch, CA, USA.

In the titanium curette (Ti-Cur) group, the instrumentation was performed for 120 s by scaling in one direction at an angle of 30 degrees on the implants.

In the ultrasonic scaler with a PEEK tip (US-PEEK) group, debridement was performed by keeping the device tip in contact with the implant surface at a 30 degree angle for 120 s. The ultrasonic scaler was used at the recommended speed for routine periodontal treatment under maximum water cooling.

In laser groups, the laser energy delivery was directed by a computer interface that dictated the selected laser tip, modes, energy, and associated water and air for each laser group. Each used laser parameter was described in Table 1.

Er:YAG laser (ErL) (2.940 nm) was used with a 90 degree-irradiation angle to the titanium surfaces according to the recommendation of the manufacturer. An R02-C handpiece was used with a 0.9 mm diameter. The spot area was calculated as 0.63 mm$^2$. To simulate clinical use, sweeping irradiation was performed in non-contact mode at approximately 1 mm [35–37].

Er: YAG very short pulse laser (ErL-VSP) parameters were: Pulse energy: 120 mJ; Pulse duration: 100 μs; Frequency: 10 Hz; resulting in an energy density per pulse of 19.04 J/cm$^2$; Air/Water output: 4/6.

Er: YAG short-pulse laser (ErL-SP) parameters were: Pulse energy: 120 mJ; Pulse duration: 300 μs; Frequency: 10 Hz; resulting in an energy density per pulse of 19.04 J/cm$^2$; Air/Water output: 4/6.

Er: YAG long-pulse laser (ErL-LP) parameters were: Pulse energy: 120 mJ; Pulse duration: 600 μs; Frequency: 10 Hz; resulting in an energy density per pulse of 19.04 J/cm$^2$; Air/Water output: 4/6.

Er, Cr:YSGG laser (ErCrL) (2.780 nm) tip was used with a 15 degree-irradiation angle to the titanium surfaces according to the recommendation of the manufacturer. To simulate clinical use, sweeping irradiation was performed in non-contact mode at approximately 1 mm. An RFPT5 radial firing fiber tip with beam divergence > 40 degrees and 0.5 mm in diameter was used in the study. The spot area was calculated as 2.5 mm$^2$ (0.025 cm$^2$) at 1 mm from the implant surfaces [38].

Er, Cr: YSGG1 laser (ErCrL-1) parameters were: Power: 1 W; Pulse energy: 100 mJ; Pulse duration: 60 µs; Frequency: 10 Hz; resulting in an energy density per pulse of 4 J/cm$^2$; Air/Water percentage output (%): 40/50.

Er, Cr: YSGG2 laser (ErCrL-2) parameters were: Power: 1.5 W; Pulse energy: 50 mJ; Pulse duration: 60 µs; Frequency: 30 Hz; resulting in an energy density per pulse of 1.8 J/cm$^2$; Air/Water percentage output (%): 40/50.

*2.3. Analyses*

2.3.1. SEM

Topographic surface alterations were assessed with SEM. The SEM observation was conducted with a FEI™ VERSA 3DLOVAC microscope (FEI, Hillsboro, OR, ABD). Analyses were performed at baseline for the control group, and before and after the interventions for the contaminated implants. Each sample was marked from the neck area to be observed from the same surface after the treatment. Each implant was scanned and photographed at five set magnifications (76×–150×–500×–1000×–2000×).

Implant Debridement Visual Index

This current index was created for the visual evaluation of SEM images. The index aims to compare the surface features of treated-contaminated implants with virgin implants by grading:

1:  Image without any contamination and resembling a positive control,
2:  Spot contamination of observed image,
3:  Image of contamination beyond spot contamination.

SEM images of the implants after debridement were assessed at (150×) magnification. The debris on the implant surface were evaluated by three blind observers. The images were randomly shown to each observer twice without specifying the sample numbers and groups.

While analyzing by statistics, 1 and 2 grades were considered as clean, and 3 as contaminated.

2.3.2. EDX

EDX analysis is utilized for quantitative evaluation and the local determination of the chemical composition. In the present study, EDX was used to measure the presence of carbon (C), titanium (Ti), oxygen (O), and nitrogen (N) elements on the implant surfaces. The spectroscopy of the emitted X-ray photons was performed by a Bruker detector with an energy resolution of about 123 eV at a working distance. The measurement was made from the 1 mm$^2$ area on the same thread determined in each implant surface.

2.3.3. Profilometry

The roughness evaluation and analysis were performed using a profilometry (Veeco Instruments Inc., Plainview, NY, USA, ABD), (Radius: 5 µm, Stylus force: 3 mg/29.4 µN, Resolution: 0.167 µm/sample, Length: 1000 µm, Duration: 20 s). Measurements were made at the same area where the identified threads in the middle third of each implant were marked. In the selected thread distance, the diamond tip of the profilometry performed measurement 20 times in the horizontal direction along the 1000 µm length and resulted in the average of the roughness.

### 2.3.4. Statistical Analysis

A power analysis demonstrated a sample size of 4 implants for each intervention group would ensure 80% power to detect the difference between the treatment methods in the morphologic features of implant surfaces with a significance level of 0.05. The data analysis was performed using SPSS v.23 software (IBM Corp., New York, NY, USA). The descriptive statistics were presented by the mean values with minimum and maximum and their standard deviations (SD). When the data were parametric, paired sampled *t*-test was used in comparing the differences of the groups. For the evaluation of the before and after treatment differences between the groups, the statistical analysis used was one-way analysis of variance (ANOVA) and Tukey HSD multiple comparison tests followed by Bonferroni post hoc testing. The data are represented as mean $\pm$ standard deviation. A value of $p < 0.05$ was considered to be statistically significant.

## 3. Results

### 3.1. SEM Analysis

Implant Debridement Visual Index (IDVI) scores of the three trained observers are given at Table 2. According to the index results, the most effective groups were ErL groups. This was followed by the US-PEEK. The debris not removed (DnR) score was higher for ErCrL groups and Ti-Cur. All three pulse settings of the ErL were found to be effective in removing hard deposits from the implant surface.

**Table 2.** Implant debridement visual index scores.

| Group | O1: DR/DnR | O2: DR/DnR | O3: DR/DnR |
|---|---|---|---|
| Ti-Cur | 0/4 | 1/3 | 1/3 |
| US-PEEK | 3/1 | 3/1 | 3/1 |
| ErL VSP | 4/0 | 4/0 | 4/0 |
| ErL SP | 4/0 | 4/0 | 3/1 |
| ErL LP | 4/0 | 4/0 | 4/0 |
| ErCrL-1 | 1/3 | 2/2 | 1/3 |
| ErCrL-2 | 2/2 | 2/2 | 2/2 |

O1: observer 1, O2: observer 2, O3: observer 3, DR: debris removed, DnR: debris not removed.

All groups were contaminated at baseline compared to the control group. The contamination was confirmed by the presence of a layer of the debris, the titanium surface features, and by the EDX lower C and Ti values.

### 3.1.1. Ti-Cur Group

Large and flat scratched areas were observed on the titanium surface after the treatment, in the Ti-Cur group. Although the honeycomb appearance [39] of the SLA implant surface was achieved in some surfaces, it was observed that the curette did not provide an effective debridement and left a residue (Figure 1a–d).

### 3.1.2. US-PEEK Group

In the ultrasonic group, clean surfaces were observed similar to the positive control group. A small amount of debris and some materials considered to be remnants of the PEEK material were observed (Figure 2a–d).

### 3.1.3. ErL-VSP Group

Although the debridement was achieved in the ErL-VSP group, delamination and deformation were also observed on the surfaces. In particular, porosity due to melting, loss of honeycomb appearance, and a relatively smooth surface with microcracks were observed (Figure 3a–d).

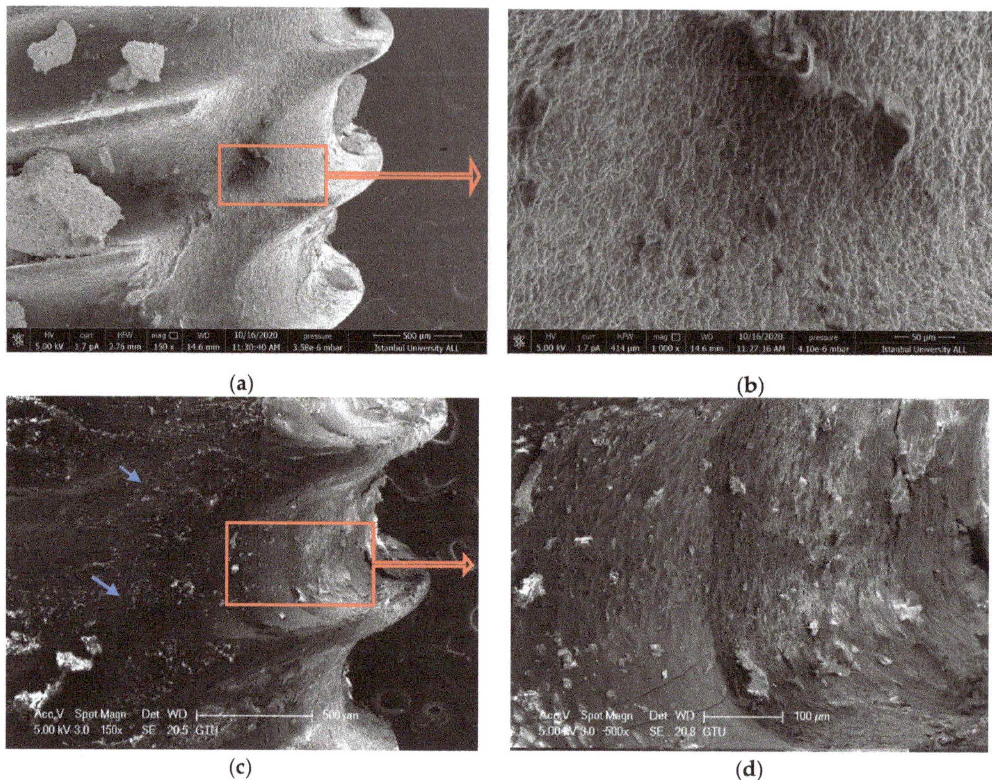

**Figure 1.** (**a,b**) SEM images showing the topography of untreated failed implant surfaces in Ti-Cur group (150× and 1000×) (**c,d**) after intervention (150× and 1000×). The blue arrows point to scratched areas. Red arrows show the further magnification area.

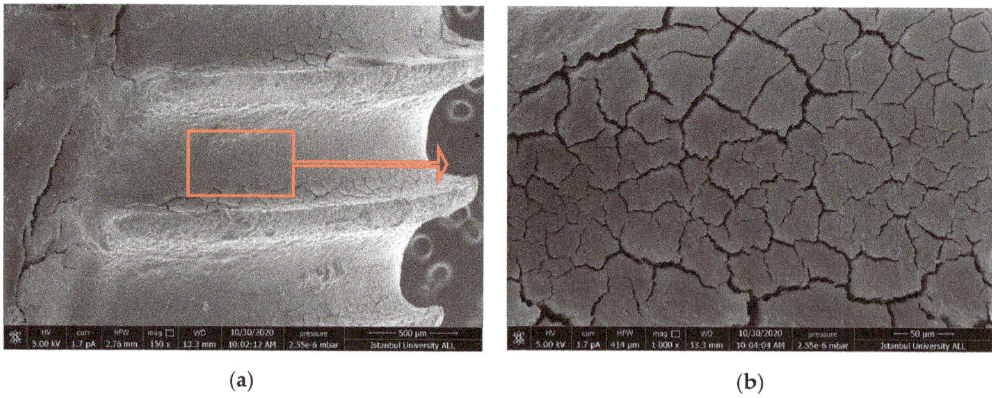

**Figure 2.** *Cont.*

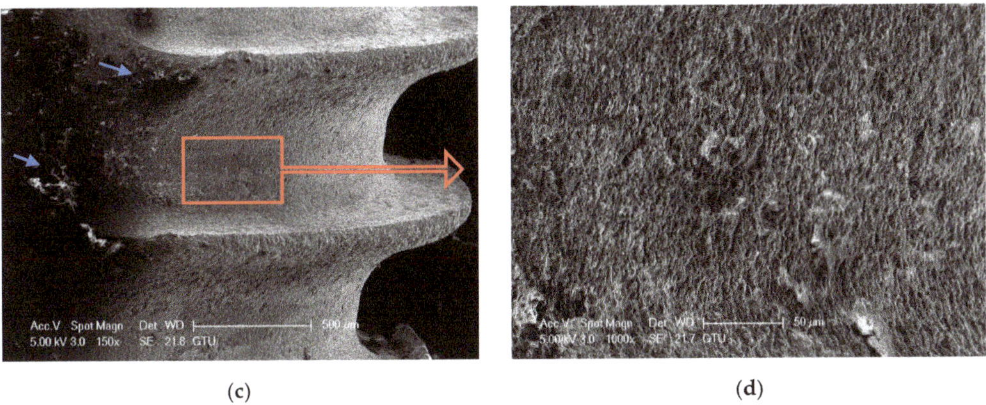

(c)                                                                 (d)

**Figure 2.** (**a,b**) SEM images showing the topography untreated failed implant surfaces in US-PEEK group (150× and 1000×). The image in (**b**) represents the contamination layer on the implant surface in all SEM images; (**c,d**) after intervention (150× and 1000×). The blue arrows point to PEEK material remnants. Red arrows show the further magnification area.

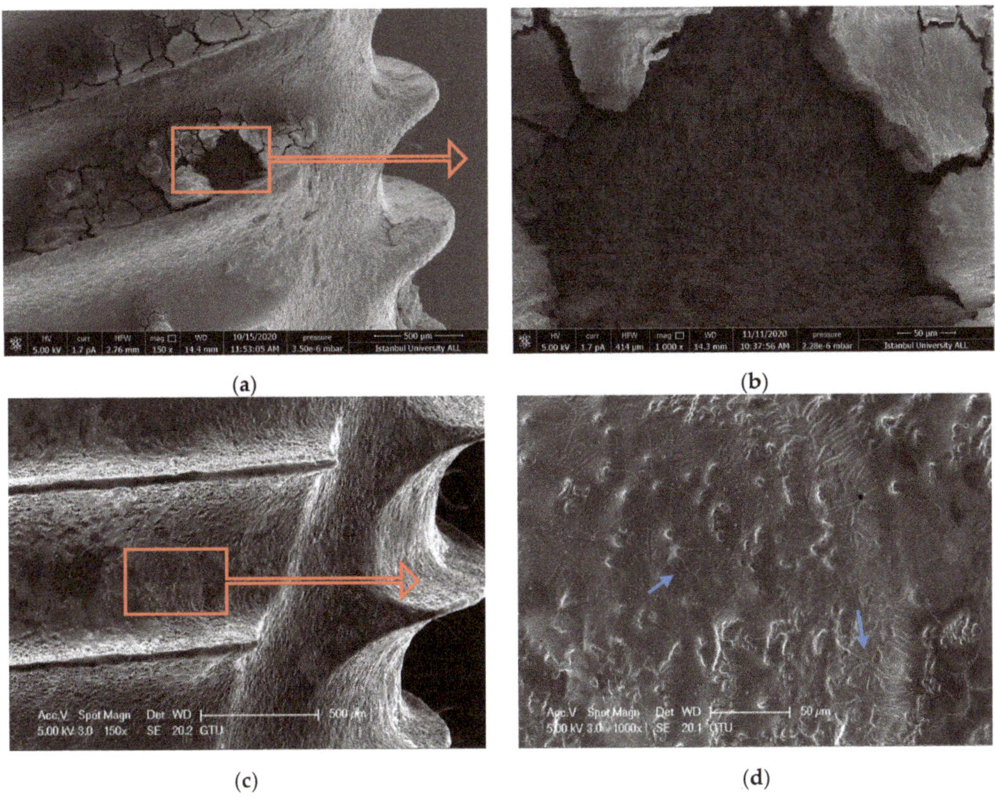

(a)                                                                 (b)

(c)                                                                 (d)

**Figure 3.** (**a,b**) SEM images showing the topography of untreated failed implant surfaces in ErL-VSP group (150× and 1000×) (**c,d**) after intervention (150× and 1000×). The blue arrows point to microcracks. Red arrows show the further magnification area.

### 3.1.4. ErL-SP Group

Although the cleaning process was completed, delamination, deformation, and melting were observed as seen in the ErL-VSP group. However, there were less undesired effects, and no microcracks were observed (Figure 4a–d).

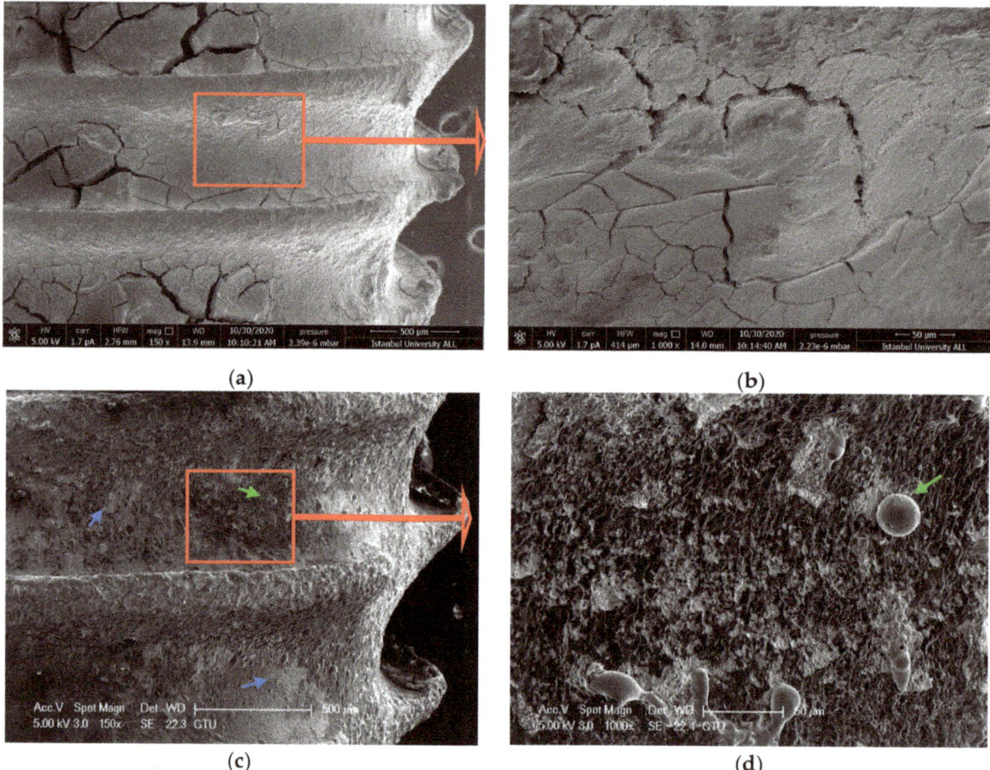

**Figure 4.** (**a**,**b**) SEM images showing the topography of untreated failed implant surfaces in ErL-SP group (150× and 1000×) (**c**,**d**) after intervention (150× and 1000×). The blue arrows indicate delamination and the green arrows indicate melting. Red arrows show the further magnification area.

### 3.1.5. ErL-LP Group

In the ErL-LP group, debridement was achieved by the typical surface appearance of virgin implants at large magnifications (Figure 5a–d). The surface topography is comparable to virgin implant surface properties. No damage was seen on the implant surfaces.

### 3.1.6. ErCrL-1 Group

In the ErCrL-1 group, it was observed that debridement was not totally achieved after laser application (Figure 6c). However, the microscopic appearance of some debrided threads was similar to the original nanomaterial surface, and no damage was observed on the surface (Figure 6d).

### 3.1.7. ErCrL-2 Group

In the ErCrL-2 group, debris was remaining as a layer as in the ErCrL-1 (Figure 7c). The areas where debris remains are observed more clearly at 1000× magnification in the debrided threads (Figure 7d).

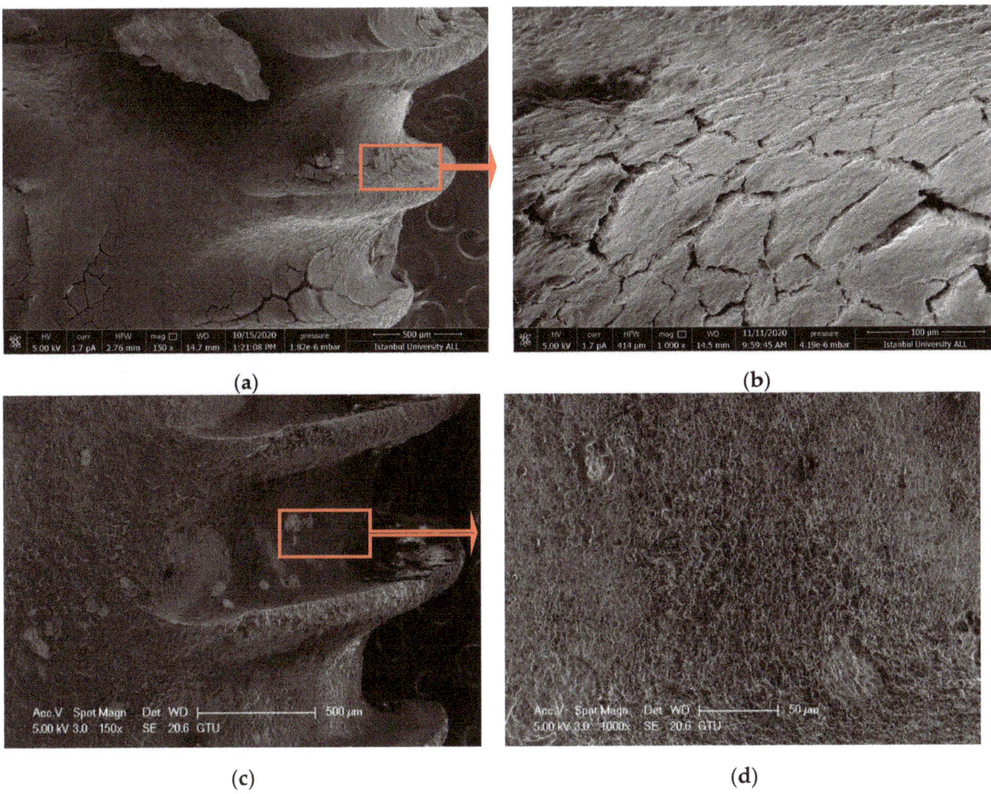

**Figure 5.** (**a**,**b**) SEM images showing the topography of untreated failed implant surfaces in ErL-LP group (150× and 1000×) (**c**,**d**) after intervention (150× and 1000×). Red arrows show the further magnification area.

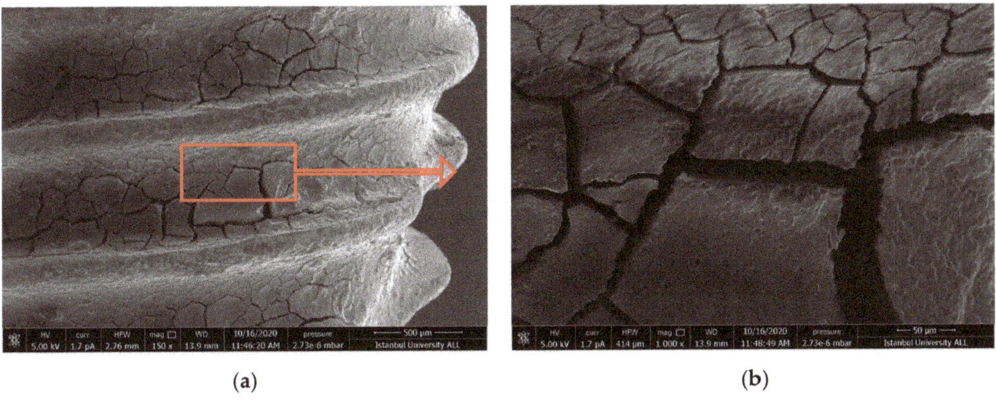

**Figure 6.** *Cont.*

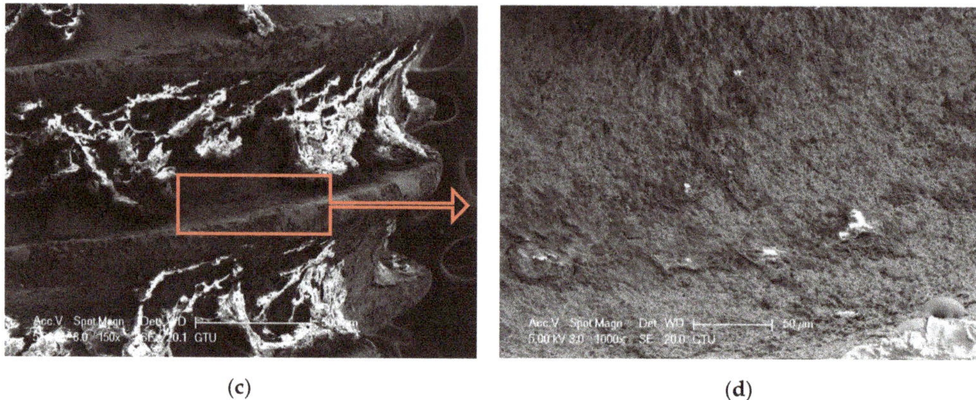

**Figure 6.** (**a**,**b**) SEM images showing the topography of untreated failed implant surfaces in ErCrL-1 group (150× and 1000×); (**c**,**d**) after intervention (150× and 1000×). Red arrows show the further magnification area.

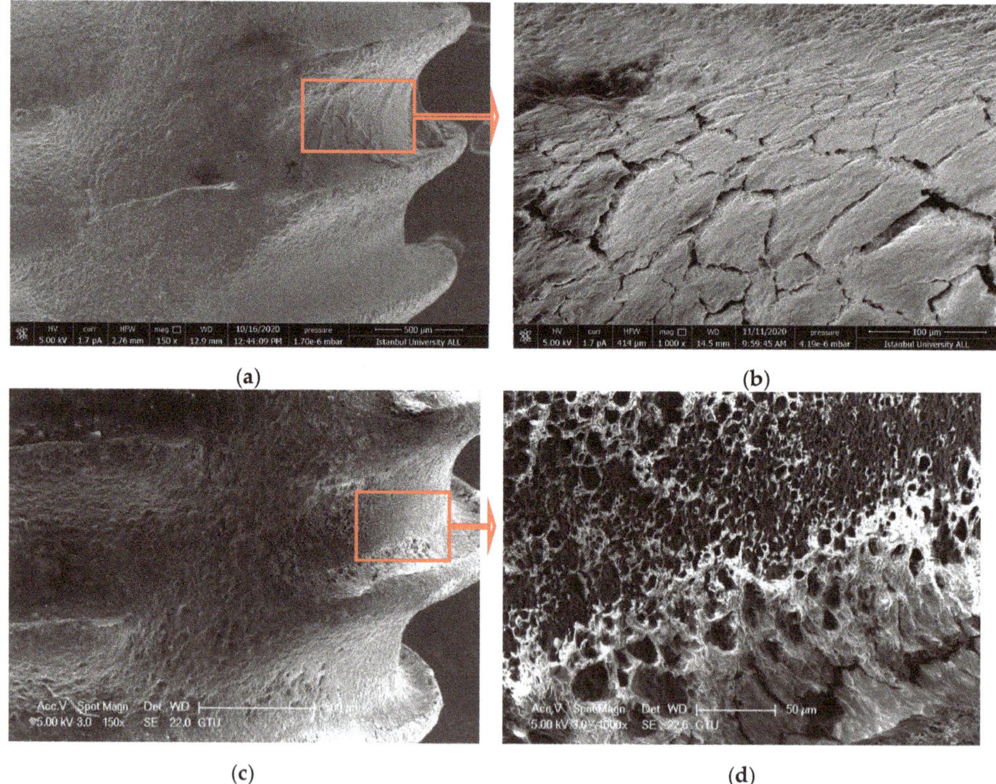

**Figure 7.** (**a**,**b**) SEM images showing the topography of untreated failed implant surfaces in ErCrL-2 group (150× and 1000× magnification); (**c**,**d**) after intervention (150× and 1000× magnification). Red arrows show the further magnification area.

### 3.1.8. Control Group

A nano-porous structure on the surface of the control titanium implants was observed. This nano-topography shows the typical micro-roughness of SLA implants (Figure 8a–d).

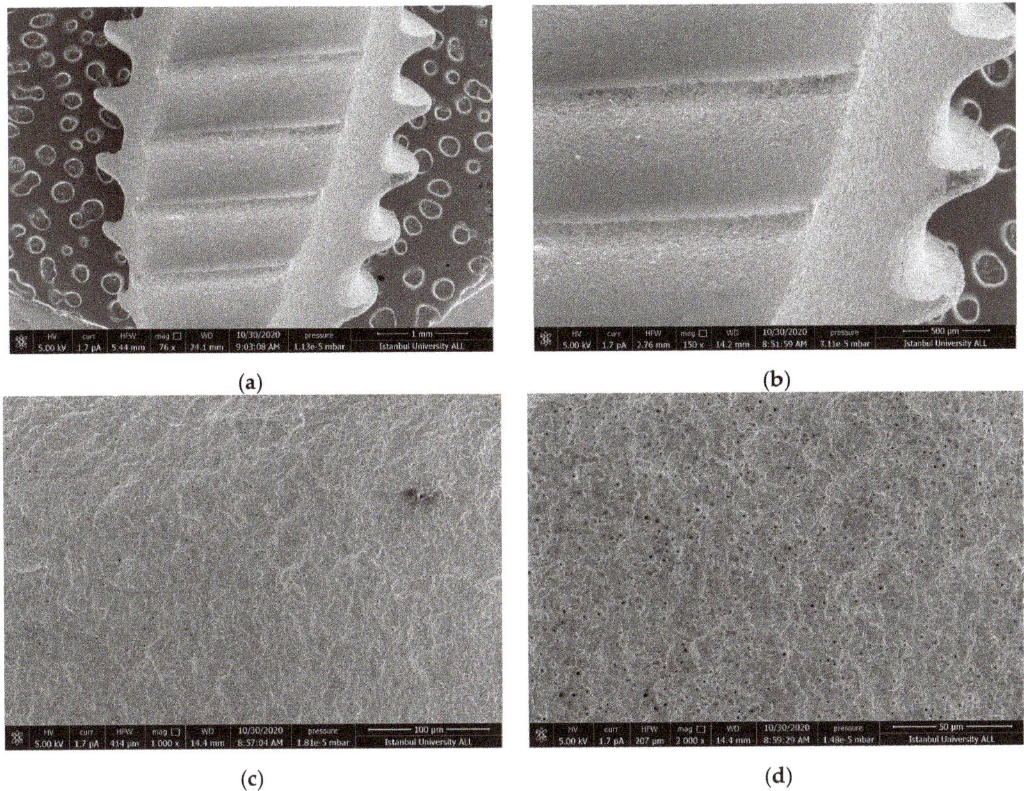

**Figure 8.** SEM images showing the topography original surface of virgin implant (**a**) 76×; (**b**) 150×; (**c**) 1000×; (**d**) 2000× magnification.

### 3.2. EDX Analysis

Higher C and lower Ti values were measured in the intervention groups at baseline [(Figure 9a,b) ($p < 0.05$, $p < 0.05$ respectively)]. All of the intervention groups were contaminated compared to the control group. The contamination was confirmed by the presence of a layer of the debris, and the titanium surface features of SEM analysis. C decreased (Figure 9a) and Ti increased (Figure 9b) as a result of debridement. While the highest percentage of C was found before the debridement of contaminated implants, the lowest C percentage was detected in virgin implants and in ErL groups after the debridement (Figure 9a). C contamination was significantly reduced after debridement procedures in all groups except ErCrL groups ($p < 0.05$).

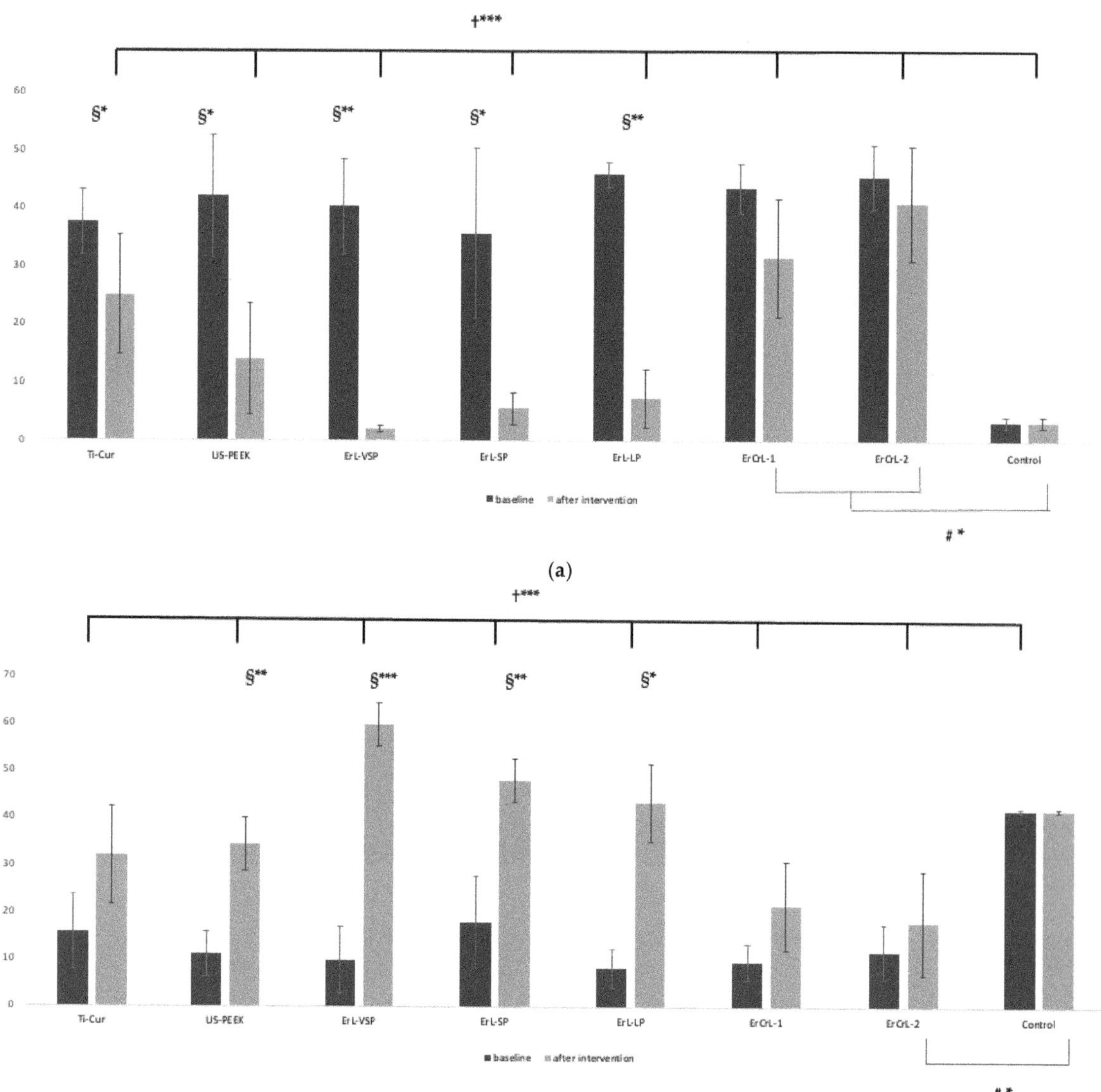

(**a**)

(**b**)

**Figure 9.** *Cont.*

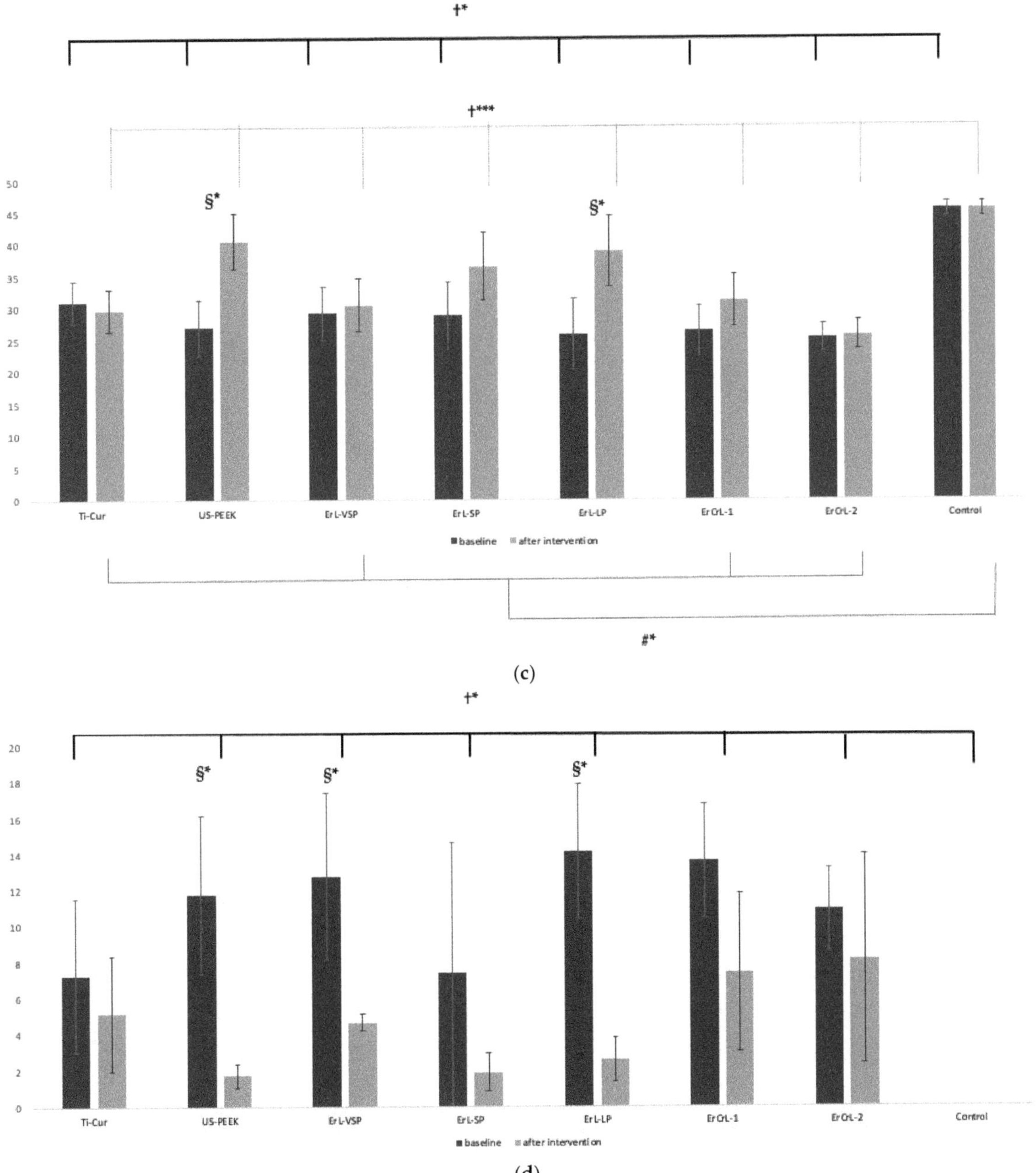

**Figure 9.** (**a**) EDX % C intragroup and intergroup comparisons; (**b**) EDX % Ti intragroup and intergroup comparisons; (**c**) EDX % O intragroup and intergroup comparisons; (**d**) EDX % N intragroup and intergroup comparisons. §: Paired Sample *t* Test †: One-Way ANOVA #: POST HOC Tukey HSD *p*-value * < 0.05 ** < 0.01 *** < 0.001.

O increased in the groups when achieving efficient debridement without surface damage (Figure 9c). According to our results, it is considered that the N also represents the contamination and decreased by debridement (Figure 9d).

### 3.3. Profilometry Analysis

Profilometry analysis results are shown in Figure 10a–p. A flattened three-dimensional topography was seen in the ErL-VSP group after intervention (Figure 10f). ErL-LP (Figure 10j) showed a surface topography similar to virgin implants (Figure 10p).

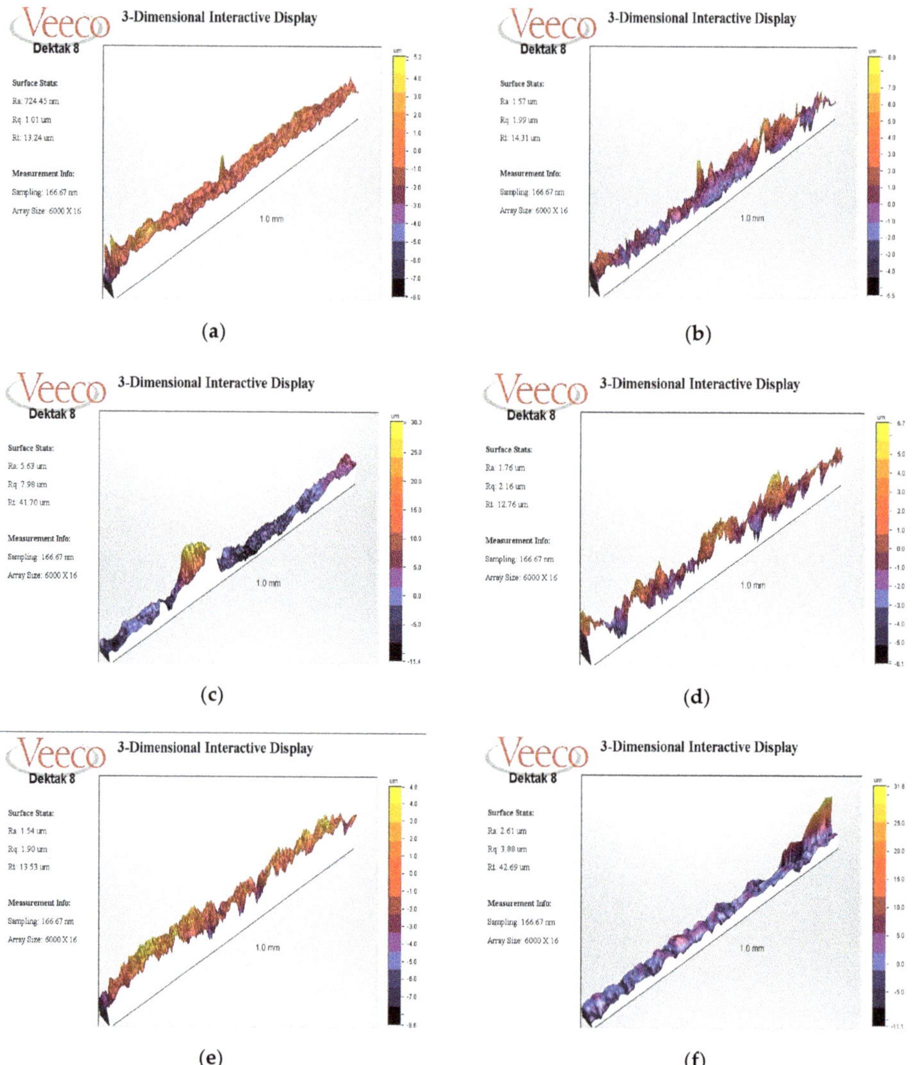

**Figure 10.** *Cont.*

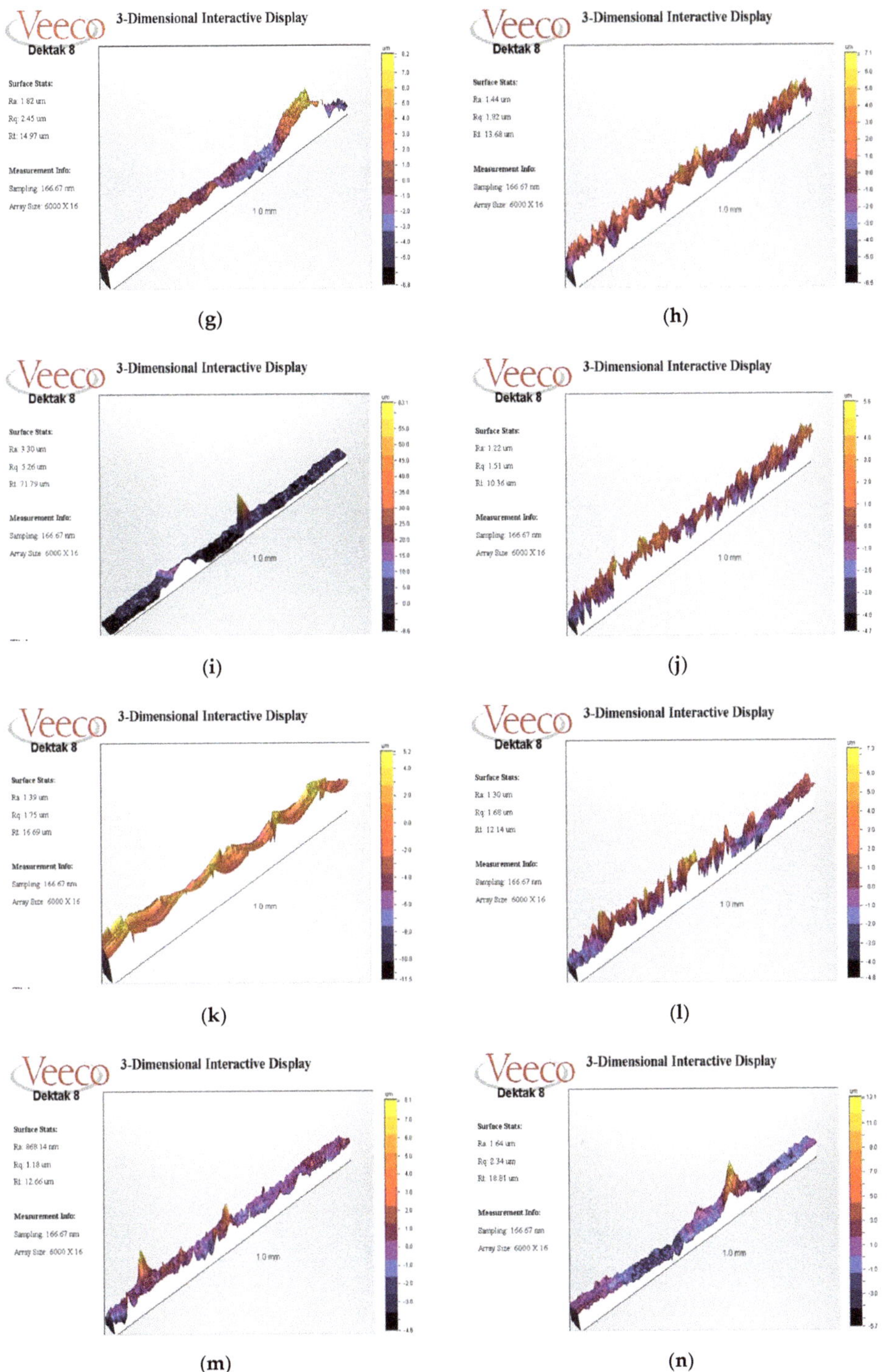

**Figure 10.** *Cont.*

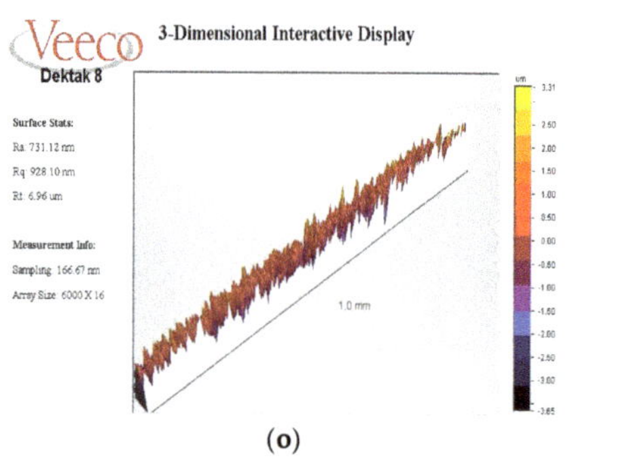
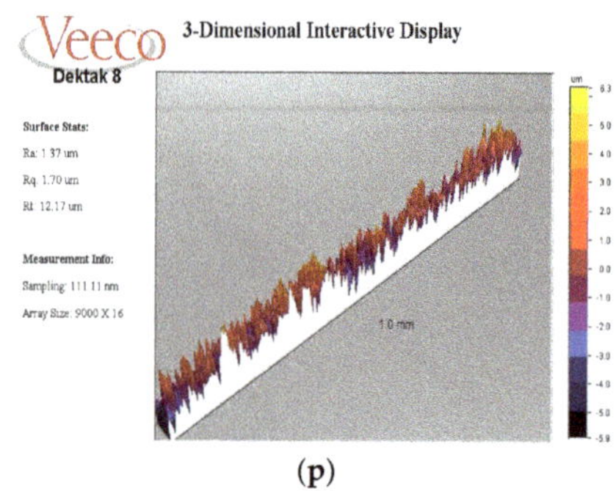

(o)　　　　　　　　　　　　　　(p)

**Figure 10.** Profilometry graphic sample of the Ti-Cur group (**a**) Ra score baseline 724.45 nm; (**b**) after intervention 1.57 μm; (**c**) The US-PEEK group Ra score baseline 5.63 μm; (**d**) after intervention 1.76 μm, In the baseline measurement, a large debris (30.3 μm) in a relatively small area was observed to increase the mean roughness (Ra 5.63 μm). Although an increase in roughness was observed in the graph after intervention, the average roughness value decreased (Ra 1.76 μm); (**e**) the ErL-VSP group Ra score baseline 1.54 μm; (**f**) after intervention 2.61 μm, Although the yellow color scale in the initial measurement graph of this sample turned into a dark color scale due to the reduction of the debris layer in the thread, debris at one point peaked in the graph; (**g**) the ErL-SP group Ra scores baseline 1.82 μm; (**h**) after intervention 1.44 μm; (**i**) The ErL-LP group Ra score baseline 3.30 μm; (**j**) after intervention 1.22 μm, in all implant samples from ErL-SP and ErL-LP groups, the roughness graph of after debridement measurement shows a closer look to the image of control implants than the initial measurement; (**k**) the ErCrL-1 group Ra score baseline 1.39 μm; (**l**) after intervention 1.30 μm; (**m**) the ErCrL-2 group Ra score baseline 868.14 nm; (**n**) after intervention 1.64 μm, in ErCrL-2 group implants, a homogeneous peak-to-valley distribution cannot be observed in the roughness graph after debridement; (**o**) the control group Ra scores 731.12 nm; (**p**) 1.37 μm.

## 4. Discussion

Our aim was to ensure that the debris-covered dirty implants salvage the surface properties and elemental composition of virgin implants without changing the surface morphology after using different debridement methods. When SEM and EDX findings were evaluated together, the groups that were more efficient in re-achieving the typical nano-surface topography by removing the debris without damaging the surface of the SLA surface were ErL-LP and US-PEEK. Although ErL-VSP debrided contaminated implant surfaces more efficiently, undesirable outcomes were seen. The ErL-SP group caused some surface changes as well, and ErCrL groups, Ti-Cur could not remove the debris.

There are several in vitro studies evaluating implant surface properties after different interventions for debridement. However, these studies have some shortcomings. Firstly, most of the studies have been done with short-term biofilm formation on discs [40,41]. Studies that removed hard tissue residue from implant surfaces are limited. Removing a layer of biofilm or hard debris involves completely different interventions. There are few studies that removed the debris layer from the surfaces of implants that were extracted due to peri-implantitis [37,42–44]. As a result, definitive protocols for laser parameters do not exist. In this study, very short pulse, short pulse, and long pulse modes of ErL groups were used for debridement. The effects of different pulses on laser energy levels transmitted to a shorter pulse can have a stronger effect on the targeted area. Therefore, in vitro studies that evaluate the settings of lasers in terms of variant pulse modes along with mJ/pulse and time are vital for clinical application.

IDVI evaluate more objectively by comparing the virgin implant surfaces with dirty implants after debridement. The nanoscale honeycomb appearance of the SLA implant surface was assessed on SEM images (greater than 150× magnifications) of the cleaned implant surfaces. Scratching, melting, and carbonization on the implant surface due to debridement methods were ignored during IDVI scoring. Observers scored only debridement effectiveness (cleanliness) for the implant surface. However, the above-mentioned undesirable effects that occurred during the intervention were noted (Figures 1d and 3d).

In our study, C, Ti, O, and, N elements were evaluated by EDX analysis. While C and N decreased after debridement, an increase was observed in Ti and O. EDX analysis around dirty implants showed that the lower percentages of C and higher Ti when the surface was cleaned [37,44]. In the studies of Scarano et al., increased surface oxide levels, decreased in porosity, and nano-roughness represented a positive change that could protect titanium against bacterial adhesion [43,45]. The study of Takagi et al. reported that C and Ca percentages on dirty surfaces decreased, while Ti percentages increased significantly after debridement with ErL and ErCrL on artificially created calcified areas. Substantial reductions in the percentage of O have also been reported. On the contrary, where there were natural calcifications on the lost implant surface, there was no substantial decrease in the O ratio in all groups [44]. In our study, the lowest O levels were seen in the ErL-VSP group, in which debridement was done thoroughly but some loss of nanostructure occurred. The absence of surface damage on SEM in the ErL-SP and ErL-LP groups made for interesting results of the study. The O ratio increased and reached levels similar to virgin implants in the two interventions. The element of O detected on implant surfaces can be attributed to the titanium dioxide ($TiO_2$) layer, which prevents corrosion of Ti and increases biocompatibility. It was also reported that the thickness of the oxide layer on the implant surface could increase three to four times after implantation compared to pre-implantation [46]. Taken together, these findings suggest that the $TiO_2$ layer remaining on the implant after ErL treatment may be important for healing in the later stages. In the SEM images of the ErL-VSP group, lower O percentages were observed in the areas where surface damage was seen. These results suggest that the measurement of elemental composition in addition to SEM images provides a quantitative assessment of titanium implant surface properties.

Hakki et al. reported that titanium curette was more effective than plastic, carbon, and titanium curettes [42]. However, when they compared the titanium curette with lasers, they reported that the curette left residue. According to the results of our study, similar to the results of Hakki and Takagi, scratches were detected in SEM images of the debridement areas (Figure 1d) [42,44]. Furthermore, the effectiveness of debridement was less than the laser's application. The scratches that occurred by using curettes or ultrasonic devices on the implant alter cell adhesion on the titanium surface and thus effect proper wound healing. Harrel et al. compared titanium, stainless steel curettes, and PEEK ultrasonic tips in terms of metal particle release during debriding titanium implant surfaces. They reported that the PEEK tip had the least amount of titanium particles removed from the surface [47]. In this present study, more scratches and debris layers were observed on the surface of the Ti-Cur compared to the US-PEEK in SEM images (Figures 1c,d and 2c,d). US-PEEK has been the most effective intervention group after the ErL in removing the hard debris layer according to the SEM and EDX analyses.

In the literature, few studies have performed debridement on failed implant surfaces that have been removed due to peri-implantitis [37,42–44]. In addition, the difficulty in comparing these research studies could be due to methodological differences as well as poorly reported laser parameters. When comparing laser devices made by different manufacturers, the optimum energy output differs between lasers. Therefore, it is very important for the clinician to fully understand the differences in the characteristics of different laser devices and to apply erbium lasers effectively and safely for the treatment of peri-implantitis.

When compared to debridement results in laser groups, ErL groups were found superior to ErCrL groups. Although the energy density of the pulse ($19.04\,\text{J}/\text{cm}^2$) in three ErL groups was the same, the applied pulse durations were different. All the ErL groups achieved debridement in the surfaces but two groups had some surface damage. ErL-LP debrided most effectively by achieving the typical surface appearance of virgin implants. It was considered that, because the ErL-LP group had the longest pulse duration (600 µs), the minor surface damages were in this group. On the other hand, although no damage was observed on the debrided threads, all the ErCrL groups failed to complete effective debridement. When the described parameters by the manufacturer were applied, it was calculated that the energy density of ErCrL-1and ErCrL-2 were 4 J/cm$^2$ and 1.8 J/cm$^2$, respectively. When comparing to ErL with ErCrL groups, the energy density parameters were lower in ErCrL groups. The inadequate energy density within the same duration of the application resulted in differing outcomes in terms of effectiveness among the interventions. Application duration is also critical for debridement. The clinical use of all debridement methods within 120 s was selected according to previously reported studies of Er:YAG lasers. Since the Er,Cr:YSGG laser is less efficient when compared to the Er:YAG laser, the Er,Cr:YSGG could give different results in a study design where the application time is not limited or longer. Another reason why Er,Cr:YSGG is less effective may be that the 2.940 μm wavelength of the Er:YAG laser matches the water absorption peak, while the Er,Cr:YSGG laser has an approximately three times lower absorption coefficient in water due to its 2.780 μm wavelength [33].

We evaluated the three-dimensional roughness data together with the SEM images by profilometry, and the Ra values were measured. There were fewer peaks and valleys before debridement procedures when a thicker debris layer masked the typical SLA surface of the implant. It was observed that Ra values measured from these areas increased after debridement procedures. On the contrary, the surface roughness decreased in implants where a relatively clean thread was selected at baseline. Since the hard attachments on the extracted implant surface due to peri-implantitis were not distributed homogeneously, the limited area examined does not represent the whole implant surface. It may be more reliable to measure with techniques that can display the entire implant surface area instead of linear values (Ra values) of a chosen spot.

## 5. Conclusions

The topographic and elemental evaluation of the surface change with SEM, EDX, and profilometry methods as a result of debridement with erbium lasers (Er:YAG and Er,Cr: YSGG) and mechanical debridement methods (titanium curette, ultrasonic device) in implants that have been removed due to peri-implantitis resulted in the following:

ErL-LP was the most efficient in debriding the implant without damaging the surface. Besides a few particles left on the implant surface, US-PEEK was effective as well. ErL-SP and ErL-VSP interventions were also efficient in terms of cleanness, but some surface damage was seen. Ti-Cur could not achieve a thorough cleaning and resulted in some surface scratching. ErCrL was ineffective in this specific application duration and energy density.

**Author Contributions:** Conceptualization, U.B., A.S.-A. and F.Y.; methodology, U.B., A.S.-A., N.A.K.-O. and G.A.-S.; validation, U.B.; formal analysis, U.B., A.S.-A.; investigation, A.S.-A. and A.G.; data curation, A.S.-A., N.A.K.-O. and G.A.-S.; writing—original draft preparation, U.B., G.A.-S., A.S.-A. and N.A.K.-O.; writing—review and editing, A.G. and F.Y.; visualization, U.B., F.Y. and A.G.; supervision, U.B., A.G. All authors have read and agreed to the published version of the manuscript.

**Funding:** This research was funded by the Scientific Research Projects Coordination Unit of Istanbul University, Grant No. TDH-2020-37077.

**Institutional Review Board Statement:** The study was conducted according to the guidelines of the Declaration of Helsinki, and approved by the Ethics Commission of Istanbul University Faculty of Dentistry (protocol code 2020/22 and 30/04/2020).

**Data Availability Statement:** Data are contained within the article.

**Acknowledgments:** The authors thank the Scientific Research Projects Coordination Unit of Istanbul University, Istanbul, Turkey.

**Conflicts of Interest:** This submission of a manuscript implies that the work described has not been published before (except in the form of abstract) and that it is not under consideration for publication elsewhere. The authors declare no conflict of interest.

# References


1. Lindhe, J.; Meyle, J.; Group D of European Workshop on Periodontology. Peri-implant diseases: Consensus Report of the Sixth European Workshop on Periodontology. *J. Clin. Periodontol.* **2008**, *35*, 282–285. [CrossRef] [PubMed]
2. Natto, Z.S.; Aladmawy, M.; Levi, P.A., Jr.; Wang, H.L. Comparison of the efficacy of different types of lasers for the treatment of peri-implantitis: A systematic review. *Int. J. Oral Maxillofac. Implant.* **2015**, *30*, 338–345. [CrossRef] [PubMed]
3. Zitzmann, N.U.; Berglundh, T. Definition and prevalence of peri-implant diseases. *J. Clin. Periodontol.* **2008**, *35*, 286–291. [CrossRef] [PubMed]
4. Albrektsson, T.; Isidor, F. Consensus Report of Session IV. In *Proceedings of the First European Workshop on Periodontology*; Lang, N.P., Karring, T., Eds.; Quintessence Publishing: London, UK, 1994; pp. 365–369.
5. Schwarz, F.; Rothamel, D.; Sculean, A.; Georg, T.; Scherbaum, W.; Becker, J. Effects of an Er:YAG laser and the Vector ultrasonic system on the biocompatibility of titanium implants in cultures of human osteoblast-like cells. *Clin. Oral Implant. Res.* **2003**, *14*, 784–792. [CrossRef]
6. Mombelli, A.; Décaillet, F. The characteristics of biofilms in peri-implant disease. *J. Clin. Periodontol.* **2011**, *38* (Suppl. 11), 203–213. [CrossRef]
7. Berglundh, T.; Armitage, G.; Araujo, M.G.; Avila-Ortiz, G.; Blanco, J.; Camargo, P.M.; Chen, S.; Cochran, D.; Derks, J.; Figuero, E.; et al. Peri-implant diseases and conditions: Consensus report of workgroup 4 of the 2017 World Workshop on the Classification of Periodontal and Peri-Implant Diseases and Conditions. *J. Clin. Periodontol.* **2018**, *45* (Suppl. 20), S286–S291. [CrossRef]
8. Trino, L.D.; Bronze-Uhle, E.S.; Ramachandran, A.; Lisboa-Filho, P.N.; Mathew, M.T.; George, A. Titanium surface biofunctionalization using osteogenic peptides: Surface chemistry, biocompatibility, corrosion and tribocorrosion aspects. *J. Mech. Behav. Biomed. Mater.* **2018**, *81*, 26–38. [CrossRef] [PubMed]
9. Mombelli, A.; Lang, N.P. Antimicrobial treatment of peri-implant infections. *Clin. Oral Implant. Res.* **1992**, *3*, 162–168. [CrossRef]
10. Renvert, S.; Roos-Jansåker, A.M.; Claffey, N. Non-surgical treatment of peri-implant mucositis and peri-implantitis: A literature review. *J. Clin. Periodontol.* **2008**, *35*, 305–315. [CrossRef] [PubMed]
11. Trejo, P.M.; Bonaventura, G.; Weng, D.; Caffesse, R.G.; Bragger, U.; Lang, N.P. Effect of mechanical and antiseptic therapy on peri-implant mucositis: An experimental study in monkeys. *Clin. Oral Implant. Res.* **2006**, *17*, 294–304. [CrossRef] [PubMed]
12. Berglundh, T.; Zitzmann, N.U.; Donati, M. Are peri-implantitis lesions different from periodontitis lesions? *J. Clin. Periodontol.* **2011**, *38* (Suppl. 11), 188–202. [CrossRef]
13. Heitz-Mayfield, L.J.; Lang, N.P. Comparative biology of chronic and aggressive periodontitis vs. peri-implantitis. *Periodontol. 2000* **2010**, *53*, 167–181. [CrossRef]
14. Renvert, S.; Lindahl, C.; Roos Jansåker, A.-M.; Persson, G.R. Treatment of peri-implantitis using an Er:YAG laser or an air-abrasive device: A randomized clinical trial. *J. Clin. Periodontol.* **2011**, *38*, 65–73. [CrossRef]
15. Subramani, K.; Wismeijer, D. Decontamination of titanium implant surface and re-osseointegration to treat peri-implantitis: A literature review. *Int. J. Oral Maxillofac. Implant.* **2012**, *27*, 1043–1054.
16. Fox, S.C.; Moriarty, J.D.; Kusy, R.P. The effects of scaling a titanium implant surface with metal and plastic instruments: An in vitro study. *J. Periodontol.* **1990**, *61*, 485–490. [CrossRef] [PubMed]
17. London, R.M.; Roberts, F.A.; Baker, D.A.; Rohrer, M.D.; O'Neal, R.B. Histologic comparison of a thermal dual-etched implant surface to machined, TPS, and HA surfaces: Bone contact in vivo in rabbits. *Int. J. Oral Maxillofac. Implant.* **2002**, *17*, 369–376.
18. Mengel, R.; Buns, C.E.; Mengel, C.; Flores-de-Jacoby, L. An in vitro study of the treatment of implant surfaces with different instruments. *Int. J. Oral Maxillofac. Implant.* **1998**, *13*, 91–96.
19. Lausmaa, J.; Linder, L. Surface spectroscopic characterization of titanium implants after separation from plastic-embedded tissue. *Biomaterials* **1988**, *9*, 277–280. [CrossRef]
20. Shibli, J.; Vitussi, T.; Garcia, R.; Zenóbio, E.; Tsuzuki, C.; Cassoni, A.; Piattelli, A.; d'Avila, S. Implant Surface Analysis and Microbiologic Evaluation of Failed Implants Retrieved From Smokers. *J. Oral Implantol.* **2007**, *33*, 232–238. [CrossRef]
21. Berglundh, T.; Abrahamsson, I.; Albouy, J.P.; Lindhe, J. Bone healing at implants with a fluoride-modified surface: An experimental study in dogs. *Clin. Oral Implant. Res.* **2007**, *18*, 147–152. [CrossRef] [PubMed]
22. Hayashi, R.; Ueno, T.; Migita, S.; Tsutsumi, Y.; Doi, H.; Ogawa, T.; Hanawa, T.; Wakabayashi, N. Hydrocarbon Deposition Attenuates Osteoblast Activity on Titanium. *J. Dent. Res.* **2014**, *93*, 698–703. [CrossRef] [PubMed]
23. Eick, S.; Meier, I.; Spoerlé, F.; Bender, P.; Aoki, A.; Izumi, Y.; Salvi, G.E.; Sculean, A. In Vitro-Activity of Er:YAG Laser in Comparison with other Treatment Modalities on Biofilm Ablation from Implant and Tooth Surfaces. *PLoS ONE* **2017**, *12*, e0171086. [CrossRef] [PubMed]
24. Koldsland, O.C.; Scheie, A.A.; Aass, A.M. Prevalence of peri-implantitis related to severity of the disease with different degrees of bone loss. *J. Periodontol.* **2010**, *81*, 231–238. [CrossRef] [PubMed]
25. Roos-Jansåker, A.M.; Lindahl, C.; Renvert, H.; Renvert, S. Nine- to fourteen-year follow-up of implant treatment. Part II: Presence of peri-implant lesions. *J. Clin. Periodontol.* **2006**, *33*, 290–295. [CrossRef]
26. Renvert, S.; Polyzois, I.; Claffey, N. Surgical therapy for the control of peri-implantitis. *Clin. Oral Implant. Res.* **2012**, *23* (Suppl. 6), 84–94. [CrossRef]
27. Romanos, G.E.; Weitz, D. Therapy of peri-implant diseases. Where is the evidence? *J. Evid. Based Dent. Pract.* **2012**, *12*, 204–208. [CrossRef]

28. Mailoa, J.; Lin, G.H.; Chan, H.L.; MacEachern, M.; Wang, H.L. Clinical outcomes of using lasers for peri-implantitis surface detoxification: A systematic review and meta-analysis. *J. Periodontol.* **2014**, *85*, 1194–1202. [CrossRef]

29. Sahrmann, P.; Ronay, V.; Hofer, D.; Attin, T.; Jung, R.E.; Schmidlin, P.R. In vitro cleaning potential of three different implant debridement methods. *Clin. Oral Implant. Res.* **2015**, *26*, 314–319. [CrossRef]

30. Kreisler, M.; Götz, H.; Duschner, H. Effect of Nd:YAG, Ho:YAG, Er:YAG, CO2, and GaAIAs laser irradiation on surface properties of endosseous dental implants. *Int. J. Oral Maxillofac. Implant.* **2002**, *17*, 202–211.

31. Schwarz, F.; Nuesry, E.; Bieling, K.; Herten, M.; Becker, J. Influence of an erbium, chromium-doped yttrium, scandium, gallium, and garnet (Er,Cr:YSGG) laser on the reestablishment of the biocompatibility of contaminated titanium implant surfaces. *J. Periodontol.* **2006**, *77*, 1820–1827. [CrossRef]

32. Taniguchi, Y.; Aoki, A.; Mizutani, K.; Takeuchi, Y.; Ichinose, S.; Takasaki, A.A.; Schwarz, F.; Izumi, Y. Optimal Er:YAG laser irradiation parameters for debridement of microstructured fixture surfaces of titanium dental implants. *Lasers Med. Sci.* **2013**, *28*, 1057–1068. [CrossRef]

33. Walsh, J.T., Jr.; Cummings, J.P. Effect of the dynamic optical properties of water on midinfrared laser ablation. *Lasers Surg. Med.* **1994**, *15*, 295–305. [CrossRef]

34. Kamel, M.S.; Khosa, A.; Tawse-Smith, A.; Leichter, J. The use of laser therapy for dental implant surface decontamination: A narrative review of in vitro studies. *Lasers Med. Sci.* **2014**, *29*, 1977–1985. [CrossRef]

35. Al-Hashedi, A.A.; Laurenti, M.; Benhamou, V.; Tamimi, F. Decontamination of titanium implants using physical methods. *Clin. Oral Implant. Res.* **2017**, *28*, 1013–1021. [CrossRef]

36. Matsuyama, T.; Aoki, A.; Oda, S.; Yoneyama, T.; Ishikawa, I. Effects of the Er:YAG laser irradiation on titanium implant materials and contaminated implant abutment surfaces. *J. Clin. Laser Med. Surg.* **2003**, *21*, 7–17. [CrossRef]

37. Nejem Wakim, R.; Namour, M.; Nguyen, H.V.; Peremans, A.; Zeinoun, T.; Vanheusden, A.; Rompen, E.; Nammour, S. Decontamination of Dental Implant Surfaces by the Er:YAG Laser Beam: A Comparative in Vitro Study of Various Protocols. *Dent. J.* **2018**, *6*, 66. [CrossRef] [PubMed]

38. Clem, D.; Heard, R.; McGuire, M.; Scheyer, E.T.; Richardson, C.; Toback, G.; Gwaltney, C.; Gunsolley, J.C. Comparison of Er,Cr:YSGG laser to minimally invasive surgical technique in the treatment of intrabony defects: Six-month results of a multicenter, randomized, controlled study. *J. Periodontol.* **2021**, *92*, 496–506. [CrossRef] [PubMed]

39. Lee, J.-B.; Jo, Y.-H.; Choi, J.-Y.; Seol, Y.-J.; Lee, Y.-M.; Ku, Y.; Rhyu, I.-C.; Yeo, I.-S.L. The Effect of Ultraviolet Photofunctionalization on a Titanium Dental Implant with Machined Surface: An In Vitro and In Vivo Study. *Materials* **2019**, *12*, 2078. [CrossRef] [PubMed]

40. Namour, M.; Verspecht, T.; El Mobadder, M.; Teughels, W.; Peremans, A.; Nammour, S.; Rompen, E. Q-Switch Nd:YAG Laser-Assisted Elimination of Multi-Species Biofilm on Titanium Surfaces. *Materials* **2020**, *13*, 1573. [CrossRef] [PubMed]

41. Yao, W.L.; Lin, J.C.Y.; Salamanca, E.; Pan, Y.H.; Tsai, P.Y.; Leu, S.J.; Yang, K.C.; Huang, H.M.; Huang, H.Y.; Chang, W.J. Er,Cr:YSGG Laser Performance Improves Biological Response on Titanium Surfaces. *Materials* **2020**, *13*, 756. [CrossRef] [PubMed]

42. Hakki, S.S.; Tatar, G.; Dundar, N.; Demiralp, B. The effect of different cleaning methods on the surface and temperature of failed titanium implants: An in vitro study. *Lasers Med. Sci.* **2017**, *32*, 563–571. [CrossRef]

43. Scarano, A.; Sinjari, B.; Di Iorio, D.; Murmura, G.; Carinci, F.; Lauritano, D. Surface analysis of failed oral titanium implants after irradiated with ErCR:YSGG 2780 laser. *Eur. J. Inflamm.* **2012**, *10*, 49–54.

44. Takagi, T.; Aoki, A.; Ichinose, S.; Taniguchi, Y.; Tachikawa, N.; Shinoki, T.; Meinzer, W.; Sculean, A.; Izumi, Y. Effective removal of calcified deposits on microstructured titanium fixture surfaces of dental implants with erbium lasers. *J. Periodontol.* **2018**, *89*, 680–690. [CrossRef]

45. Scarano, A.; Lorusso, F.; Inchingolo, F.; Postiglione, F.; Petrini, M. The Effects of Erbium-Doped Yttrium Aluminum Garnet Laser (Er: YAG) Irradiation on Sandblasted and Acid-Etched (SLA) Titanium, an In Vitro Study. *Materials* **2020**, *13*, 4174. [CrossRef] [PubMed]

46. Textor, M.S., C.; Frauchiger, V.; Tosatti, S.; Brunette, D.M. Properties and Biological Significance of Natural Oxide Films on Titanium and Its Alloys. In *Titanium in Medicine*; Springer: Berlin/Heidelberg, Germany, 2001; pp. 171–230.

47. Harrel, S.K.; Wilson, T.G., Jr.; Pandya, M.; Diekwisch, T.G.H. Titanium particles generated during ultrasonic scaling of implants. *J. Periodontol.* **2019**, *90*, 241–246. [CrossRef] [PubMed]

*Article*

# Ipriflavone-Loaded Mesoporous Nanospheres with Potential Applications for Periodontal Treatment

**Laura Casarrubios [1], Natividad Gómez-Cerezo [2,3], María José Feito [1], María Vallet-Regí [2,3,*], Daniel Arcos [2,3,*] and María Teresa Portolés [1,3,*]**

[1] Departamento de Bioquímica y Biología Molecular, Facultad de Ciencias Químicas, Universidad Complutense de Madrid, Instituto de Investigación Sanitaria del Hospital Clínico San Carlos (IdISSC), 28040 Madrid, Spain; laura.casarrubios.molina@gmail.com (L.C.); mjfeito@ucm.es (M.J.F.)

[2] Departamento de Química en Ciencias Farmacéuticas, Facultad de Farmacia, Universidad Complutense de Madrid, Instituto de Investigación Sanitaria Hospital 12 de Octubre i+12, Plaza Ramón y Cajal s/n, 28040 Madrid, Spain; magome21@ucm.es

[3] CIBER de Bioingeniería, Biomateriales y Nanomedicina, CIBER-BBN, 28040 Madrid, Spain

* Correspondence: vallet@ucm.es (M.V.-R.); arcosd@ucm.es (D.A.); portoles@quim.ucm.es (M.T.P.)

Received: 3 December 2020; Accepted: 18 December 2020; Published: 21 December 2020

**Abstract:** The incorporation and effects of hollow mesoporous nanospheres in the system $SiO_2$–CaO (nanoMBGs) containing ipriflavone (IP), a synthetic isoflavone that prevents osteoporosis, were evaluated. Due to their superior porosity and capability to host drugs, these nanoparticles are designed as a potential alternative to conventional bioactive glasses for the treatment of periodontal defects. To identify the endocytic mechanisms by which these nanospheres are incorporated within the MC3T3-E1 cells, five inhibitors (cytochalasin B, cytochalasin D, chlorpromazine, genistein and wortmannin) were used before the addition of these nanoparticles labeled with fluorescein isothiocyanate (FITC–nanoMBGs). The results indicate that nanoMBGs enter the pre-osteoblasts mainly through clathrin-dependent mechanisms and in a lower proportion by macropinocytosis. The present study evidences the active incorporation of nanoMBG–IPs by MC3T3-E1 osteoprogenitor cells that stimulate their differentiation into mature osteoblast phenotype with increased alkaline phosphatase activity. The final aim of this study is to demonstrate the biocompatibility and osteogenic behavior of IP-loaded bioactive nanoparticles to be used for periodontal augmentation purposes and to shed light on internalization mechanisms that determine the incorporation of these nanoparticles into the cells.

**Keywords:** endocytosis; ipriflavone; mesoporous nanospheres; nanoparticles; oxidative stress; pre-osteoblasts

---

## 1. Introduction

Bioactive glasses are a group of bioceramics that exhibit bone regeneration properties. Since their discovery in 1971, over 1.5 million patients have been treated with Bioglass 45S5, the original four-component Bioglass composition (45 wt % $SiO_2$, 24.5 wt % CaO, 24.5 wt % $Na_2O$, 6 wt % $P_2O_5$). In addition to orthopedic surgery as bone graft substitutes, bioactive glasses applications in dentistry involve their use as dental restorative materials, mineralizing agents, coating material for dental implants, pulp capping and root canal treatment [1]. The first particulate form of Bioglass, trademark PerioGlass®, in 1993, is still sold for the treatment of periodontal defects and has become a standard for the treatment of these types of clinical defects [2].

The research developed during the subsequent decades has resulted in new materials that significantly differs from the original melt-derived Bioglass 45S5. The use of the sol–gel process in the 1990s [3,4], the preparation of bioactive star gels [5] and the development of mesoporous bioactive glasses (MBG) revealed new potential applications in the field of bone tissue regeneration and drug delivery platforms [6–8]. Compared to conventional bioactive glasses, MBGs exhibit higher surface area and porosity, which give them excellent drug-loading ability and superior bone-forming capacity [9–11]. These characteristics make MBGs very attractive as bone-graft material to be used in the regeneration of periodontal bone defects since they can augment the height and bone volume of the alveolar ridge for the insertion of dental implants, whereas they can deliver antibiotic or antiosteoporotic drugs to prevent infection or promote bone healing in the case of patients with diminished bone-forming capability, respectively.

The advances in nanomedicine have opened new research lines involving the synthesis and development of nanoparticles, including carbon-based nanomaterials, hydroxyapatite, iron oxide, zirconia, silica, silver or titania, among others [12]. Thus, nanodentistry is a consequence of the progress in nanomaterials, tissue engineering and nanomedicine, being very beneficial for diagnostic procedures, treatment and prevention of oral and dental diseases. Currently, the use of nanoparticles in dentistry comprises dental filling, reinforcement of dental implants, polishing of enamel surface, prevention of caries, teeth whitening and anti-sensitivity agents [13]. In this sense, recent advances have been made with gold nanoparticles as a biomaterial in dentistry due to their antifungal and antibacterial activity, mechanical properties and availability of different sizes and concentrations [14]. However, the studies focused on the use of bioactive nanoparticles for periodontal bone augmentation are very scarce, and most of them have been carried out with hydroxyapatite nanoparticles [15,16]. In this context, the recent developments in the preparation of mesoporous bioactive glass nanoparticles could provide a very interesting alternative for this purpose [17–20]. On the other hand, the coupling of osteogenesis and angiogenesis is crucial in periodontal tissue regeneration and biomaterials loaded with different agents that act synergistically on both processes are very recently being designed to achieve periodontal regeneration [21].

One of the most interesting strategies to promote bone regeneration under osteoporotic conditions consists of loading bioactive materials with different drugs to treat osteoporotic bone by either promoting the osteogenesis process or inhibiting the activity of osteoclasts, or both [22,23]. Among the drugs used for this purpose, it has been shown that ipriflavone (IP) prevents osteoporosis by inhibiting bone resorption [24]. On the other side, oral administration of IP (1200 mg daily) to subjects diagnosed of primary hyperparathyroidism indicated that this drug has great potential in the therapy of metabolic bone pathologies in which there is high bone turnover [25]. As a nanotherapeutic strategy, different inorganic nanoparticles have been designed for drug incorporation and intraosseous administration in osteoporosis and regenerative therapies for bone diseases [26,27]. This type of administration, with nanoparticles loaded with drugs that will be released inside the bone cells, allows significantly reducing the quantity of drug required to carry out the desired effect.

In the present work, we have evaluated the effects of mesoporous bioactive nanospheres (nanoMBGs) loaded with IP on MC3T3-E1 osteoprogenitor cells, the most relevant model of *in vitro* osteogenesis [28], as a nanotherapeutic strategy to promote bone regeneration. The rationale behind this selection is the osteogenic potential and drug delivery capabilities of nanoMBGs, which could provide an excellent strategy as a bone graft for periodontal defects and also for the treatment of infections and inflammatory processes such as those that occur in periodontitis. These nanospheres are synthesized in the ternary system $SiO_2$–$CaO$–$P_2O_5$ and have shown excellent *in vitro* bioactive behavior in previous studies [29]. Since the effectiveness of treatment with nanoparticles designed for intracellular drug release depends on their efficient incorporation into cells, we have investigated the mechanisms of incorporation of these nanospheres into pre-osteoblasts. Thus, to identify the endocytic mechanisms by which these nanoMBGs are incorporated within the MC3T3-E1 cells, five inhibitors (cytochalasin B, cytochalasin D, chlorpromazine, genistein and wortmannin) were used before the addition of these nanoparticles labeled with fluorescein isothiocyanate (FITC–nanoMBGs). On the other hand, to assess the intracellular action of the drug, the effects of unloaded and IP-loaded nanospheres (nanoMBG–IPs) on MC3T3-E1 pre-osteoblasts were evaluated in a comparative study by analyzing the following cellular parameters: cell viability, apoptosis, cell cycle, intracellular content of reactive oxygen species, intracellular content of $Ca^{2+}$, production of interleukin 6, alkaline phosphatase activity and matrix mineralization. The study of all these parameters is focused on testing the absence of cytotoxicity of nanoMBG–IPs and their potential as a nanotherapeutic strategy for the intracellular delivery of ipriflavone to promote osteogenesis in the periodontal defects. The final aim of this study is to demonstrate the biocompatibility and osteogenic behavior of nanoMBG–IP and to shed light on the mechanisms that rule the incorporation of these nanoparticles into the cells.

## 2. Materials and Methods

### 2.1. Preparation, Characterization and Labeling of Mesoporous SiO₂–CaO Nanospheres

Mesoporous $SiO_2$–CaO–$P_2O_5$ nanospheres (nanoMBGs) were synthesized following the method described in previous work [30]. This method consists of the preparation of an O/W emulsion where two different templates are dissolved. Briefly, poly(styrene)-block-poly(acrylic acid) (PS-b-PAA) with average Mw = 38,000, was dissolved in tetrahydrofuran (THF) and poured on a hexadecyltrimethylammonium bromide (CTAB) water solution. Then, the appropriated amounts of $Ca(NO_3)_2 \cdot 4H_2O$, triethyl phosphate (TEP) and tetraethyl orthosilicate (TEOS) were added dropwise dissolved in water and ethanol, respectively. After 24 h stirring, the product was collected by centrifugation, dried and calcined at 550 °C to remove the organic templates (see Supplementary Materials for a detailed description of the synthesis).

Scanning electron microscopy (SEM) and transmission electron microscopy (TEM) images were collected with a JEOL F-6335 microscope and a JEOL-1400 microscope (JEOL, Tokyo, Japan), respectively.

Textural properties were studied by means of nitrogen adsorption analysis using an ASAP 2020 equipment (Micromeritics, Norcross, GA, USA). For this aim, nanoMBGs were degassed at 150 °C for 15 h. Fourier-transform infrared spectroscopy (FT-IR) was carried out using a Nicolet Magma IR 550 spectrometer (Nicolet Instruments, Madison, WI, USA). In order to collect more information from the surface of the nanoparticles, the spectra were collected by means of the attenuated total reflectance (ATR) sampling technique. Thermogravimetric analysis (TGA) was performed using a TG/DTA Seiko SSC/5200 thermobalance (SEIKO instruments, Chiba, Japan). The samples were heated from 50 to 600 °C at a heating rate of 1 °C min$^{-1}$, using $\alpha$-$Al_2O_3$ as reference.

For fluorescein isothiocyanate (FITC)-labeling, aminopropyl triethoxysilane (APTES) was dissolved in ethanol. Subsequently, 0.6 mg of fluorescein isothiocyanate was added and stirred for 5 h. This solution was added dropwise on the nanoMBG particle suspension, and the labeled particles were washed and collected by centrifugation (see Supplementary Materials for a detailed description of the labeling).

### 2.2. Antiosteoporotic Drug Loading

Ipriflavone (IP) was chosen as an antiosteoporotic drug for this study. For this aim, 300 mg of IP (7-isopropoxy-3-phenyl-4H-1-benzopyran-4-one) were dissolved in 6 mL of acetone as previously reported [31]. Subsequently, 80 mg of nanoMBGs were poured on this solution and stirred in a rotatory incubator at 100 rpm for 24 h. Ipriflafone-loaded nanoparticles (nanoMBG–IP) were filtered and washed with acetone and water, thus removing the excess of IP physically adsorbed on the external surface.

### 2.3. Cell Culture of MC3T3-E1 Pre-Osteoblasts for FITC–NanoMBG Incorporation. Evaluation of the Endocytic Mechanisms for FITC-NanoMBG Cell Entry

Since MC3T3-E1 osteoprogenitor cells are the most relevant model of in vitro osteogenesis [28], this cell line was chosen to investigate the entry mechanisms of these mesoporous bioactive nanospheres labeled with FITC in undifferentiated osteoblasts. This cell line was kindly provided by Dr. B.T. Pérez-Maceda (CIB, CSIC, Madrid, Spain). On the other hand, in this study, we have analyzed the effects of these nanospheres loaded with ipriflavone on the differentiation of pre-osteoblasts into mature osteoblasts, as explained below. For FITC–nanoMBG incorporation studies, MC3T3-E1 pre-osteoblasts ($10^5$ cells/mL) were seeded in 24 well culture plates with Dulbecco's Modified Eagle's Medium (DMEM, Sigma Chemical Company, St. Louis, MO, USA) with fetal bovine serum (FBS, Gibco, BRL, 10% *vol/vol*), 1 mM L-glutamine (BioWhittaker Europe, Verviers, Belgium) and antibiotics (200 µg penicillin and 200 µg streptomycin per mL, BioWhittaker Europe, Verviers, Belgium). Cells were cultured for 24 h in a 5% $CO_2$ incubator at 37 °C, and different doses of FITC–nanoMBGs (10, 30 and 50 µg/mL) were added afterward into the culture medium and maintained several times. Cells were harvested with trypsin-EDTA (0.25%), and FITC–nanoMBG incorporation was quantified through flow cytometry. The FITC–nanoMBG fluorescence was detected in a FACScalibur Becton Dickinson flow cytometer with a 530/30 filter, exciting the sample at 488 nm. The data acquisition and flow cytometric analysis conditions were set through negative and positive controls using the CellQuest Program of Becton Dickinson and maintained for all measurements. A total of $10^4$ cells were analyzed in each sample in order to ensure a correct statistical significance.

To identify the endocytic mechanisms by which these FITC–nanoMBG nanospheres are incorporated within the MC3T3-E1 cells, the inclusion in the culture medium of several specific endocytosis inhibitors was carried out before adding the nanoparticles, maintaining the cells 2 h under these conditions. The endocytosis inhibitors were: 20 µM cytochalasin B (MP Biomedicals, Eschwege, Germany), 4 µM cytochalasin D (MP Biomedicals, Eschwege, Germany), 30 µM chlorpromazine (Enzo Life Sciences, Barcelona, Spain), 3.7 µM genistein (Enzo Life Sciences, Barcelona, Spain), and 23 µM wortmannin (Enzo Life Sciences, Barcelona, Spain). Then, the culture medium was changed by a fresh medium containing 50 µg/mL FITC–nanoMBGs and cells were maintained for 2 h at 37 °C in a 5% $CO_2$ incubator. Finally, cells were collected with trypsin-EDTA (0.25%) and the FITC–nanoMBG incorporation in each case was quantified by flow cytometry as stated above. All the analyses were compared with their respective controls without inhibitors.

*2.4. Cell Size and Complexity Analysis*

To study the cell size and complexity, forward angle (FSC) and side angle (SSC) scatters were detected, respectively, in a FACScalibur Becton Dickinson flow cytometer. A total of $10^4$ cells were analyzed in each sample in order to ensure a correct statistical significance.

*2.5. Cell Viability Studies*

Cell viability was measured by adding 0.005% (wt/vol) propidium iodide (PI) in PBS (Sigma-Aldrich, St. Louis, MO, USA) into the samples to stain the dead cells. The PI exclusion indicates the plasma membrane integrity. PI fluorescence was detected in a FACScalibur Becton Dickinson flow cytometer (Becton Dickinson, San Jose, CA, USA) with a 530/30 filter, exciting the sample at 488 nm. A total of $10^4$ cells were analyzed in each sample in order to ensure a correct statistical significance.

*2.6. Cell-Cycle Analysis and Apoptosis Detection by Flow Cytometry*

Cells in 0.5 mL of PBS were mixed with 4.5 mL of ethanol 70% and maintained overnight at 4 °C. Cell suspensions were then centrifuged for 10 min at 310× $g$ and resuspended in 0.5 mL of RNAsa solution containing 0.1% Triton X-100, 20 µg/mL of IP and 0.2 mg/mL of RNAsa (Sigma-Aldrich, St. Louis, MO, USA). After 30 min of incubation at 37 °C, PI fluorescence was detected in a FACScan Becton Dickinson flow cytometer with a 585/42 filter, exciting the sample at 488 nm. The CellQuest Program of Becton Dickinson was used to calculate the percentage of cells in each cycle phase: $G_0/G_1$ (growth), S (DNA synthesis) and $G_2/M$ (growth and mitosis). To quantify the cell apoptosis, the SubG$_1$ fraction (cells with fragmented DNA) was evaluated. A total of $10^4$ cells were analyzed in each sample in order to ensure a correct statistical significance.

### 2.7. Intracellular Reactive Oxygen Species (ROS) Content

Cell suspensions were incubated for 30 min at 37 °C with 100 µM of 2′,7′-dichlorofluorescein diacetate (DCFH/DA, Serva, Heidelberg, Germany). DCFH/DA can penetrate the cells and can be hydrolyzed by cytosolic esterases, producing DCFH, which is instantly oxidized by ROS to DCF, highly fluorescent and whose fluorescence intensity depends directly on the intracellular content of reactive oxygen species (ROS). DCF fluorescence was measured in a FACScalibur Becton Dickinson flow cytometer with a 530/30 filter, exciting the sample at 488 nm. A total of $10^4$ cells were analyzed in each sample in order to ensure a correct statistical significance.

### 2.8. Confocal Microscopy Studies

Cells were cultured on circular glass coverslips with 50 µg/mL FITC–nanoMBGs in the culture medium for 24 h. Afterward, cells were fixed with p-formaldehyde (3.7%) and permeated, adding 500 µL of Triton-X100 (0.1% in PBS). After 20 min of incubation with BSA (1% in PBS), samples were stained with 100 µL of rhodamine-phalloidin 1:40, washed with PBS and stained with 100 µL of 4′,6-diamidino-2-phenylindole ($3 \times 10^{-6}$ M in PBS, DAPI, Molecular Probes, Inc., Eugene, OR, USA). Finally, samples were observed through a Leica SP2 confocal laser scanning microscope. The fluorescence of rhodamine and DAPI were excited at 540 and 405 nm, respectively, and detected at 565 and 420/586 nm, respectively.

### 2.9. Intracellular Calcium Content

After incubation of cell suspensions for 30 min with the probe Fluo4-AM (5 µM, Thermo Fisher Scientific, Madrid, Spain), which can penetrate the cells and be hydrolyzed by cytosolic esterases, Fluo4 fluorescence was measured in a FACScan Becton Dickinson flow cytometer with a 530/30 filter, exciting the sample at 488 nm. Finally, to check the assay sensitivity, A-23,187 ionophore (5 µM, Sigma-Aldrich, St. Louis, MO, USA) was added to each sample. A total of $10^4$ cells were analyzed in each sample in order to ensure a correct statistical significance.

### 2.10. Alkaline Phosphatase Activity

A total of $2 \times 10^4$ cells/mL were cultured in 24 well plates and maintained for 24 h in a 5% $CO_2$ incubator at 37 °C, with 1 mL/well of culture medium (DMEM with 10% FBS, 1 mM L-glutamine 1 mM and antibiotics), supplemented with 10 mM L-ascorbic acid and 50 µg/mL β-glycerolphosphate in order to promote cell differentiation. To evaluate the nanomaterial effects on alkaline phosphatase (ALP) activity, as a key indicator of osteoblast phenotype expression, 50 µg/mL of nanoMBGs with or without ipriflavone were added into the wells and cells were maintained for 11 days in a 5% $CO_2$ incubator at 37 °C, refreshing the culture medium every 4 days. ALP activity was detected using Reddi and Huggins′ method (Reddi and Huggins, 1972, SpinReact S.A., Girona, Spain), and the obtained values were normalized with respect to total cell protein content, measured using Bradford's method with bovine serum album (BSA) as standard.

### 2.11. Mineralization Assay

A total of $2 \times 10^4$ cells/mL were seeded in 12 well plates and maintained for 24 h in a 5% $CO_2$ incubator at 37 °C, with 1.5 mL/well of culture medium (DMEM with 10% FBS, 1 mM L-glutamine 1 mM and antibiotics), supplemented with 10 mM L-ascorbic acid and 50 µg/mL β-glycerophosphate in order to promote cell differentiation. Then, 50 µg/mL of nanoMBGs with or without ipriflavone were added into the wells and cells were maintained for 11 days in a 5% $CO_2$ incubator at 37 °C, refreshing the culture medium every 4 days. Afterward, the culture medium was removed, and the cell cultures were treated with glutaraldehyde (10%) as a fixer for 1 h. Then, cells were stained with 40 mM Alizarin Red at pH 4.2 for 45 min in order to analyze the matrix mineralization. Finally, the stained extracellular deposits were dissolved with cetylpyridinium chloride (10% at pH 7), and the absorbance of the supernatants was measured at 620 nm.

### 2.12. Interleukin 6 (IL-6) Detection

The concentration of IL-6 secreted to the culture medium by $2 \times 10^4$ cells/mL, after treatment with 50 µg/mL of nanoMBGs with or without ipriflavone, was measured using an ELISA IL-6 kit (Gen-Probe, Diaclone). This method is based on a sandwich ELISA where plates are pre-coated with a capture antibody highly specific for IL-6 and, after the incubation with the samples, a biotinylated secondary antibody is added, and the correct unions are revealed with streptavidin-avidin conjugated with horseradish peroxidase in a colorimetric reaction which is quantified in an ELISA Plate Reader at 450 nm, with a sensitivity of 10 pg/mL and an inter-assay variation coefficient <10%. Recombinant cytokine was adopted as standard.

### 2.13. Statistics

The results obtained appear as means of three replicate experiments plus their standard deviations, analyzed with the 22nd version of Statistical Package for the Social Sciences (SPSS). Statistical comparisons were carried out with the analysis of variance (ANOVA), and Scheffé and Games–Howell test was employed for post hoc analysis of differences between study groups, considering $p < 0.005$ as statistically significant.

## 3. Results and Discussion

### 3.1. Characterization of Mesoporous Nanospheres

Prior to any biological assay, the main physic-chemical features of the nanoparticles must be determined. For this purpose, electron microscopy (SEM and TEM) experiments, textural properties determination and FTIR analysis before and after drug-loading was carried out. Figure 1a shows an SEM image of nanoMBGs, pointing out that this material is made of non-aggregated spheres ranging in size between 150 and 250 nanometers. The spheres show porosity accessible to the external surface. TEM image (Figure 1b) provides more detailed information about the porous structure of nanoMBG spheres. The TEM image evidence that our spheres are composed of an inner cavity of about 100 nm in diameter, surrounded by a shell that exhibits a radial porosity. These two types of porosity are clearly reflected in the nitrogen adsorption/desorption isotherm shown in Figure 1c. The adsorption isotherm corresponds to a highly porous material with high surface area (see Table 1) and with a wide hysteresis loop type H2, characteristic of ink bottle-like pore as a clear reflection of the wide central cavity connected to the narrow necks of the radial pores of the shell. Finally, FTIR spectra evidence the presence of ipriflavone after the loading process (Figure 1d) with the characteristic absorption band of this compound (see Figure S1 in Supplementary Materials). Thermogravimetric analysis indicated 18% in weight of ipriflavone-load (see Figure S2 in Supplementary Materials), and the decrease of the textural parameters also evidence that the drug is filling or even occluding the pores of the spheres (Table 1).

### 3.2. Effects of NanoMBGs and NanoMBG-IPs on Size, Complexity, Apoptosis and Cell Cycle of MC3T3-E1 Pre-Osteoblasts

Once the main physic-chemical characteristics of nanoMBGs were determined, we proceeded to assess the potentially deleterious effects that these nanoparticles could exert on pre-osteoblast in terms of cell size, complexity, apoptosis or harmful variations in the cell cycle. No changes in pre-osteoblast size and complexity (FSC and SSC, respectively) were observed after the intracellular incorporation of nanoMBGs or nanoMBG–IPs (see Figure S3 in Supplementary Materials). In this context, we have observed in previous studies with MC3T3-E1 pre-osteoblasts that the incorporation of another type of nanoparticles, such as graphene oxide nanosheets, produced alterations as the increase in cell size (FSC) without changes in cell complexity (SSC) [32]. However, previous studies with RAW 264.7 macrophages and nanoMBGs evidenced a significant increase of macrophage complexity (SSC) after the treatment with these nanospheres due to their uptake by macrophages [29]. It is well known that these cell parameters, FSC and SSC, depend on different factors such as the cell surface and some organelles (lysosomes, mitochondria, nucleus or pinocytic vesicles) as well as on the presence of granulated material within the cell [33].

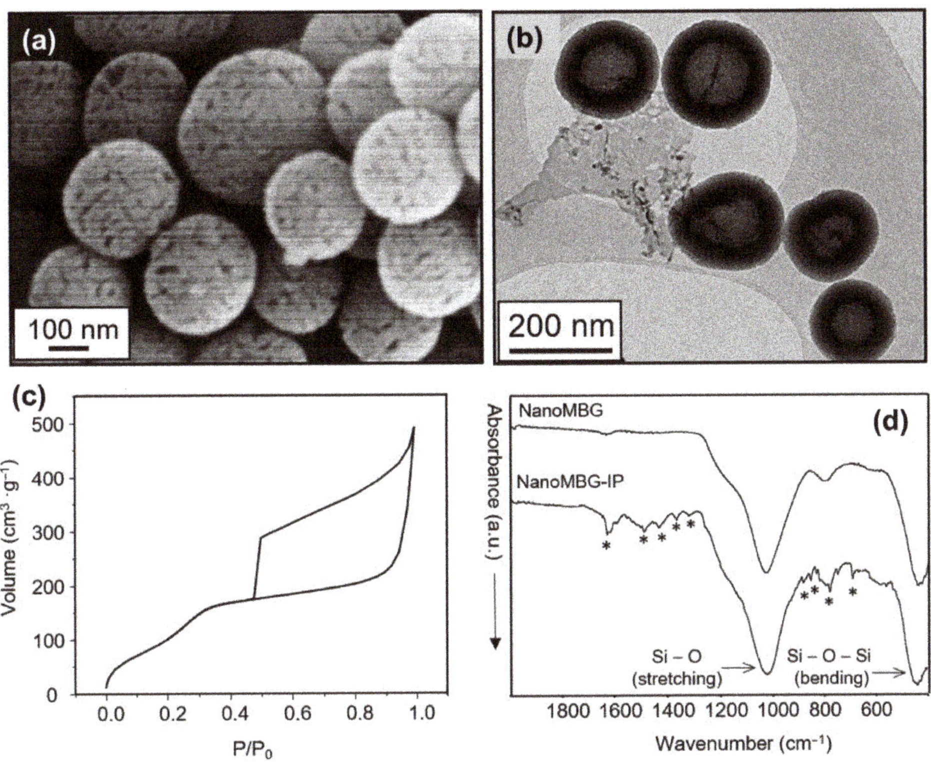


**Figure 1.** Characterization of mesoporous nanospheres. (**a**) scanning electron micrograph of hollow mesoporous nanospheres in the system $SiO_2$–CaO (nanoMBG) spheres. (**b**) Transmission electron image of nanoMBG spheres. (**c**) Nitrogen adsorption/desorption isotherm of nanoMBG spheres. (**d**) FTIR spectra of nanoMBG and IP-loaded nanospheres (nanoMBG–IP) spheres (* indicates the absorption bands corresponding to ipriflavone).

**Table 1.** Textural properties for nanoMBG and nanoMBG–IP spheres measured by $N_2$ adsorption.

| Sample | Surface Area $(m^2 \cdot g^{-1})$ | Pore Volume $(cm^3 \cdot g^{-1})$ | Pore Size (nm) |
|---|---|---|---|
| **nanoMBG** | 543.6 | 0.435 | −2.5 nm |
| **nanoMBG–IP** | 14.4 | 0.057 | NA |

The effects of nanoMBG and nanoMBG–IP nanospheres on cell cycle phases ($G_0/G_1$, S and $G_2/M$) of MC3T3-E1 pre-osteoblast and the percentage of cells in apoptosis (SubG$_1$ fraction) were analyzed. Figure 2 shows that the treatment with 50 μg/mL of nanospheres without ipriflavone for 24 h did not induce alterations on $G_0/G_1$, S and $G_2/M$ phases. In the same way, nanoMBG–IPs did not induce changes in $G_0/G_1$ and $G_2/M$ phases. Nevertheless, a significant increment ($p < 0.005$) of the synthesis phase (S) was observed after the incubation of the MC3T3-E1 pre-osteoblasts with 50 μg/mL of nanoMBG–IPs, thus evidencing the protective impact of the ipriflavone into these cells. Moreover, nanoMBG and nanoMBG–IPs did not induce apoptosis on MC3T3-E1 pre-osteoblasts, detected as SubG1 fraction, in comparison with control cultures.

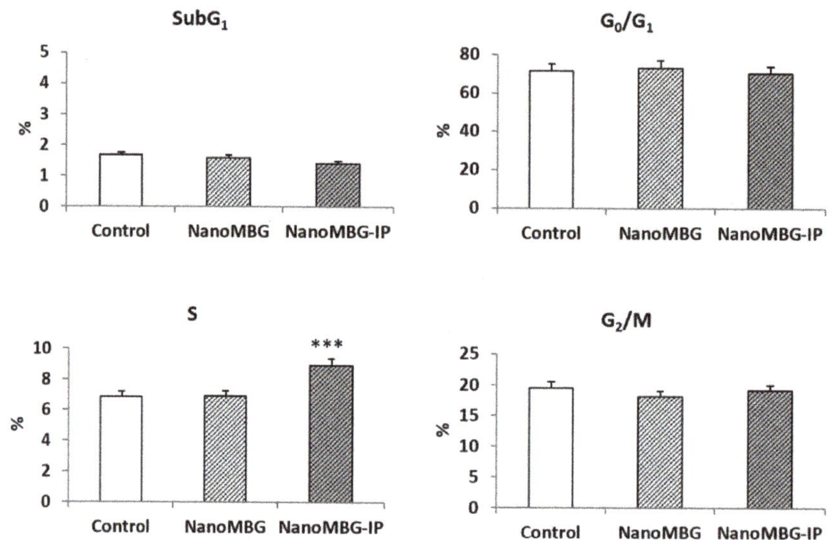


**Figure 2.** Effects of nanoMBGs and nanoMBG–IPs on cell cycle phases of MC3T3-E1 pre-osteoblasts and apoptosis percentage (Sub G$_1$ fraction) after 24 h of treatment with 50 μg/mL of nanospheres. Control conditions without nanospheres were performed at the same time. Statistical significance: *** $p < 0.005$.

### 3.3. Effects of NanoMBGs and NanoMBG-IPs on Viability, Intracellular Reactive Oxygen Species (ROS) and Calcium Content of MC3T3-E1 Pre-Osteoblasts

Although no adverse effects on cell cycle were observed and IP evidenced a protective impact, the incorporation of nanoparticles could trigger an increment of the intracellular content of reactive oxygen species (ROS), oxidative stress, a decrease of cell viability and toxicity mechanisms [34]. On the other hand, bioactive mesoporous materials exhibit a high capability for releasing $Ca^{2+}$ and other ions such as soluble silicate that can stimulate the proliferation and differentiation of osteoblasts [35–37], inducing bone regeneration due to the release of these two ions [38]. Considering all these facts, in the present work, we have evaluated the cell viability, intracellular content of ROS and cytosolic calcium of MC3T3-E1 pre-osteoblasts after treatment with nanoMBGs and nanoMBG–IPs. Control conditions without nanospheres were performed at the same time. Figure 3 shows the obtained results. The fluorescence profiles of control cells, cells with Fluo4 and cells with Fluo4 plus A23187 ionophore are also shown in the lower-left figure. The fluorescence increase observed after the addition of A23187 ionophore to the cells demonstrates the sensitivity of the assay. No viability changes but significant decreases of both intracellular ROS and calcium content were observed after incubation with 50 μg/mL of nanoMBGs and nanoMBG–IPs. These results evidence the absence of oxidative stress or toxicity caused by these nanospheres in MC3T3-E1 pre-osteoblasts after their uptake.

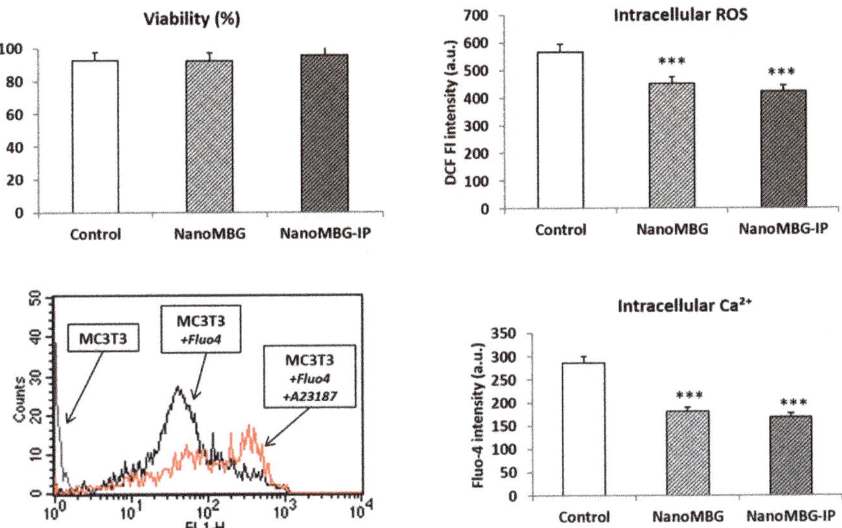

**Figure 3.** Effects of 50 µg/mL of nanoMBG and nanoMBG–IP nanospheres on viability, intracellular content of reactive oxygen species (ROS) and cytosolic calcium of MC3T3-E1 pre-osteoblasts, after 24 h of incubation. Control conditions without nanospheres were performed at the same time. Fluorescence profiles of control cells, cells with Fluo4 and cells with Fluo4 plus A23187 ionophore are shown in the lower-left figure. Statistical significance: *** $p < 0.005$.

### 3.4. Effects of NanoMBGs and NanoMBG-IPs on Differentiation of MC3T3-E1 Pre-Osteoblasts

The set of results obtained and described in Sections 3.2 and 3.3 evidence the excellent behavior in terms of cell viability and the absence of cytotoxicity of nanoMBGs and nanoMBG–IP. However, the application as osteoregenerative material requires the capability to stimulate the differentiation of the pre-osteoblasts toward the osteoblastic phenotype. The differentiation process of the MC3T3-E1 pre-osteoblasts includes three successive phases: (a) initial stage with active cell proliferation, but without expression of differentiation markers such as alkaline phosphatase (ALP) or mineral depositions; (b) intermediate stage with the maturation of the matrix and a high expression of ALP; and (c) final stage with matrix mineralization characterized by the presence of mineral depositions due to ALP activity [39,40]. On the other hand, ipriflavone is a synthetic drug that prevents osteoporosis by inhibiting bone resorption and maintaining bone thickness [24]. Thus, the use of nanoMBG–IPs for intracellular delivery of this drug could be a nanotherapeutic strategy to promote bone regeneration. In this context, we evaluate the impact of nanoMBG–IPs on MC3T3-E1 pre-osteoblast differentiation as a prototype of in vitro osteogenesis through the measurement of ALP activity and the quantification of matrix mineralization as key markers of MC3T3-E1 cell differentiation after 11 days of treatment with different doses of these nanospheres. Controls without nanospheres and with nanoMBGs, but without ipriflavone were performed at the same time.

Figure 4 shows that the cell incorporation of nanoMBG without ipriflavone induced a decrease of ALP activity compared to control cells after 11 days of incubation with 10 and 50 µg/mL. However, significant increases of ALP activity were observed after treatment with 5, 10 and 50 µg/mL of these nanospheres loaded with ipriflavone (nanoMBG–IPs), thus indicating the efficient intracellular release of IP and its positive in vitro effect on osteogenesis. The effect of the highest dose (50 µg/mL) of nanoMBG–IPs was lower than the obtained with 5 and 10 µg/mL of nanoMBG–IPs, evidencing the convenience of using lower doses than 50 µg/mL.

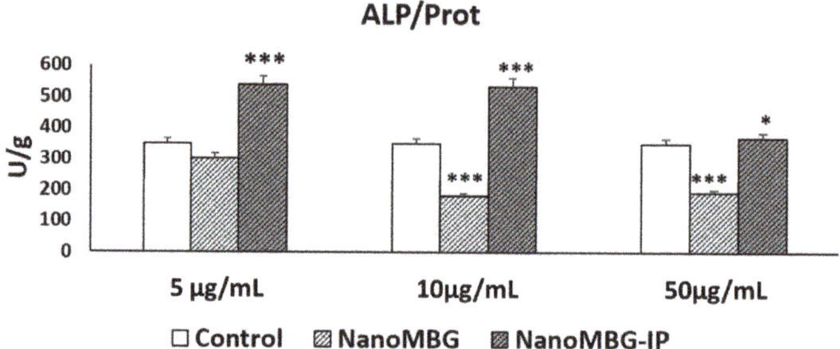

**Figure 4.** Effects of several doses of nanoMBGs and nanoMBG–IPs on MC3T3-E1 pre-osteoblast differentiation after 11 days, evaluated through the measurement of alkaline phosphatase (ALP) activity. Control conditions without nanospheres were performed at the same time. Statistical significance: *** $p < 0.005$, * $p < 0.05$.

Increases of matrix mineralization were detected after the incubation with 50 μg/mL of nanoMBGs and nanoMBG–IPs for 11 days, but these effects were not statistically significant (Figure S4, Supplementary Materials), probably as a result of the lower precision and sensitivity of this test.

The ALP activity results demonstrate the efficient intracellular release of the drug from the nanoMBG–IPs and suggest their potential application as intracellular drug delivery systems in a nanotherapeutic strategy to promote bone regeneration.

Regarding the effects of other nanoparticles on MC3T3-E1 cell differentiation, in previous studies with this cell type and graphene oxide (GO) nanosheets, we observed that the treatment with 40 μg/mL of 400 nm PEG-GO for 3 days did not affect the differentiation process 12 days after the intracellular uptake of the nanomaterial [32].

*3.5. Effects of NanoMBGs and NanoMBG-IPs on Interleukin 6 (IL-6) Production by MC3T3-E1 Pre-Osteoblasts*

Despite having demonstrated the absence of cytotoxicity of these nanospheres and having observed their capability to promote pre-osteoblast differentiation, the inflammatory response that any kind of nanoparticles could elicit should be evaluated. In this sense, the detection *in vitro* of inflammatory cytokines provides valuable information about these potential clinical complications. IL-6 is produced by many cells, including osteoblasts, monocytes, macrophages and bone marrow mononuclear cells [41]. In bone, this cytokine induces osteoclast differentiation [42,43]. On the other hand, recent studies in a murine model have shown the IL-6 is related to the processes of revascularization and bone formation after ischemic osteonecrosis [44]. In the present work, we have quantified the levels of IL-6 secreted by cultured MC3T3-E1 pre-osteoblasts after incubation with nanoMBGs and nanoMBG–IPs. Figure 5 suggests that no significant changes of IL-6 secretion were detected after treatment with these nanospheres.

With respect to IL-6, it is important to note that this cytokine and tumor necrosis factor-alpha (TNF-α) play a key role in the inflammatory response, infection and stress [45]. In the present work, no significant changes of *in vitro* IL-6 secretion by pre-osteoblasts were detected after nanoMBG and nanoMBG–IP treatment, thus indicating that the local nanomaterial administration *in vivo* would not trigger the production of this pro-inflammatory cytokine and would not activate the innate immune system. These results agree with the switch of the M1 pro-inflammatory macrophage phenotype to the M2 reparative phenotype previously observed [29].

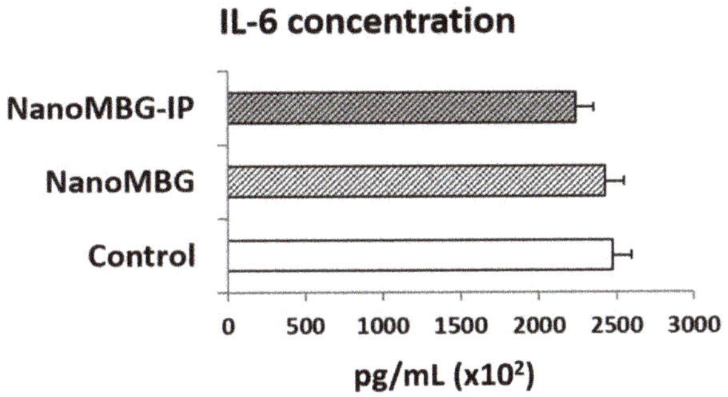

## IL-6 concentration

$$\text{pg/mL (x10}^2)$$

**Figure 5.** Effects of nanoMBGs and nanoMBG–IPs on interleukin 6 (IL-6) production by cultured MC3T3-E1 pre-osteoblasts after treatment with 50 µg/mL of nanospheres for 24 h. Control conditions without nanospheres were performed at the same time.

The results obtained so far evidence not only the excellent biocompatibility of nanoMBG–IP but also their capability to promote pre-osteoblasts differentiation towards osteoblast phenotype, thus confirming the osteogenic potential of nanoMBG–IP. In this sense, the intracellular release of the drug seems to play an important role in this process. The following experiments were carried out to shed some light on the mechanism that rules the incorporation of these nanoparticles within cells.

### 3.6. Uptake of NanoMBGs by MC3T3-E1 Pre-Osteoblasts

In order to evaluate the nanoparticles uptake by pre-osteoblast cells, nanoMBG nanospheres were labeled with FITC. As a first approach, MC3T3-E1 pre-osteoblasts were cultured for 15, 30 and 60 min with 10, 30 and 50 µg/mL of FITC–nanoMBG. The cells were then detached, and the amount of cell-associated fluorescence was detected by flow cytometry as a measure of the intracellular uptake of these nanospheres. As can be observed in Figure 6A, the fluorescence intensity of osteoprogenitor cells after each treatment reveals a fast and dose-dependent FITC–nanoMBG uptake after 15 min. On the other hand, a decrease in fluorescence related to the intracellular content of these nanospheres was observed after 60 min of treatment with all the doses used (Figure 6A). This fact indicates that, after FITC–nanoMBG uptake, the exocytosis of this nanomaterial also occurs, according to the process described for other nanoparticles in mammalian cells [46].

For confocal microscopy studies, the dose of 50 µg/mL of FITC–nanoMBGs and 24 h time were chosen to observe if the intracellular uptake of this nanomaterial by MC3T3-E1 pre-osteoblasts could damage the cytoskeleton structure in these conditions of high dose and longer treatment time. Control cultures without this nanomaterial were performed at the same time. Figure 6B shows the abundance of nanospheres in the cytoplasm of the pre-osteoblasts and the integrity of their morphology. The results evidence that the incorporation of FITC–nanoMBGs did not induce changes in the pre-osteoblast cytoskeleton.

### 3.7. Endocytic Mechanisms for FITC–NanoMBG Entry into MC3T3-E1 Pre-Osteoblasts

Five specific endocytosis inhibitors were added into the culture wells before the nanomaterial addition in order to identify the endocytic mechanisms by which these FITC–nanoMBG nanospheres are incorporated within the MC3T3-E1 cells. This indirect method consists of pretreating the cells with different inhibitors that specifically block a certain mechanism of endocytosis. In this way, when the inhibitor used reduces the entry of the nanospheres, we can know that this mechanism that has been blocked constitutes an entry route. On the contrary, if the inhibitor does not decrease the entry of the nanospheres, we will know that the mechanism that is blocking the inhibitor is not involved in the entry of the nanospheres. Figure 7 shows a scheme of the assay, a table with the mechanism affected by each inhibitor (its specific action and the corresponding reference) and a graph with the effects of these agents on the FITC–nanoMBG uptake. Previous studies allowed us to choose the dose of the different inhibitors [47–51]. The results showed two incorporation mechanisms for FITC–nanoMBG entry into MC3T3-E1 pre-osteoblasts.

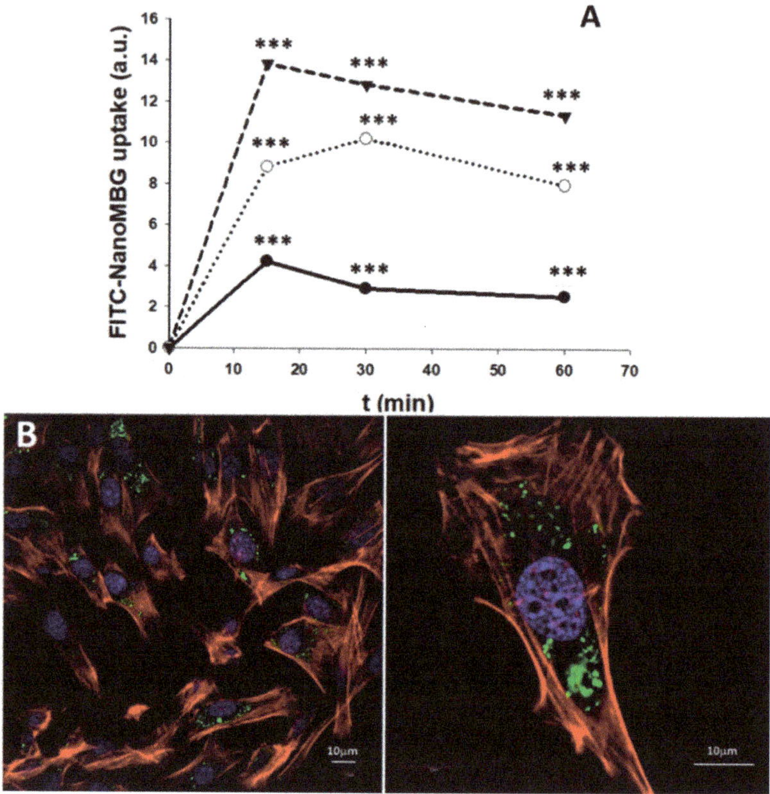

**Figure 6.** Intracellular uptake of nanoMBG nanospheres labeled with FITC by MC3T3-E1 pre-osteoblasts. (**A**) flow cytometric analysis of fluorescence intensity of cells with intracellular FITC–nanoMBG nanospheres after incubation with 10 (●), 30 (○) and 50 µg/mL (▼) for different times (15, 30 and 60 min). Statistical significance: *** $p < 0.005$. (**B**) Confocal microscopy images of MC3T3-E1 pre-osteoblasts after 24 h of incubation with 50 µg/mL of nanoMBG nanospheres labeled with fluorescein isothiocyanate (FITC). Nuclei were stained with DAPI (blue), F-actin filaments were stained with rhodamine-phalloidin (red), and FITC–nanoMBGs are observed in green.

Cytochalasins B and D, which block actin polymerization and inhibit macropinocytosis, reduce the FITC–nanoMBG incorporation by pre-osteoblasts, although only the effect of Cytochalasin B was significant ($p < 0.05$, Figure 3). Chlorpromazine is an inhibitor of clathrin-dependent mechanisms, and this agent produced a very pronounced diminution ($p < 0.005$) of FITC–nanoMBG incorporation by MC3T3-E1 cells, thus indicating that the clathrin-dependent endocytic mechanism is the main route implicated in the entry of these nanospheres into pre-osteoblasts. Previous studies with nanosheets of graphene oxide and Saos-2 osteoblasts evidenced that these nanosheets can enter in mature osteoblasts through pathways dependent on microtubules [51]. It is important to note that the mechanisms of entry of nanomaterials into cells depend on the cell type and the characteristics of the nanoparticles. In the present study, the treatment with either wortmannin or genistein did not trigger significant changes on FITC–nanoMBG incorporation by pre-osteoblasts. Wortmannin blocks the activity of phosphoinositide 3-kinase (PI3K) and phosphoinositide 4-kinase (PI4K) [47], with key roles in cell development and growth as adhesion, apoptosis, cytoskeletal organization, motility, proliferation, thus preventing phagocytosis mechanisms [52]. Genistein blocks Src tyrosine kinases and the dynamics of caveolae [50], and no differences were observed in the uptake of these nanospheres when it was present in the cell culture. Since wortmannin and genistein did not reduce FITC–nanoMBG uptake by pre-osteoblasts, we can conclude that neither phagocytosis nor caveolae-mediated incorporation is routes implicated in the *in vitro* uptake of these nanospheres by MC3T3-E1 cells.

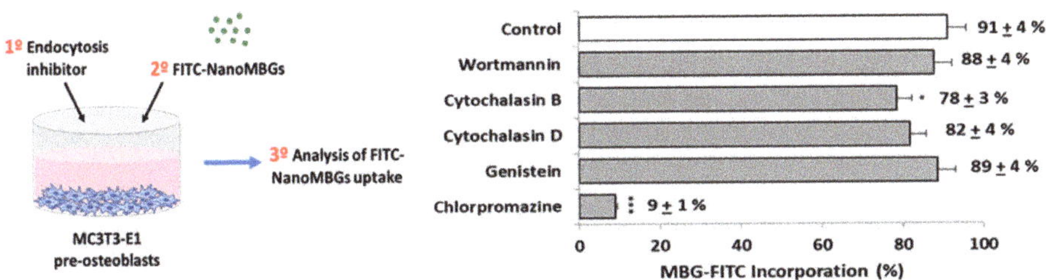


| INHIBITOR | AFFECTED MECHANISM | ACTION | REFERENCE |
|---|---|---|---|
| **WORTMANNIN** | Phagocytosis | Irreversible inhibition of phosphatidylinositol 3-kinase (PI3K) | Bandmann, V. *et al.* 2012 |
| **CYTOCHALASIN B** | Macropinocytosis | Inhibition of actin polymerization preventing microfilaments action | Sato, K. *et al.* 2009 |
| **CYTOCHALASIN D** | Macropinocytosis | Inhibition of actin polymerization preventing microfilaments action and other endocytosis pathways | Mäger, I. *et al.* 2012 |
| **GENISTEIN** | Clathrin-independent endocytosis | Inhibition of Src tyrosine kinases and caveolae dynamics | Schulz, W.L. *et al.* 2012 |
| **CHLORPROMAZINE** | Clathrin-dependent endocytosis | Inhibition of clathrin-coat assembly and alteration of membrane fluidity | Mäger, I. *et al.* 2012 |

**Figure 7.** Inhibitory effects of several endocytosis inhibitors on FITC–nanoMBG uptake by MC3T3-E1 pre-osteoblasts. Cells were incubated with each inhibitor for 2 h, the medium was then removed, and the cultures were treated with 50 µg/mL FITC–nanoMBGs for 2 h. Statistical significance: * $p < 0.05$, *** $p < 0.005$.

## 4. Conclusions

The novelty of this work is the knowledge of the effects of ipriflavone-loaded mesoporous nanospheres on the differentiation of bone-forming cells. In previous studies, the effects of these nanoparticles on already differentiated osteoblasts in coculture with osteoclasts were analyzed [25], but until now, their effects on osteoprogenitor cells were unknown. Another of the novel objectives of the present work was to understand the mechanisms by which these nanoparticles are incorporated into osteoprogenitor cells. The obtained results demonstrate active incorporation of nanoMBG–IPs by MC3T3-E1 pre-osteoblasts that stimulates their differentiation into mature osteoblast phenotype with increased alkaline phosphatase activity, thus indicating the efficient intracellular release of the drug and its positive *in vitro* effect on osteogenesis. The main mechanism by which FITC-Nano-MBGs enter pre-osteoblasts is the clathrin-dependent route, although these nanospheres can also enter through micropinocytosis. The present work reveals the absence of cytotoxicity of nanoMBG–IPs and their great potential as a nanotherapeutic strategy for the intracellular delivery of ipriflavone to promote osteogenesis in the periodontal defects. On the other hand, having demonstrated the intracellular incorporation of these nanospheres and their effective intracellular release of ipriflavone, this study represents the starting point for the use of these nanospheres as carriers of very diverse drugs (antibiotics, anti-inflammatory, antiresorptive and osteogenic drugs) not only for periodontal defects but also for infections and inflammatory processes such as those that occur in periodontitis.

**Supplementary Materials:** The following are available online at http://www.mdpi.com/2079-4991/10/12/2573/s1, Figure S1: FTIR spectrum of ipriflavone; Figure S2: Thermogravimetric analysis before (nanoMBG) and after loading with ipriflavone (nanoMBG–IP); Figure S3: Effects of nanoMBGs and nanoMBG–IPs (50 µg/mL) on cell size and complexity of MC3T3-E1 pre-osteoblasts after 24 h of treatment with 50 µg/mL of nanospheres. Control conditions without nanospheres were performed at the same time; Figure S4: Effects of nanoMBGs and nanoMBG–IPs on matrix mineralization by MC3T3-E1 pre-osteoblasts after 11 days of treatment with 50 µg/mL of nanospheres by Alizarin Red staining. Control conditions without nanospheres were performed at the same time.

**Author Contributions:** Conceptualization, D.A. and M.T.P.; methodology, L.C., N.G.-C., D.A., M.J.F. and M.T.P.; validation, D.A. and M.T.P.; formal analysis, L.C.; investigation, L.C., N.G.-C., D.A., M.J.F. and M.T.P.; resources, M.T.P., D.A. and M.V.-R.; data curation, D.A. and M.T.P.; writing—original draft preparation, M.T.P. and D.A.; writing—review and editing, L.C., N.G.-C., M.J.F., D.A., M.T.P. and M.V.-R.; visualization, N.G.-C., D.A. and M.T.P.; supervision, M.T.P, D.A., and M.V.-R.; project administration, M.T.P., D.A. and M.V.-R.; funding acquisition, M.T.P., D.A. and M.V.-R. All authors have read and agreed to the published version of the manuscript.

**Funding:** This research was funded by Ministerio de Economía y Competitividad, Agencia Estatal de Investigación (AEI) and Fondo Europeo de Desarrollo Regional (FEDER) (MAT2016-75611-R AEI/FEDER, UE to D.A. and M.T.P.); European Research Council (Advanced Grant VERDI; ERC-2015-AdG Proposal No. 694160 to M.V.-R.).

**Acknowledgments:** The realization of these studies was possible thanks to research grants from Ministerio de Economía y Competitividad, Agencia Estatal de Investigación (AEI) and Fondo Europeo de Desarrollo Regional (FEDER) (MAT2016-75611-R AEI/FEDER, UE). MVR acknowledges financing from the European Research Council (Advanced Grant VERDI; ERC-2015-AdG Proposal No. 694160). LC is grateful to the Universidad Complutense de Madrid for an UCM fellowship. The authors thank the staff of the ICTS Centro Nacional de Microscopía Electrónica (Spain) and the Centro de Citometría y Microscopía de Fluorescencia (Universidad Complutense de Madrid (Spain) for the support in the studies of electron microscopy, flow cytometry and confocal microscopy.

**Conflicts of Interest:** The authors declare no conflict of interest.

# References

1. Skallevold, H.E.; Rokaya, D.; Khurshid, Z.; Zafar, M.S. Bioactive Glass Applications in Dentistry. *Int. J. Mol. Sci.* **2019**, *20*, 5960. [CrossRef] [PubMed]

2. Greenspan, D.C. *Bioglass and Bioactivity: A Brief Look Back in Bioactive Glasses: Properties, Composition and Recent Applications*; Arcos, D., Vallet-Regí, M., Eds.; Novascience Publishers: New York, NY, USA, 2020.

3. Li, R.; Clark, A.E.; Hench, L.L. An investigation of bioactive glass powders by sol-gel processing. *J. Appl. Biomater.* **1991**, *2*, 231–239. [CrossRef] [PubMed]

4. Arcos, D.; Vallet-Regí, M. Sol–gel silica-based biomaterials and bone tissue regeneration. *Acta Biomater.* **2010**, *6*, 2874–2888. [CrossRef] [PubMed]

5. Manzano, M.; Arcos, D.; Delgado, M.R.; Ruiz, E.; Gil, F.J.; Vallet-Regí, M. Bioactive Star Gels. *Chem. Mater.* **2006**, *18*, 5696–5703. [CrossRef]

6. Gómez-Cerezo, M.N.; Lozano, D.; Arcos, D.; Vallet-Regí, M.; Vaquette, C. The effect of biomimetic mineralization of 3D-printed mesoporous bioglass scaffolds on physical properties and *in vitro* osteogenicity. *Mater. Sci. Eng. C* **2020**, *109*, 110572. [CrossRef]

7. Gómez-Cerezo, N.; Casarrubios, L.; Saiz-Pardo, M.; Ortega, L.; De Pablo, D.; Díaz-Güemes, I.; Fernández-Tomé, B.; Enciso, S.; Sánchez-Margallo, F.; Portolés, M.T.; et al. Mesoporous bioactive glass/ε-polycaprolactone scaffolds promote bone regeneration in osteoporotic sheep. *Acta Biomater.* **2019**, *90*, 393–402. [CrossRef]

8. Polo, L.; Gómez-Cerezo, N.; García-Fernández, A.; Aznar, E.; Vivancos, J.-L.; Arcos, D.; Vallet-Regí, M.; Martínez-Máñez, R. Mesoporous bioactive glasses equipped with stimuli-responsive molecular gates for the controlled delivery of levofloxacin against bacteria. *Chem. Eur. J.* **2018**, *24*, 18944. [CrossRef]

9. Yan, X.; Yu, C.; Zhou, X.; Tang, J.; Zhao, D. Highly Ordered Mesoporous Bioactive Glasses with Superior *In Vitro* Bone-Forming Bioactivities. *Angew. Chem. Int. Ed.* **2004**, *43*, 5980–5984. [CrossRef]

10. Wu, C.; Chang, J. Mesoporous bioactive glasses: Structure characteristics, drug/growth factor delivery and bone regeneration application. *Interface Focus* **2012**, *2*, 292–306. [CrossRef]

11. Wu, C.; Chang, J. Multifunctional mesoporous bioactive glasses for effective delivery of therapeutic ions and drug/growth factors. *J. Control. Release* **2014**, *193*, 282–295. [CrossRef]

12. Priyadarsini, S.; Mukherjee, S.; Mishra, M. Nanoparticles used in dentistry: A review. *J. Oral Biol. Craniofacial Res.* **2018**, *8*, 58–67. [CrossRef] [PubMed]

13. Bapat, R.A.; Joshi, C.P.; Bapat, P.; Chaubal, T.V.; Pandurangappa, R.; Jnanendrappa, N.; Gorain, B.; Khurana, S.; Kesharwani, P. The use of nanoparticles as biomaterials in dentistry. *Drug Discov. Today* **2019**, *24*, 85–98. [CrossRef] [PubMed]

14. Bapat, R.A.; Chaubal, T.V.; Dharmadhikari, S.; Abdulla, A.M.; Bapat, P.; Alexander, A.; Dubey, S.K.; Kesharwani, P. Recent advances of gold nanoparticles as biomaterial in dentistry. *Int. J. Pharm.* **2020**, *586*, 119596. [CrossRef] [PubMed]

15. Zakrewski, W.; Dobrzynski, M.; Rybak, Z.; Szymonowicz, M.; Wiglusz, R.J. Selected nanomaterials' application enhaced with the use of stem cells in acceleration of alveolar bone regeneration during augmentation process. *Nanomaterials* **2020**, *10*, 1216. [CrossRef]

16. Hannig, C.; Hannig, M. Natural enamel wear—A physiological source of hydroxylapatite nanoparticles for biofilm management and tooth repair? *Med. Hypotheses* **2010**, *74*, 670–672. [CrossRef]

17. Lee, J.-H.; Kang, M.-S.; Mahapatra, C.; Kim, H.-W. Effect of Aminated Mesoporous Bioactive Glass Nanoparticles on the Differentiation of Dental Pulp Stem Cells. *PLoS ONE* **2016**, *11*, e0150727. [CrossRef]

18. Zheng, K.; Balasubramanian, P.; Paterson, T.E.; Stein, R.; MacNeil, S.; Fiorilli, S.; Vitale-Brovarone, C.; Shepherd, J.; Boccaccini, A.R. Ag modified mesoporous bioactive glass nanoparticles for enhanced antibacterial activity in 3D infected skin model. *Mater. Sci. Eng. C* **2019**, *103*, 109764. [CrossRef]

19. Huang, W.; Yang, J.; Feng, Q.; Shu, Y.; Liu, C.; Zeng, S.; Guan, H.; Ge, L.; Pathak, J.L.; Zeng, S. Mesoporous Bioactive Glass Nanoparticles Promote Odontogenesis and Neutralize Pathophysiological Acidic pH. *Front. Mater.* **2020**, *7*, 241. [CrossRef]

20. Neščáková, Z.; Zheng, K.; Liverani, L.; Nawaz, Q.; Galusková, D.; Kaňková, H.; Michálek, M.; Galusek, D.; Boccaccini, A.R. Multifunctional zinc ion doped sol–gel derived mesoporous bioactive glass nanoparticles for biomedical applications. *Bioact. Mater.* **2019**, *4*, 312. [CrossRef]

21. Shang, L.; Liu, Z.; Ma, B.; Shao, J.; Wang, B.; Ma, C.; Ge, S.-H. Dimethyloxallyl glycine/nanosilicates-loaded osteogenic/angiogenic difunctional fibrous structure for functional periodontal tissue regeneration. *Bioact. Mater.* **2021**, *6*, 1175–1188. [CrossRef]

22. Portal-Núñez, S.; Mediero, A.; Esbrit, P.; Sánchez-Pernaute, O.; Largo, R.; Herrero-Beaumont, G. Unexpected Bone Formation Produced by RANKL Blockade. *Trends Endocrinol. Metab.* **2017**, *28*, 695–704. [CrossRef] [PubMed]

23. Luhmann, T.; Germershaus, O.; Groll, J.; Meinel, L. Bone targeting for the treatment of osteoporosis. *J. Control. Release* **2012**, *161*, 198–213. [CrossRef] [PubMed]

24. Reginster, J.-Y.L. Ipriflavone: Pharmacological properties and usefulness in postmenopausal osteoporosis. *Bone Miner.* **1993**, *23*, 223–232. [CrossRef]

25. Mazzuoli, G.; Romagnoli, E.; Carnevale, V.; Scarda, A.; Scarnecchia, L.; Pacitti, M.T.; Rosso, R.; Minisola, S. Effects of ipriflavone on bone remodeling in primary hyperparathyroidism. *Bone Miner.* **1992**, *19* (Suppl. 1), S27–S33. [CrossRef]

26. Duan, H.; Diao, J.; Zhao, N.; Ma, Y. Synthesis of hollow mesoporous bioactive glass microspheres with tunable shell thickness by hydrothermal-assisted self-transformation method. *Mater. Lett.* **2016**, *167*, 201–204. [CrossRef]

27. Gera, S.; Sampathi, S.; Dodoala, S. Role of Nanoparticles in Drug Delivery and Regenerative Therapy for Bone Diseases. *Curr. Drug Deliv.* **2017**, *14*, 904–916. [CrossRef] [PubMed]

28. Addison, W.; Nelea, V.; Chicatun, F.; Chien, Y.-C.; Tran-Khanh, N.; Buschmann, M.D.; Nazhat, S.; Kaartinen, M.T.; Vali, H.; Tecklenburg, M.; et al. Extracellular matrix mineralization in murine MC3T3-E1 osteoblast cultures: An ultrastructural, compositional and comparative analysis with mouse bone. *Bone* **2015**, *71*, 244–256. [CrossRef]

29. Casarrubios, L.; Gómez-Cerezo, N.; Feito, M.J.; Vallet-Regí, M.; Arcos, D.; Portolés, M.T. Incorporation and effects of mesoporous $SiO_2$–CaO nanospheres loaded with ipriflavone on osteoblast/osteoclast cocultures. *Eur. J. Pharm. Biopharm.* **2018**, *133*, 258–268. [CrossRef]

30. Li, Y.; Bastakoti, B.P.; Yamauchi, Y. Smart Soft-Templating Synthesis of Hollow Mesoporous Bioactive Glass Spheres. *Chem. A Eur. J.* **2015**, *21*, 8038–8042. [CrossRef]

31. López-Noriega, A.; Arcos, D.; Vallet-Regí, M. Functionalizing Mesoporous Bioglasses for Long-Term Anti-Osteoporotic Drug Delivery. *Chem. Eur. J.* **2010**, *16*, 10879–10886. [CrossRef]

32. Cicuéndez, M.; Silva, V.S.; Hortigüela, M.J.; Matesanz, M.C.; Vila, M.; Portolés, M.T. MC3T3-E1 pre-osteoblast response and differentiation after graphene oxide nanosheet uptake. *Colloids Surf. B Biointerfaces* **2017**, *158*, 33–40. [CrossRef] [PubMed]

33. Udall, J.N.; Moscicki, R.A.; Preffer, F.I.; Ariniello, P.D.; Carter, E.A.; Bhan, A.K.; Bloch, K.J. Flow Cytometry: A New Approach to the Isolation and Characterization of Kupffer Cells. *Adv. Exp. Med. Biol.* **1987**, *216*, 821–827. [CrossRef]

34. Matesanz, M.-C.; Vila, M.; Feito, M.-J.; Linares, J.; Gonçalves, G.; Vallet-Regí, M.; Marques, P.A.A.P.; Portolés, M.-T. The effects of graphene oxide nanosheets localized on F-actin filaments on cell-cycle alterations. *Biomaterials* **2013**, *34*, 1562–1569. [CrossRef] [PubMed]

35. Mao, L.; Xia, L.; Chang, J.; Liu, J.; Jiang, L.; Wu, C.; Fang, B. The synergistic effects of Sr and Si bioactive ions on osteogenesis, osteoclastogenesis and angiogenesis for osteoporotic bone regeneration. *Acta Biomater.* **2017**, *61*, 217–232. [CrossRef]

36. Beck, G.R.; Ha, S.-W.; Camalier, C.E.; Yamaguchi, M.; Li, Y.; Lee, J.-K.; Weitzmann, M.N. Bioactive silica-based nanoparticles stimulate bone-forming osteoblasts, suppress bone-resorbing osteoclasts, and enhance bone mineral density in vivo. *Nanomed. Nanotechnol. Biol. Med.* **2012**, *8*, 793–803. [CrossRef]

37. Zhang, X.; Zeng, D.; Li, N.; Wen, J.; Jiang, X.; Liu, C.; Li, Y. Functionalized mesoporous bioactive glass scaffolds for enhanced bone tissue regeneration. *Sci. Rep.* **2016**, *6*, srep19361. [CrossRef]

38. Gómez-Cerezo, N.; Casarrubios, L.; Morales, I.; Feito, M.; Vallet-Regí, M.; Arcos, D.; Portolés, M.T. Effects of a mesoporous bioactive glass on osteoblasts, osteoclasts and macrophages. *J. Colloid Interface Sci.* **2018**, *528*, 309–320. [CrossRef]

39. Quarles, L.D.; Yohay, D.A.; Lever, L.W.; Caton, R.; Wenstrup, R.J. Distinct proliferative and differentiated stages of murine MC3T3-E1 cells in culture: An in vitro model of osteoblast development. *J. Bone Miner. Res.* **1992**, *7*, 683–692. [CrossRef]

40. Stein, G.S.; Lian, J.B. Molecular Mechanisms Mediating Proliferation/Differentiation Interrelationships During Progressive Development of the Osteoblast Phenotype. *Endocr. Rev.* **1993**, *14*, 424–442. [CrossRef]

41. Ara, T.; Declerck, Y.A. Interleukin-6 in bone metastasis and cancer progression. *Eur. J. Cancer* **2010**, *46*, 1223–1231. [CrossRef]

42. Udagawa, N.; Takahashi, N.; Katagiri, T.; Tamura, T.; Wada, S.; Findlay, D.M.; Martin, T.J.; Hirota, H.; Taga, T.; Kishimoto, T.; et al. Interleukin (IL)-6 induction of osteoclast differentiation depends on IL-6 receptors expressed on osteoblastic cells but not on osteoclast progenitors. *J. Exp. Med.* **1995**, *182*, 1461–1468. [CrossRef] [PubMed]

43. Horwood, N.J.; Elliott, J.; Martin, T.J.; Gillespie, M. Osteotropic Agents Regulate the Expression of Osteoclast Differentiation Factor and Osteoprotegerin in Osteoblastic Stromal Cells. *Endocrinol.* **1998**, *139*, 4743. [CrossRef]

44. Kuroyanagi, G.; Adapala, N.S.; Yamaguchi, R.; Kamiya, N.; Deng, Z.; Aruwajoye, O.; Kutschke, M.; Chen, E.; Jo, C.; Ren, Y.; et al. Interleukin-6 deletion stimulates revascularization and new bone formation following ischemic osteonecrosis in a murine model. *Bone* **2018**, *116*, 221–231. [CrossRef] [PubMed]

45. Mariani, E.; Lisignoli, G.; Borzì, R.M.; Pulsatelli, L. Biomaterials: Foreign Bodies or Tuners for the Immune Response? *Int. J. Mol. Sci.* **2019**, *20*, 636. [CrossRef] [PubMed]

46. Oh, N.; Park, J.H. Endocytosis and exocytosis of nanoparticles in mammalian cells. *Int. J. Nanomed.* **2014**, *9*, 51–63. [CrossRef]

47. Bandmann, V.; Müller, J.D.; Köhler, T.; Homann, U. Uptake of fluorescent nano beads into BY2-cells involves clathrin-dependent and clathrin-independent endocytosis. *FEBS Lett.* **2012**, *586*, 3626–3632. [CrossRef] [PubMed]

48. Sato, K.; Nagai, J.; Mitsui, N.; Yumoto, R.; Takano, M. Effects of endocytosis inhibitors on internalization of human IgG by Caco-2 human intestinal epithelial cells. *Life Sci.* **2009**, *85*, 800–807. [CrossRef] [PubMed]

49. Mäger, I.; Langel, K.; Lehto, T.; Eiríksdóttir, E.; Langel, Ü. The role of endocytosis on the uptake kinetics of luciferin-conjugated cell-penetrating peptides. *Biochim. Biophys. Acta* **2012**, *1818*, 502–511. [CrossRef]

50. Schulz, W.L.; Haj, A.K.; Schiff, L.A. Reovirus Uses Multiple Endocytic Pathways for Cell Entry. *J. Virol.* **2012**, *86*, 12665–12675. [CrossRef]

51. Linares, J.; Matesanz, M.C.; Vila, M.; Feito, M.J.; Gonçalves, G.; Vallet-Regí, M.; Marques, P.A.A.P.; Portolés, M.T. Endocytic Mechanisms of Graphene Oxide Nanosheets in Osteoblasts, Hepatocytes and Macrophages. *ACS Appl. Mater. Interfaces* **2014**, *6*, 13697–13706. [CrossRef]

52.    Cantley, L.C. The Phosphoinositide 3-Kinase Pathway. *Science* **2002**, *296*, 1655–1657. [CrossRef] [PubMed]

**Publisher's Note:** MDPI stays neutral with regard to jurisdictional claims in published maps and institutional affiliations.

MDPI

St. Alban-Anlage 66

4052 Basel

Switzerland

Tel. +41 61 683 77 34

Fax +41 61 302 89 18

www.mdpi.com

*Nanomaterials* Editorial Office

E-mail: nanomaterials@mdpi.com

www.mdpi.com/journal/nanomaterials